Color Atlas of

MUSCLE PATHOLOGY

W. J. K. Cumming BSc, MD, FRCPI, FRCP
Consultant Neurologist
University Hospital of South Manchester, Withington;
Honorary Lecturer, School of Biological Sciences
University of Manchester;
Director, North West Regional Neuromuscular Unit
Manchester, UK

J. J. Fulthorpe MPhil, FIBMS
Senior Chief Medical Laboratory Scientific Officer
Muscular Dystrophy Research Laboratories
Newcastle General Hospital
Newcastle upon Tyne, UK

P. Hudgson MB, FRCP, FRACP
Senior Lecturer in Neurology
University of Newcastle upon Tyne;
Consultant Neurologist
Clinical Director (Adult Neuromuscular Diseases)
Muscular Dystrophy Research Laboratories
Newcastle General Hospital,
Newcastle upon Tyne, UK

M. Mahon BSc
Lecturer in Anatomy
School of Biological Sciences
University of Manchester;
Research Director
North West Regional Neuromuscular Unit
Manchester, UK

Mosby-Wolfe

London Baltimore Bogotá Boston Buenos Aires Caracas Carlsbad, CA Chicago Madrid Mexico City Milan Naples, FL New York Philadelphia St. Louis Sydney Tokyo Toronto Wiesbaden

Published in 1994 by Mosby Wolfe, an imprint of Times Mirror International Publishers Limited

Printed in Italy by Imago

ISBN 0 7234 2016 5

For full details of all Times Mirror International Publishers Limited titles, please write to Times Mirror International Publishers Limited, Lynton House, 7–12 Tavistock Square, London WC1H 9LB, England.

A CIP catalogue record for this book is available from the British Library.

Library of Congress Cataloging-in-Publication Data: applied for

Contents

Foreword		iv
Preface		v
Acknowledgements		vi
1	Introduction	1
2	Preparation of the biopsy	3
3	Normal muscle structure and analysis of the biopsy	15
4	Pathological reactions of the muscle fibre	45
5	X-linked recessive myopathies (Xp21.2 myopathies)	91
6	Spinal muscular atrophies	105
7	Inflammatory myopathies	113
8	Congenital myopathies	135
9	Metabolic disorders	147
10	Miscellaneous disorders	169
11	Interpreting the biopsy	177
Appendices:		
A	Glossary of abbreviations	181
B	Micrograph labels	182
C	Histological and histochemical procedures	183
D	Immunocytochemical procedures	189
E	Electron microscopical procedures	191
F	Neuromuscular disorders – Gene location	193
References		196
Bibliography		198
Index		200

Foreword

Perhaps more than any other branch of medicine, the diagnosis and treatment of muscle disease depends upon close collaboration and understanding between physician and pathologist. So much so, in fact, that many of the most significant advances in human muscle pathology have come from neurologists or from those who have joint neurology/neuropathology practices.

In the 1950s, Adams in the U.S.A. and Walton in Britain laid a firm foundation for the clinical pathology of muscle disease. A decade later, with the introduction of electron microscopical and histochemical techniques for the diagnosis of muscle disease, clinicians such as Dubowitz and Engel were still in the forefront. The rapid growth of molecular biology and immunocytochemistry has further broadened the scope for diagnosis and management of muscle disease, but has left many of us with a bewildering and complex maze of information and techniques.

One of the most important questions to answer is 'How do we grasp and assimilate the plethora of data derived from complex investigations so that the needs of patients, with rare but devastating muscle disease, are met in the most cost-effective manner?'. This *Color Atlas of Muscle Pathology*, in effect, goes a long way towards answering that question. It is an intensely practical book, concentrating initially on the preparation and analysis of muscle biopsies and later on their interpretation. By the extensive use of colour and monochrome photographs, the reader is guided through the major histological, histochemical and electron microscopical appearances of normal muscle, the general pathological reactions of muscle, and the pathology of specific muscle diseases. In addition, the atlas looks forward through the well-developed molecular biology of the X-linked dystrophies to the further characterisation of other muscular dystrophies, and the increasing use of immunocytochemistry in the localisation of gene products.

The authors are to be congratulated upon the clear and concise presentation of the text and illustrations in the atlas. Its title is perhaps rather modest, as the contents also effectively bridge the gap between the clinical presentation and the pathology of muscle disease. As such, it will be of enormous value to both the experienced and the uninitiated in the field of muscle pathology and clinical muscle disease.

Professor R. O. Weller BSc MD PhD FRCPath
Department of Pathology (Neuropathology)
Southampton General Hospital

Preface

The literature on neuromuscular pathology has expanded rapidly during the last forty years, and, because of this, one has to examine the need for yet another contribution with care, if not scepticism. Nevertheless, the experience of the four authors (two laboratory scientists and two clinicians) is unusual in its breadth and depth, and encompasses most of the epoch referred to above.

In addition, it is only too painfully clear that most of the findings published in peer-reviewed journals and monographs on neuromuscular disease during this period have been ignored by clinicians (including paediatricians and neurologists) and most particularly by general pathologists.

In light of this, we felt well and truly justified in recording our personal experiences in the hope of producing a 'bench book' which is of practical application for clinicians and pathologists alike.

With the above (we hope laudable) objective in mind, we have 'gone back to basics' and set out the text in traditional fashion, commencing with normal morphology (both gross and microscopic) and the technical procedures necessary to produce interpretable data in what is acknowledged to be a 'difficult' area.

With respect to the latter, we cannot emphasise too strongly the need for these studies to be carried out in centres of reference, and ***not*** in local hospitals, if only because the requisite technology and expertise is unlikely to be available.

Turning to the interpretation of myopathological data, we have done our best to ensure that the closest possible correlations between clinical syndromes/disorders and their morphological substrates are established in a step-by-step manner. It has to be said that most of the advances in our understanding of the natural history of the myopathies (and their pathologies) have been made by clinicians with a penchant for gazing down microscopes, and in collaboration with scientists interested in clinical disease.

WJKC & MM Manchester
JJF & PH Newcastle Upon Tyne
1994

Acknowledgements

We wish to thank primarily our patients, whose biopsies are represented in this text. Without them, none of this work would have been possible.

We are grateful to colleagues who have allowed us to modify material already published in their name and to their Publishers for their cooperation.

WJKC and MM are particularly grateful to Mrs. Milly Taylor for all her help and patience throughout the years in the preparation of muscle biopsies, and we are all grateful to a succession of Research Students and Research Registrars who have passed through the Unit, whose comment and criticism has helped focus our attention on the clinical pathological correlates of neuromuscular disease.

We are indebted for secretarial assistance to Gill Hennessey and Mrs. Pam Watts who, over the years, have coped with our frequent requests for assistance with great patience and humour.

We would like to thank the following for allowing us to reproduce figures with modification:

Prof. F.G.I. Jennekens and Churchill Livingstone for **3.6**
Dr K.P. Campbell and John Wiley & Sons for **5.1**
Michel Fardeau and Churchill Livingstone for **8.1**
Prof. V. Dubrowitz and Pergammon Press Ltd for the table in Appendix F

1

Introduction

Muscle biopsy is an essential component of the investigation of patients with neuromuscular disorders, and is the deciding factor for diagnosis in most instances. Certainly it is the criterion on which management decisions are made in the majority of cases.

The techniques involved in producing material which is capable of reliable interpretation are now much more complex than those employed when the first definitive textbook of myopathology was published (Adams, Denny-Brown and Pearson, 1953). These authors relied almost exclusively on formalin and paraffin embedding. Although their illustrations were good, if not excellent in some cases, they themselves admitted that artefacts produced by fixation were a serious problem and could introduce major difficulties in interpretation. These problems are discussed and illustrated in detail in Chapters 2 and 3.

The introduction of the cryostat and the development of enzyme histochemistry has revolutionised our approach to the investigation of neuromuscular disorders. In particular, it has made it possible to diagnose chronic denervating disorders of central origin, congenital myopathies, and metabolic muscle disease. (The coexistence of the latter two is not infrequent.) In turn, this has led to ongoing clarification of the clinical nosology of the neuromuscular disorders, and has drastically revised the way in which we analyse their clinical presentations. Most importantly, it has led to the realisation that the traditional concepts of limb girdle and facioscapulohumeral dystrophies are hopelessly outmoded. These are now recognised as clinical syndromes, each with several widely differing pathological substrates (*Table 1.1*).

Sharp differences of opinion exist as to the best way of providing the highly sophisticated and expensive services and investigations (e.g. electrophysiological, biochemical, molecular genetic and morphological) that are required to deal with patients suffering from neuromuscular disorders. Although Cumming, Hudgson and Lecky (1995) have discussed this controversy elsewhere, we feel that it is worthwhile re-stating the view that routine histopathological laboratories in District General Hospitals are not necessarily the best places to operate a muscle and nerve biopsy service.

The excitable nature of the membranes of nerve and muscle tissue complicates their handling. The tissues are more likely to shrink or contract than tissue from liver or gastrointestinal mucosa, for instance, when placed in orthodox fixatives. To ensure the production of reliable, high quality material, the technical staff of the laboratories that handle muscle biopsies require thorough training in the techniques of preparing cryostat and ultrathin sections.

It is of the utmost importance for the reporting morphologists, whatever their training or experience, to have the closest possible liaison with both clinical colleagues and those providing electrophysiological and

Table 1.1 Neuromuscular disorders

Disorder	Examples	Chapter	Disorder	Examples	Chapter
Myopathic disorders					
• Dystrophies			• Secondary		
X-linked recessive(Xp21.2)	DMD, BMD	5	Toxic	Alcohol	10
Emery-Dreifuss		5	Drug induced	Steroids	10
Myotonic		10	Endocrine	Hypothyroid	10
Congenital		8		Acromegaly	10
• Inflammatory	PM, DM, IBM	7	Trauma	Burns	10
• Metabolic					
Glycolytic	Types II, IV and V	9	**Neurogenic disorders**		
Lipid storage	CPT, carnitine	9	• Central	SMA	6
Mitochondrial	CPEO, Kearn-Sayre's	9		MND	10
• Ion channel			• Junctional	MyG/LEMS	10
Periodic paralysis (Na)	Hypo/hyperkalaemia	10	• Peripheral	HMSN	-
Myotonia congenita (Cl)	Thomsen's disease	-			
Malignant hyperthermia			**Syndromes**		
• Congenital myopathies	Centronuclear	8	• Limb-girdle		5–9
	CFTD	8	• Facioscapulohumeral		5–9
	Nemaline	8	• Scapuloperoneal		5–9
	Core	8	• Distal		5–10

biochemical support. In practice, these highly desirable ends are best achieved in a centralised Regional Reference Service, if only because of the relative rarity of many of the pathological conditions to be investigated. Details of the technical problems involved in the handling and preparation of the muscle biopsy for microscopical examination are discussed in depth in Chapter 2 and in Appendix C.

The traditional histopathological reliance on one or two poorly orientated and often badly fixed haematoxylin and eosin (H&E) stained sections for making a diagnosis is totally unsatisfactory, and will lead to serious errors of management. Without going into extensive forensic detail, it is worth quoting some personal experiences which underline this point.

One of the authors had to deal with a female patient who had been told, on the basis of a technically unsatisfactory biopsy, that she was suffering from Duchenne muscular dystrophy. Another was asked to give a second opinion on material taken from a patient who suffered from motor neurone disease. Again, the material was technically unsatisfactory. It clearly showed 'grouped' atrophy with numerous nuclear clumps. However, the patient had been treated with corticosteroids for two years because the nuclear clumps had been mistakenly 'read' as an inflammatory infiltrate, thus leading to a firm, ***but incorrect***, diagnosis of polymyositis.

With reference to establishing a satisfactory muscle biopsy service, the following basic principles should be observed:

- The material should be obtained by an experienced rather than an 'occasional' operator. This operator should have immediate contact with the laboratory staff while performing the procedure.
- Selection of the muscle to be biopsied should be done jointly by the clinician in charge of the case and the surgeon. If needle biopsy is the only technique available, choice is limited to the larger limb muscles. If an open biopsy is to be carried out, the authors choose the quadriceps, deltoid and biceps, in that order. However, grossly weak and wasted muscles should be avoided, as insufficient muscle will be found to permit satisfactory histological analysis. (Sections from such muscles tend to show end-stage pathology, and contain only fat and fibrous connective tissue, with a few scattered, extremely atrophic muscle fibres.)
- One of the authors of this book has a personal preference for open muscle biopsies, and another for needle biopsies. Each technique conveys certain advantages in particular clinical situations, and needle biopsy is the only viable technique for obtaining serial samples in, for example, the monitoring of the response of inflammatory myopathy to therapy. Satisfactory histological material can be obtained by either technique, and we do not consider that dogmatic adherence to one or the other is either necessary or desirable.
- The techniques of performing muscle biopsies are detailed in Chapter 2. Suffice it to say at this point that, immediately after removal, the samples obtained should be dealt with by an experienced member of the laboratory staff. The critical period for the tissue sample is from its removal from the patient until its freezing, or its fixation for electron microscopy. Thereafter, a more leisurely and reflective approach can be taken in preparing the individual sections.
- The reporting morphologist should be well aware of the clinical background of the case before attempting to report the biopsy. Pathogenesis of most neuromuscular disorders is an interplay between sex, age, symptoms and the disorder's duration. This argues for a close relationship between the referring clinician and the morphologist.
  In the particular instance of acute rhabdomyolysis, an expedient biopsy report can exclude other inflammatory processes that are amenable to treatment. However, if rhabdomyolysis is due to a metabolic problem of some kind, a subsequent biopsy will certainly be necessary as glycolytic enzyme activity, mitochondrial structure and function, and lipid deposits cannot be identified reliably in necrotic muscle.
- It is regrettable, but true, that little or no exposure to nerve and muscle biopsy material is included in the routine training of morbid histologists. This applies even to those preparing for a career in neuropathology. It is perhaps not surprising therefore that nerve and muscle reports emanating from many laboratories tend to be idiosyncratic and bear little or no relationship to the pathological process involved. This emphasises the importance for the patient that this service is undertaken in specialised laboratories by experienced scientific and medical staff. It also suggests that training schedules for pathologists should be revised to include exposure to this increasingly active field of interventionist investigation.

As far as reporting the muscle biopsy itself is concerned, we trust that our experience over a period of more than 25 years will be of some value to those involved in the field. Considerable help may be gained from statistical and morphometric data, particularly the relative proportions of the major histochemical fibre types and their size and spatial distributions.

Normal data for the muscles most commonly biopsied are available in the literature (Chapters 2 and 3), and are classified according to the age and sex of patients. These should form an essential part of the observers' armamentarium when analysing the visual data. However, personal experience is often a major factor in arriving at a diagnosis based upon the histological and morphological appearances of a biopsy.

2

Preparation of the biopsy

Since the procedures described here involve the handling of fresh, unfixed human tissue, precautions must be taken to minimise the risk to laboratory staff of infection with HIV, hepatitis B and other agents.

A protocol should be adopted in which protective clothing is worn, waterproof dressings are applied to cuts and abrasions, and all instruments are sterilised after use in activated glutaraldehyde. The working area should be covered with disposable material, and should be wiped down with hypochlorite solution after use. In addition, all debris from such material, especially from the cryostat, must be disposed of in an appropriate manner.

Once the necessity for a muscle biopsy has been established, close cooperation between the referring physician and the laboratory-based muscle morphologist must follow, as emphasised in Chapter 1.

To obtain the maximum information from a muscle biopsy it should ***not*** be treated as a general pathology specimen, with formalin fixation and paraffin embedding. Not only does the laboratory need to be forewarned of intended muscle biopsies (in order to have liquid nitrogen available), but laboratory staff may suggest the need for further specimens, which could be suitably prepared for additional investigative procedures.

Information accompanying the biopsy should include the patient's name and hospital number, their sex and age, and a brief clinical summary concerning the distribution and duration of the symptoms. Data on other investigations, such as CK levels and EMG findings, or nerve conduction velocities, should also be included. A record of the exact site of the biopsy is essential, together with a provisional diagnosis of the disorder.

In practice, a special muscle biopsy request form (**2.1**) ensures consistent records (especially when inexperienced individuals are providing biopsies), and permits the laboratory to select the relevant histochemical and immunocytochemical investigations for accurate diagnosis. Data

2.1 A muscle biopsy request form.

Regional Neuromuscular Unit – Muscle Biopsy Request		
Biopsy No:	Date of Biopsy: Date of Birth: Referring Consultant:	Hospital: Hospital No: Research Code:
Patient's Surname:		Sex:
Patient's Firstname:		Age:
Address:		
Previous Biopsy:		
Biopsy Site:	Left / Right:	Needle / Open / PM:
Clinical Details: MRC Rating:		
Provisional Diagnosis:		Final Diagnostic Code:
Lm:	Em:	Mes:
CK:	Bioch:	EMG:
Biopsy Appearance:		
Comments:		
Signed:		Date:

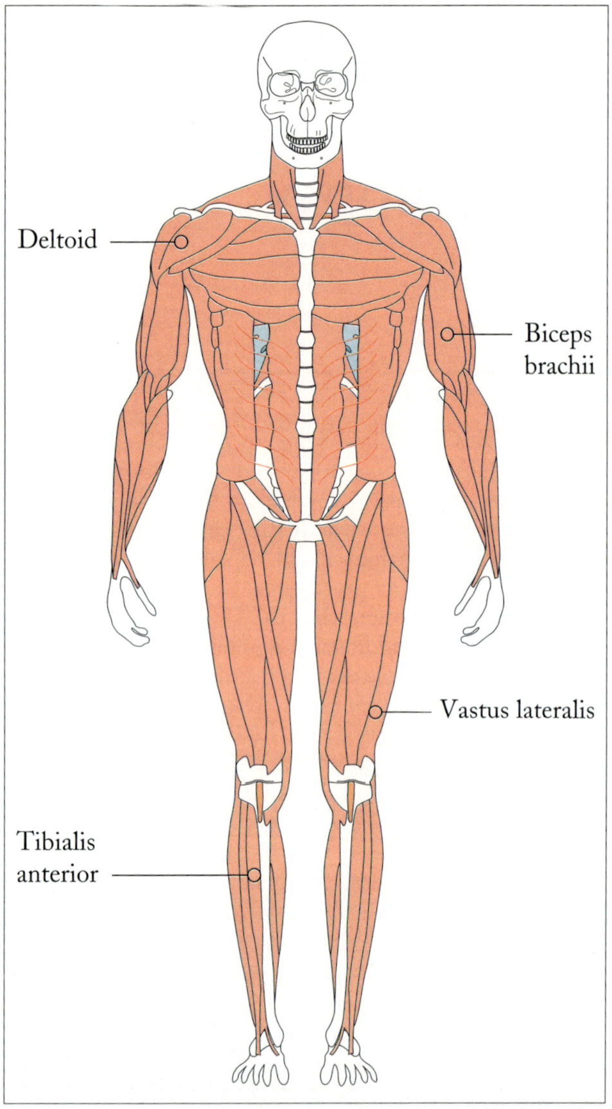

2.2 Preferred sites for muscle biopsy.

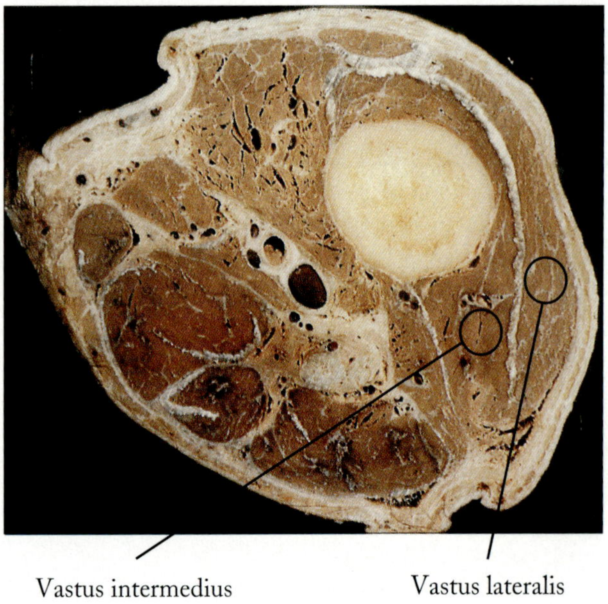

2.3 Transverse section of a cadaveric thigh at the level of the most commonly used biopsy site, showing the disposition of the muscles.

organised in this format allows for easier storage, retrieval and cross-correlation of clinical and biopsy data by computer.

Once the biopsy specimen has been removed from the patient, it is vital that clinicians and other hospital staff follow the correct handling procedures. An unlabelled, undocumented piece of 'muscle' arriving unexpectedly in the laboratory with an attached message reading 'Send report to Dr. X', is hardly likely to justify the patient's discomfort and inconvenience. It also runs the risk of being confused with other similarly presented specimens!

From the referring physician's (and the patient's) point of view, the worst reports are those that state 'The specimen contains only fatty connective tissue.' or 'Fixation has rendered this specimen unsuitable for diagnosis'. Attention to the correct procedures for obtaining and handling specimens ***must*** be adhered to in order to avoid these problems.

Biopsy site

The importance of selecting the appropriate muscle for biopsy was mentioned in Chapter 1. ***The muscle with maximal clinical signs should not be selected***, because it is likely that the specimen will present an 'end stage' appearance of mostly fat and fibrous connective tissue. Conversely, samples from muscles with mild clinical involvement usually reveal primary changes, which may help elucidate the nature or cause of the disorder.

Sites of previous biopsy, drug injection or trauma, should be avoided, and EMG needle investigations are preferably carried out on the contralateral limb, to avoid induced inflammatory changes. The choice of site (**2.2**) may also depend on the availability of normative data, and on which technical procedures (e.g. open, needle or conchotome biopsy) are involved.

Specimens are obtained most frequently from the thigh, at a site in the lateral portion of quadriceps femoris approximately a hand's breadth proximal to the superior border of the patella. This region is relatively free of large blood vessels and nerves. The disposition of the muscles at this site is shown in **2.3**, where it may be noted that samples are likely to be derived from either vastus lateralis or vastus intermedius.

Other sites which may be chosen, dependant on clinical presentation, are deltoid, biceps brachii or tibialis anterior.

Open muscle biopsy

This is the preferred technique in some centres. It may be carried out when needle biopsy has proved difficult or provided insufficient material for analysis, or when motor point biopsies are required for endplate investigations. The large specimens obtained by the open technique allow portions to be processed separately for biochemistry, immunoblotting, and light and electron microscopy. Material is obtained under general or local anaesthesia, although care must be taken with the latter not to infiltrate the underlying muscle (*see below*).

Using sterile technique, an incision parallel to the muscle belly and approximately 3–4 cm in length is made through the skin and fascia. A cylinder of the underlying muscle is removed with the aid of a specially designed clamp (**2.4–2.5**) that assists orientation of the muscle fibres and helps prevent the formation of contraction-induced artefacts.

The clamped ends should be discarded from the material examined microscopically, but may be used for immunoblotting and biochemistry (**2.6–2.7**). Similar methods may be used for autopsy samples.

Disadvantages of the open technique are the large volume of muscle removed, the possibility of scar tissue being obtained should a subsequent biopsy be required, and the inconvenience and potential problems associated with general anaesthesia. It is, however, the preferred technique when muscles other than the quadriceps are to be biopsied.

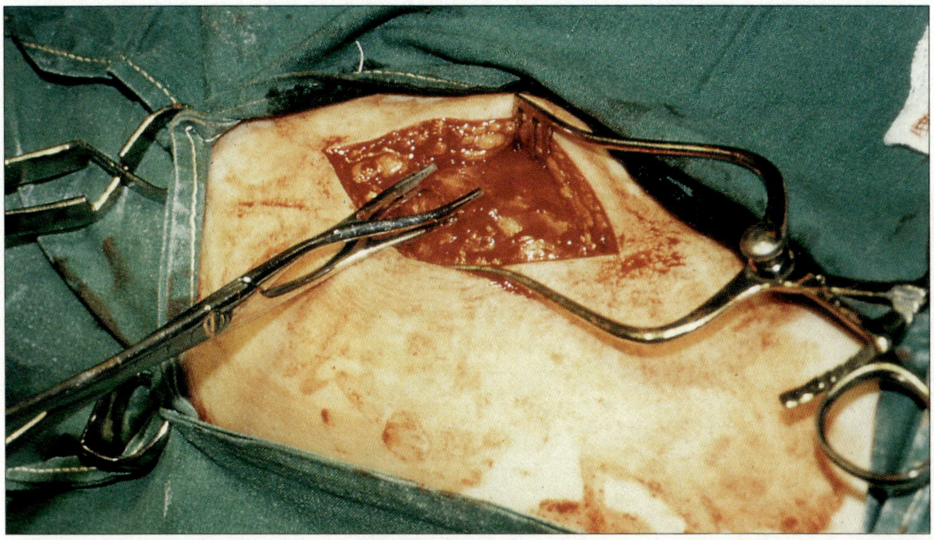

2.4 This specially designed biopsy clamp is used to obtain an accurately orientated sample of muscle in the open biopsy technique.

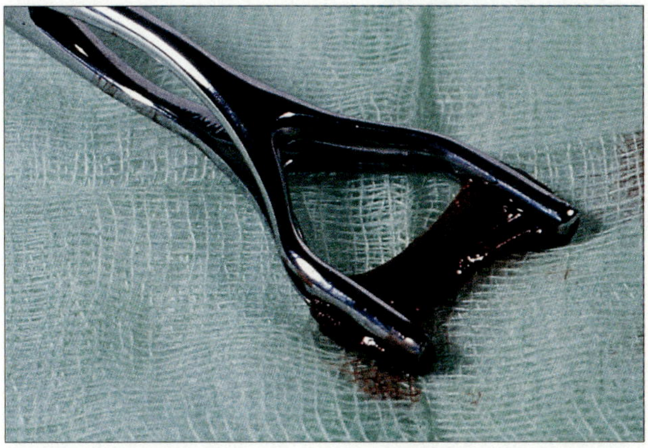

2.5 The open biopsy specimen is shown held between the jaws of the clamp. The muscle fibres run parallel to the long axis of the sample. Immediately after removal, a small bundle of fibres are dissected from the edge of the original biopsy and placed in the electron microscope clamp under slight tension.

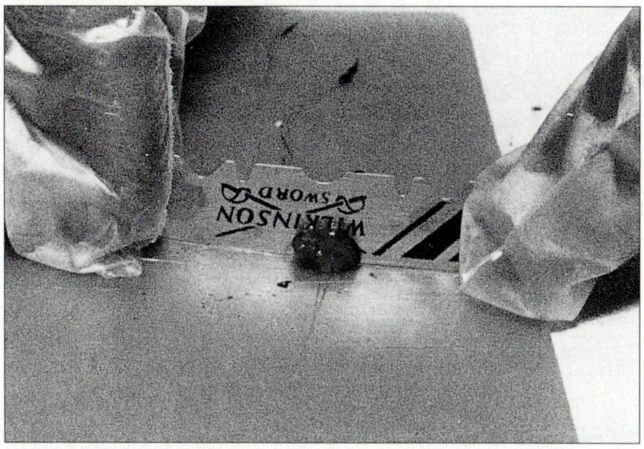

2.6 Following release of the specimen from the clamp, the crushed ends are discarded and the central portion cut into slices 3–4 mm thick.

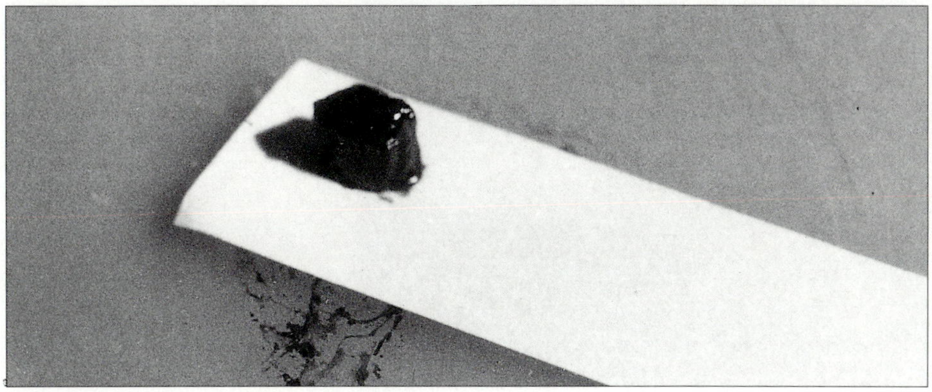

2.7 Each slice of muscle is attached, with the fibres standing upright, to one end of a pre-numbered strip of filter paper. The filter paper acts both as a label and as a means of handling the specimen when freezing and attaching it to the microtome.

Needle muscle biopsy

This was the earliest method for obtaining muscle biopsies, as descriptions of Duchenne's 'harpoon' reveal (Duchenne, 1868). More recently, the need for the histochemical and electron microscopical investigation of sequential, fresh specimens of an athlete's muscle, has meant that the needle method became more regularly invoked (Bergstrom, 1962), and was then applied to the clinical situation (Edwards, 1971).

Nowadays it is often the preferred method, especially in children, patients with respiratory problems, and for follow-up in patients where repeat biopsy may provide information on disease progression or response to treatment regimes. It is also less time-consuming and more cost-effective than open biopsy, since it does not involve an anaesthetist or theatre staff, and can be performed as an outpatient procedure. The small scar left by this technique is obviously more cosmetically acceptable to the patient than that left by an open biopsy.

The major disadvantage is the small size of the specimen (compare **2.8**, **2.9**), which is occasionally difficult to orientate. Furthermore, focal lesions are more likely to be missed, and there may be insufficient material for morphometric or biochemical analysis. However, multiple samples may be obtained from the same muscle when required. Complications of this procedure are minimal provided clotting factors are checked prior to biopsy (Cumming and Mahon, 1984).

The biopsy needle employed (**2.10–2.12**) is similar to that modified by Young *et al.* (1978). It consists of an

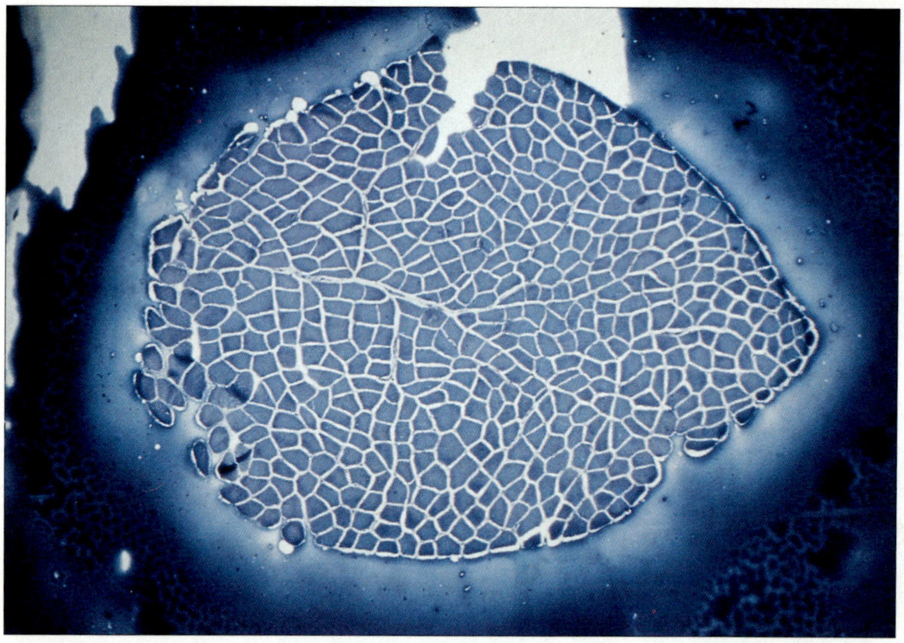

2.8 Transverse section of muscle sample, obtained by the needle biopsy technique. (Toluidine blue, ×60)

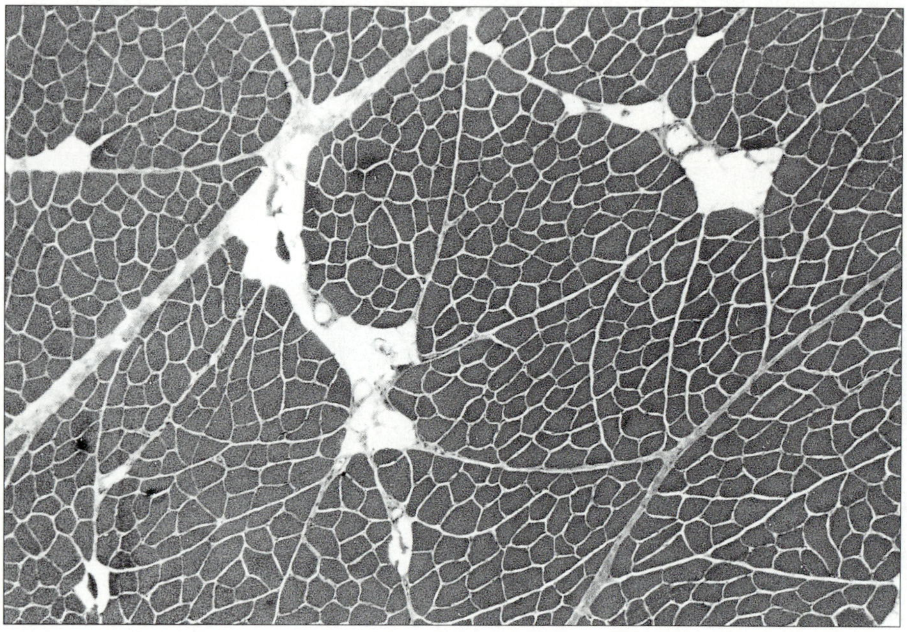

2.9 Transverse section of part of an open muscle biopsy sample. (H&E, ×80)

expensive (but re-useable), hollow-bored cylinder about 4.5mm in diameter with a cut-out window at one end. An inner cylinder, with a sharpened distal end, is used to excise tissue samples protruding through the window. Inside this is a solid rod, which is used for pushing the samples out of the needle cavity. A narrower version is available for paediatric use.

The assembled needle should be presented in the usual hospital sterile pack. The procedure is carried out under sterile conditions. The skin and subcutaneous tissue at the biopsy site are anaesthetised by the local infiltration of 1% lignocaine. Care must be taken not to introduce anaesthetic into the underlying deep fascia and muscle, as degenerative changes occur rapidly.

After anaesthesia, a small scalpel incision, approximately 5mm long (using a size 15 blade), is made in the skin and deep fascia to facilitate introduction of the needle. The needle, minus the innermost rod, is inserted with pressure into the muscle bulk as shown in **2.11**, so that a small piece of tissue can be pressed with the free hand into the needle window. Ramming the inner cylinder down the needle bore cuts off a small sample.

This latter process may be repeated to obtain two or three 'bites' of the muscle.

Upon removal of the needle from the muscle, the wound is closed over with firm pressure and skin Elastoplast. The muscle specimens should be removed carefully with the aid of the central rod and an unused hypodermic needle or watchmaker's forceps.

It is important to have the inner cylinder regularly resharpened to ensure that a clean cut is obtained, which is essential for high quality specimens.

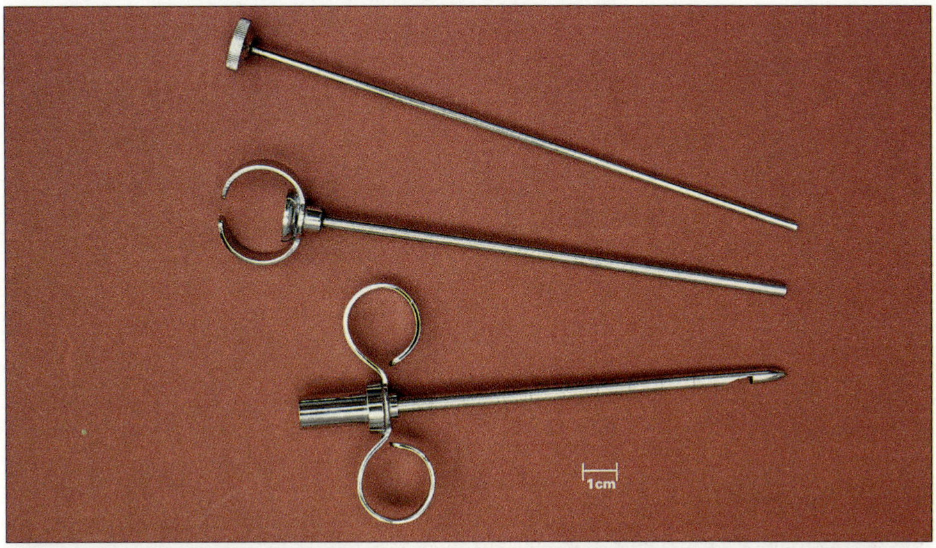

2.10 A muscle biopsy needle consisting of an outer portion with finger grips, an inner cutting blade, and an innermost rod for removing the specimen.

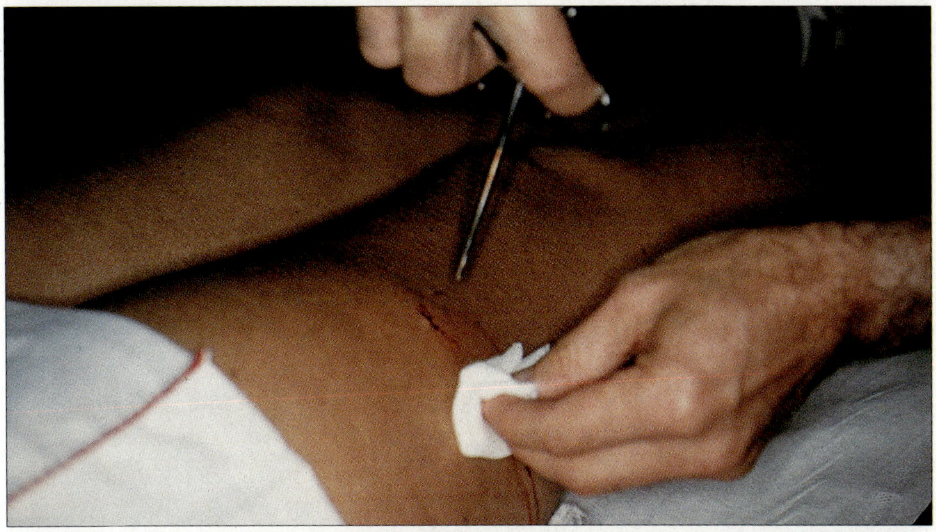

2.11 A muscle biopsy needle inserted into the thigh.

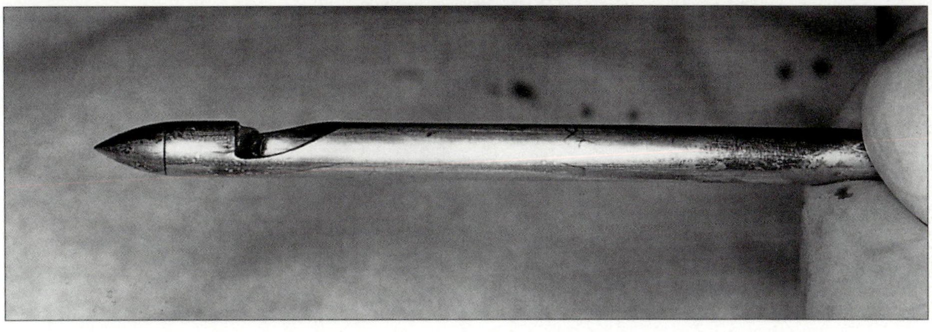

2.12 Detail of the muscle biopsy needle, showing the cut-out window near the tip into which a portion of muscle to be sampled is pressed.

The sample(s) obtained are relatively small (about the size of a match head) (**2.13–2.14**), weigh around 20–40 mg, and contain between 100–700 muscle fibres (**2.8**), although this varies with the skill of the operator. Some workers advocate the use of a suction mechanism attached to the needle during biopsy to obtain larger specimens.

The use of the chonchotome biopsy technique ('alligator forceps') has also been advocated (Coakley *et al.*, 1986), particularly for sampling the superficial portions of anterior leg muscles. The authors have no personal experience of this method which will not therefore be considered further.

Specimen handling

After removal from the patient, specimens should be dealt with by an experienced member of the laboratory staff. In the cases of biochemical analysis and immunoblotting, the sample should be immediately snap frozen in liquid nitrogen.

Preparation for light microscopy

Tissue for light microscopy should be placed on a non-absorbent surface in an airtight container, preferably

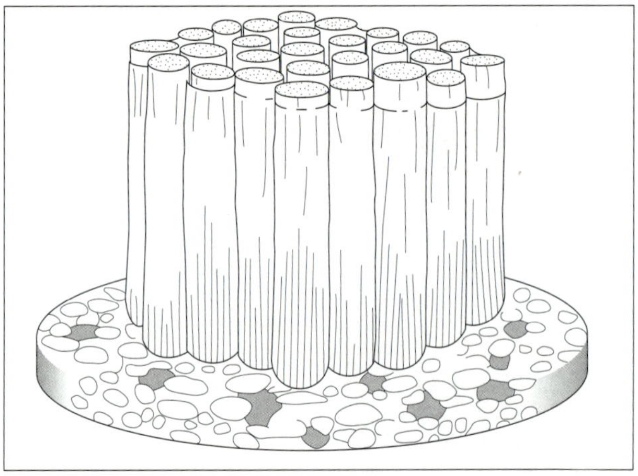

2.13 The optimum orientation of fibres for freezing and transverse sectioning in a needle biopsy sample.

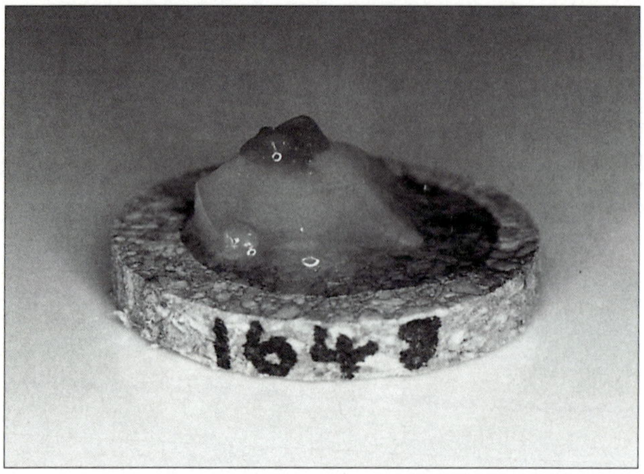

2.14 A specimen from a needle muscle biopsy, mounted in agar gel on a labelled cork disc, ready for freezing and microtomy.

Table 2.1 Histological and histochemical procedures.

This table indicates the range of procedures that are available to investigate the muscle biopsy. One procedure from each group is **mandatory**; the preferred procedure in each group is marked *. Frequently, several procedures will be required to completely elucidate any given problem. Detailed procedures are given in Appendix C.

Category	Method	Use
Morphology	Toluidine blue	General appearance and orientation
	Haematoxylin and eosin*	Muscle fibre architecture, nuclear position and number, degeneration and regeneration, hyaline fibres, fibrosis
	Van Giesson	Fibrosis
	Masson's trichome	Fibrosis
	Modified Gomori trichome	Nemaline rods and 'ragged-red' fibres
Fibre Typing	Myofibrillar ATPase*	Types I, IIa, IIb, IIc
	(pH 4.3, 4.6, 10.2)	Type grouping
Enzymes - Oxidative	NADH-Tr*	Mitochondrial and tubular aggregates
	Succinate dehydrogenase	Mitochondria
	Cytochrome oxidase	Mitochondria, enzyme deficiency
Enzymes - Glycolytic	Phosphorylase*	Enzyme deficiency
	Phosphofructokinase	Enzyme deficiency
Enzymes - Hydrolytic	Acid phosphatase	Inflammation, lysosomes
	Non-specific esterase*	Inflammation, lysosomes
	Acetylcholinesterase	Neuromuscular and myotendinous junctions
Storage	PAS*	Glycogen, carbohydrate
	Alcian blue	Mucopolysaccharide
	Sudan black B	Lipid
	Oil red O*	Lipid
Other	Methyl green pyronine	RNA
	Acridine orange	RNA
	Von Kossa	Calcium

moisturised with gauze soaked in isotonic saline. Small specimens may be prevented from excess drying by wrapping in metal foil. ***Under no circumstances should the specimen be placed in fixative.***

Ideally, the specimens should be left to relax for 10–15 minutes prior to freezing, which should itself be performed within an hour of removal. It is advisable to keep material cool in a refrigerator (not a freezer), if any delay is unavoidable. Tissue culture fluid provides an ideal transport medium in such circumstances. Packing in dry-ice is not advised since partial freezing and subsequent ice damage will occur.

Preparation for histology, histochemistry and morphometry

The basic aims are the optimal preservation of tissue architecture with minimal artefact formation, and the retention of enzyme activity and intracellular inclusions. Standard histological procedures (involving tissue fixation and paraffin embedding) are therefore ***not*** recommended for muscle tissue, since:

- Fibres may shrink by as much as 30% during processing.
- Enzyme activity is abolished or reduced.
- Certain storage products (e.g. glycogen and lipids) are not retained.

Bearing in mind that ***all*** methods are subject to the risk of artefacts appearing during sectioning, it is the inability to type muscle fibres in specimens prepared in this way that is the most serious drawback. This alone makes the process totally unacceptable in the light of currently available techniques.

Freezing the tissue for cryostat sectioning is therefore the preferred method of preparing muscle biopsy specimens. Since orientation of cryostat specimens is more difficult than that of conventional wax blocks, more care is required on initial orientation of the tissue for freezing. The maximum amount of useful information is obtained from muscle when sections are taken transversely to the long axis of the fibres (**2.8–2.9**, **2.13–2.14**).

For large, open biopsy specimens, the orientation of muscle fascicles is readily visible. The muscle can be dealt with relatively easily, by freezing them in a side-room off the surgical suite.

For the smaller, needle biopsy specimens, it is most convenient to orientate the tissue in the laboratory under a medium-power dissecting microscope, where the orientation of muscle fibres can be recognised. In practice, one places the specimens in a hollowed-out space in a block of 1% agar gel or chilled bovine liver, mounted on a labelled cork disc. The adherence of the cork to the gel, or directly to large specimens, can be assisted with viscous embedding compounds such as OCT (Tissue-Tek). Care must be taken not to envelop the tissue, as this insulates it from the freezing agent.

The mounted, labelled block, is then rapidly frozen by plunging it quickly and completely, into a good heat transfer agent (e.g. dichlorodifluoromethane, Arcton 12 ICI, or CFC-free alternative) for about 30 seconds. The cryogen should have been previously cooled to around -160°C in liquid nitrogen, or to -70°C using isopentane. Freezing is best achieved by pre-freezing the Arcton and using it when about two thirds of it has melted. Arcton is available in pressurised containers, but can be re-used if stored below its boiling point of -30°C.

Rapid freezing is essential for dense tissues such as muscle. It avoids the problem of ice crystal artefacts (**3.22**), which may form if samples are frozen directly in liquid nitrogen or cooled in isopentane chilled in dry ice. Neither the use of spray coolants, nor freezing by attachment to pre-cooled chucks is to be recommended, since the rate of cooling is too slow to avoid tissue damage due to the formation of ice crystals.

Frozen specimens are stored at -70°C until required for sectioning. For longer term storage they are best heat-sealed into labelled, lay-flat plastic tubing, to prevent them drying out through sublimation. They should be stored in a deep freeze or liquid nitrogen refrigerator that is fitted with an appropriate alarm system. Enzyme activity is preserved for many years if specimens are stored in this way.

When transferring specimens between freezers or to the cryostat, ***utmost care is required*** to avoid minor thawing or damage to the delicate cutting face of the tissue block. Thawing and refreezing causes extensive damage to the tissue due to the formation of ice crystal artefacts. If large specimens are available, it may be prudent to store some additional material permanently in the form of fixed, paraffin embedded tissue blocks, or for methacrylate embedding to provide thin (1μm) sections.

Sectioning of muscle is carried out in a cryostat, optimally at -22°C, although slightly warmer temperatures are recommended for very fatty specimens. About 20 minutes is sufficient to allow deep frozen blocks to adjust to cryostat temperature before cutting. During the cutting, the block may be attached to the chuck with OCT. Spray coolants (e.g. Lipfreeze) are best avoided, since they often thaw the specimen and introduce tiny ice crystals.

Approximately twenty, serial, 8–10 μm thick transverse sections are sufficient from each block to carry out the battery of staining techniques recommended. Sections should be taken at the start and part way through the series and stained immediately with toluidine blue or H&E (Appendix C) to check on tissue orientation, tissue preservation and the presence of relevant material (**2.8–2.9**).

It is preferable to collect specimens on clean, pre-labelled, numbered glass microscope slides rather than cover slips. Slides are more robust, more convenient to handle and, if the sections are mounted at one end, equally economical in the use of reagents. For immunocytochemical staining, sections should be mounted on slides subbed with chrome-alum, gelatine or poly-L-lysine.

Some of the more common artefacts induced by poor sectioning technique and tissue handling are illustrated in Chapter 3 (**3.21–3.25**). The histological and histochemical methods used for human skeletal muscle specimens are given in *Table 2.1*, and detailed in Appendix C.

The methods listed in *Table 2.1* are sufficient to demonstrate:

- Muscle fibre architecture.
- Connective tissue.
- Muscle fibre types and their distributions.
- The presence of stored material and defects.
- Deficiencies of muscle enzymes.

They also permit the identification and classification of inflammatory cells. The use of serial, numbered sections, allows individual muscle fibres to be 'followed' through the various staining techniques.

If the clinical notes or preliminary observations suggest, for example, a metabolic disorder, further histochemical methods are employed (*Table 2.1*, Appendix C).

Attention to details (e.g. molarity, pH, drying and incubation time) is important during the histochemical procedures. Glassware must be clean, as traces of fixative or other contaminants may inhibit the enzyme reactions. Every batch of muscle biopsy slides for histochemistry should include a normal, control 'marker' specimen, incubated in parallel; this is especially important in confirming enzyme deficiencies.

Furthermore, it is always useful to retain a few unstained sections in a separate deep freeze to be used for additional procedures, to send to other laboratories, or as an insurance in case the tissue blocks become irrevocably damaged.

Techniques that could be carried out on these serial sections are immunocytochemistry, immunofluorescence, phase contrast, polarising or electron microscopy.

Finally, it is important to be aware that certain stain reaction products, for example phosphorylase, fade rapidly, and the preparation must be examined and the results recorded immediately. ATPase preparations may also fade, though somewhat variably, over a few weeks or months, depending on the mountant employed. A photographic record is therefore recommended. In autopsy specimens phosphorylase activity and glycogen are usually absent, although other enzymes, including ATPase activity, are often well preserved up to 48 hours postmortem (Eriksson *et al.*, 1980).

Preparation for electron microscopy

Skeletal muscle can be prepared for electron microscopy using fixatives and embedding media that are in common use for other mammalian tissues. Double fixation, with glutaraldehyde followed by osmium tetroxide, is generally recognised as providing the optimal preservation of the ultrastructure of both normal and pathological tissue.

Muscle, being a highly organised and excitable tissue with a unique cytoplasmic structure, must be handled with particular care and attention to enable it to be examined in both longitudinal and transverse planes (**3.26–3.27**). If placed directly into fixative, muscle fibres will contract and distort (**3.41**).

To avoid this disorientation, the muscle is best fixed at its resting length by holding the fibres under slight tension. Allowing the specimen to relax for a few minutes in a moist chamber prior to fixation helps in this respect. Often, the time taken to transport the biopsy from the ward or theatre to the laboratory is sufficient. If possible, this time should be standardised. Care must be taken not to allow the surface of the specimen to become dry (**3.40**).

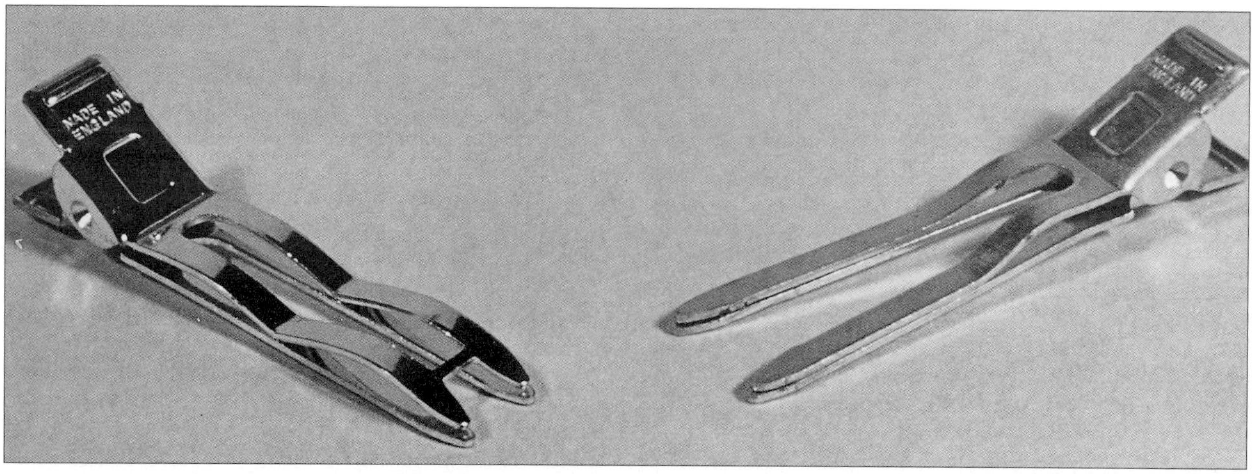

2.15 A modified hairgrip that can be used to hold the sample during fixation for electron microscopy. The bridge is cut out and the prongs are flattened and separated slightly.

A suitable clamp to hold the muscle during fixation can be made by the modification of a 'hairgrip' as illustrated in **2.15**. A small bundle of fibres is detached from the biopsy sample by cutting one end free and then inserting it into the clamp under slight tension (**2.16a**). The bundle is then cut free and placed in fixative (**2.16b**). It is just as important to avoid overstretching the fibres by applying too much tension, as it is to apply tension in the first place.

Where the sample is too small to be clamped, as, for example, when obtained using a biopsy needle, the muscle can be left to relax on a saline-moistened filter paper in a covered container for 15–20 minutes. The fibres can then be orientated under a dissecting microscope and attached to a small piece of card for support before immersion in fixative.

After primary fixation in glutaraldehyde (or in paraformaldehyde for ultrastructural immunocytochemistry) (**2.16b**, Appendix D) the muscle is freed from the clamp, the crushed ends are discarded, and the central portion cut into small strips about 1×3 mm, in the long axis of the fibres (**2.17**).

Following secondary fixation in osmium tetroxide, the specimen is dehydrated and impregnated with resin. The strips of muscle are embedded in a flat embedding mould in such a way that sections can be cut in either the

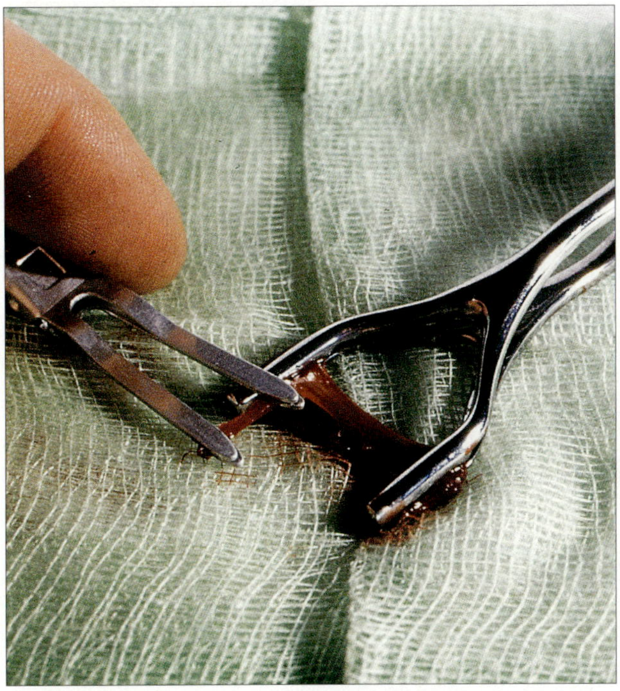

2.16a A small bundle of muscle fibres is freed from the biopsy (see **2.5**) and gripped in the clamp.

b) The sample is then immersed in glutaraldehyde fixative.

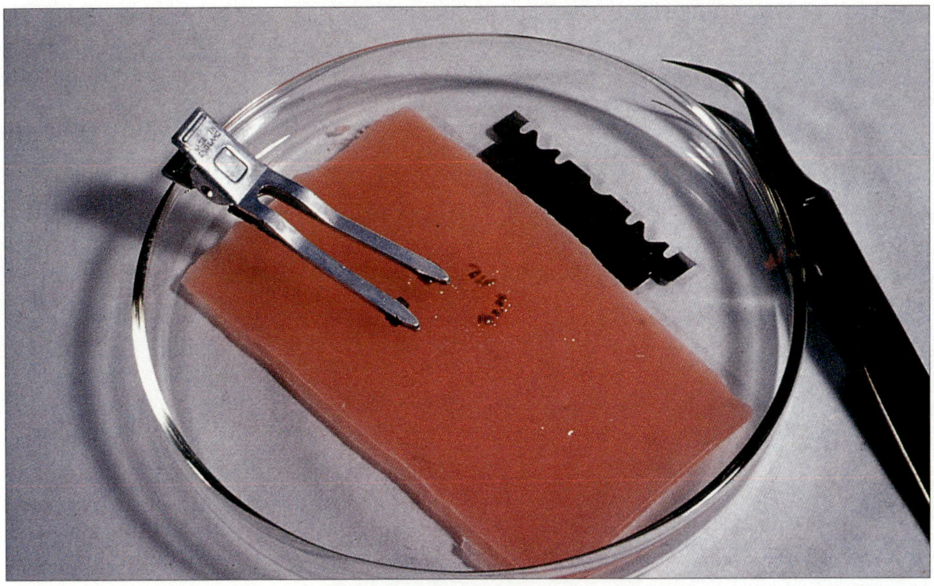

2.17 The fixed tissue is trimmed into small strips prior to dehydration and impregnation with resin. Tissue crushed by the clamp must be discarded.

longitudinal or transverse planes of the fibres (**2.18**). Survey sections are cut at 0.5–1.0 μm thickness, and stained on a hotplate with 1% toluidine blue in 1% borax.

Examination of these unmounted sections using an optical microscope equipped with objectives that are not corrected for use with a cover slip, allows rapid screening of the specimen for areas of particular interest (**2.19–2.20**). The sections for electron microscopy are cut at about 60 nm and double stained with uranyl acetate and lead citrate.

2.18 Small strips of muscle are embedded in epoxy resin to allow sections to be cut in either transverse or longitudinal planes.

2.19 Longitudinal sections of resin-embedded muscle, from which appropriate areas are selected for electron microscopy.
a) Normal.
(1 μm section, toluidine blue, ×4200);

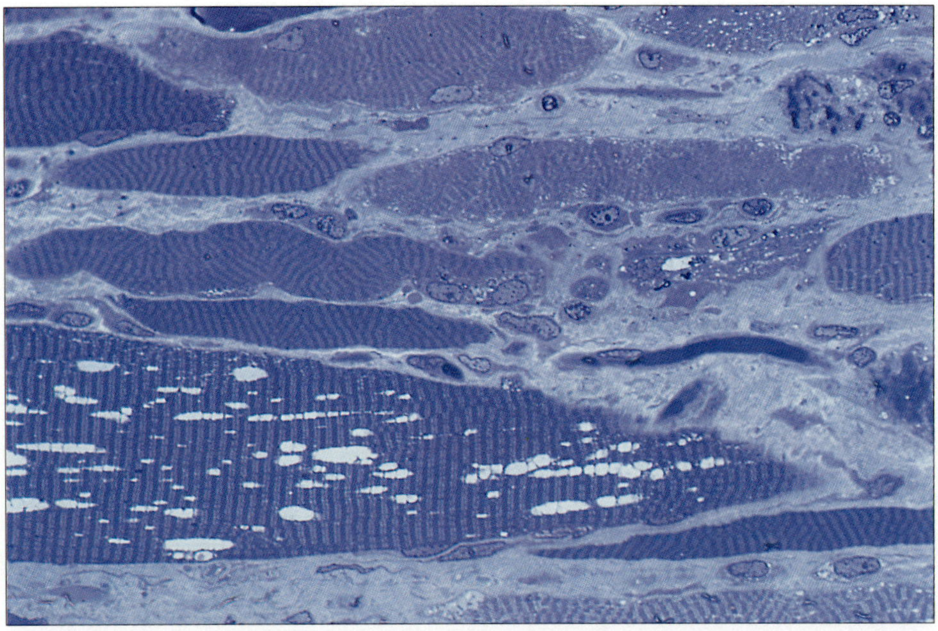

b) Pathological material viewed by light micrograph.
(1 μm section, toluidine blue, ×3850).

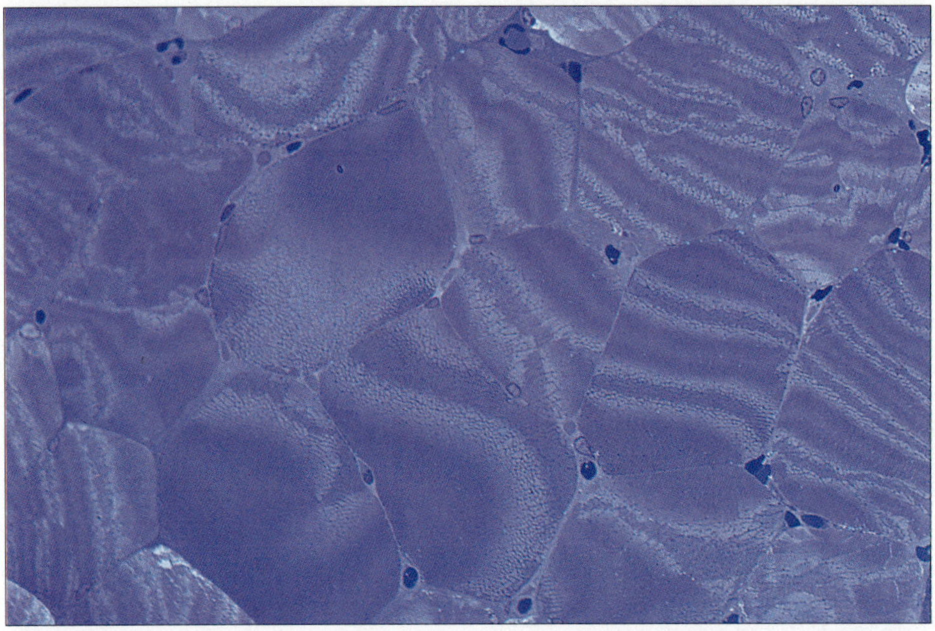

2.20 Transverse sections of muscle viewed by optical microscopy.
a) Normal.
(1 μm section, toluidine blue, ×4200);

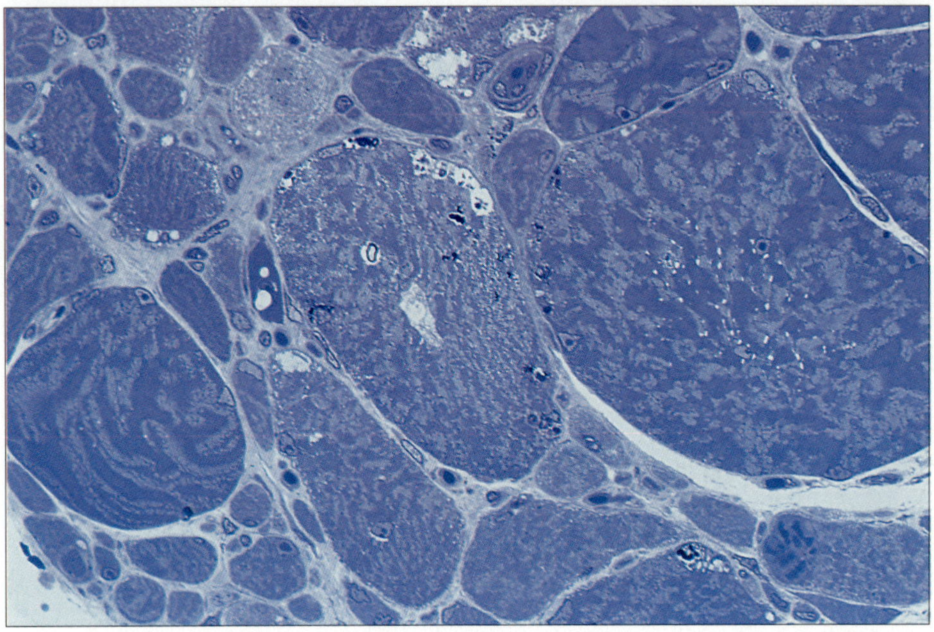

b) Pathological.
(1 μm section, toluidine blue, ×4200).

The use of immunocytochemistry

The evaluation of Xp21.2 myopathies (formerly Duchenne and Becker muscular dystrophy) ***requires*** the demonstration of an abnormality of the expression of dystrophin (see Chapter 5). Hence, immunocytochemical techniques, which were previously the province of the muscle research laboratory, are now mandatory in the 'routine' laboratory (see Appendix D).

The commercial availability of monoclonal antibodies that distinguish leucocyte subsets and MCH antigens permits further elucidation of the inflammatory myopathies. Antibodies to myosin isoforms and other cytoskeletal proteins are also becoming increasingly iportant in the analysis of the neuromuscular disorders.

Whereas monoclonal antibodies are of most use with fresh material, some polyclonal antibodies may reveal useful information in the re-analysis of previously stored, fixed (including paraffin fixed) material.

Other methods

Detailed preparation of muscle tissue for other specialised but less commonly used methods, such as biochemical analysis, immunoblotting, cell culture and *in situ* hybridisation, is beyond the scope of this book. The reader is referred to specialised texts for further information. These methods are currently the tools of the research worker and reference laboratories, and their employment in a routine clinical biopsy service is rare on account of difficulty in standardisation, interpretation of results and expense.

The appearance of light microscopical and ultrastructural muscle biopsy specimens from normal healthy individuals is described in the following chapter, where the qualitative and quantitative methods for their analysis are discussed. It must, however, be emphasised again, that care in preparation and interpretation of the muscle biopsy sample in relation to clinical data is vital in the proper diagnosis of neuromuscular disorders.

3

Normal muscle structure and analysis of the biopsy

Except where stated, all of the specimens shown in this chapter are taken from the quadriceps of normal healthy adults, and were taken by needle biopsy before being frozen.

Normal muscle structure

In describing the normal histological, histochemical and ultrastructural appearances of human skeletal muscle, it is important to bear in mind throughout that many aspects of muscle structure differ both qualitatively and quantitatively between:

- Individuals, e.g. varying with age, sex or activity.
- Different muscles from the same subject, according to the function of a particular muscle.
- Different regions within an individual muscle.
- Adjacent muscle fibres according to their motor unit characteristics.

Furthermore, since the course of many neuromuscular diseases follow preferred anatomical and temporal distributions, the importance of knowing both the exact site of muscle biopsy and the relevant clinical details before drawing conclusions regarding clinicopathological diagnoses cannot be overemphasised.

Normal muscle histology

In limb muscles, the muscle fibres are arranged in bundles or fascicles, surrounded by perimysial connective tissue. Some intermuscular variation in fascicular size, organisation and amount of connective tissue may occur, depending on the exact topography (i.e. parallel or pennate arrangement) of a particular muscle, which in turn depends on the forces required of it during normal function (for review, Williams *et al.,* 1989).

The muscle architecture and muscle fibre morphology are shown to best advantage in transverse cryostat sections, stained with H&E or trichrome (**3.1–3.3**). The detail shown in **3.1–3.4** is from near-transverse, serial sections of frozen specimens obtained by needle biopsy.

Within fascicles, muscle fibres lie closely packed with little intervening endomysial connective tissue. Both layers of connective tissue should be virtually free of fat and leucocytes. Neurovascular bundles are located in the perimysium, between fascicles, and capillaries are found in the endomysium surrounding muscle fibres.

In the authors' study of quadriceps femoris muscle samples from 80 healthy individuals (aged 1–84 years), the total connective tissue content (including blood vessels

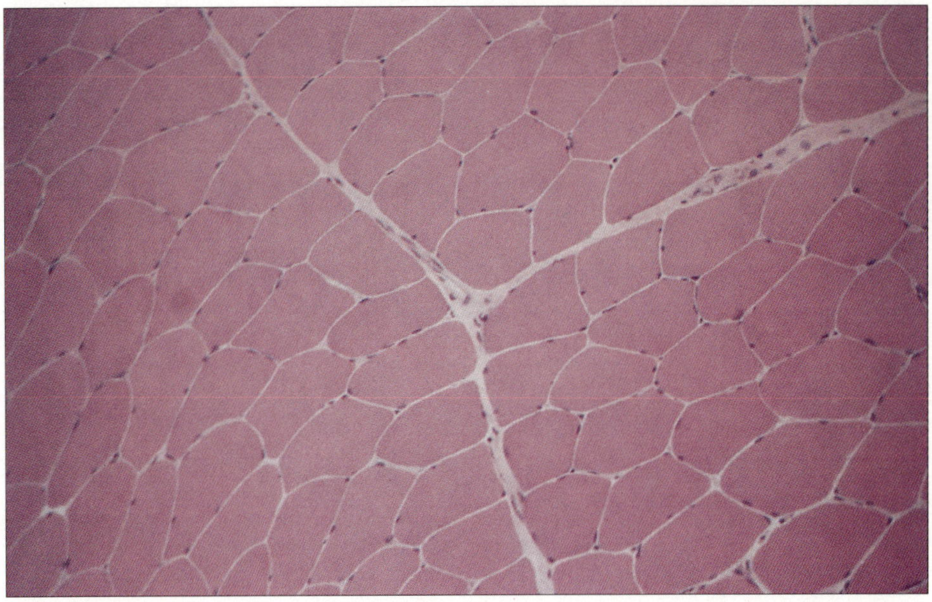

3.1 Normal muscle architecture shows closely packed polygonal muscle fibre profiles. There is little variation in muscle fibre size or shape. Cytoplasmic staining is uniform, and small, peripherally located nuclei are abundant. (H&E, ×140)

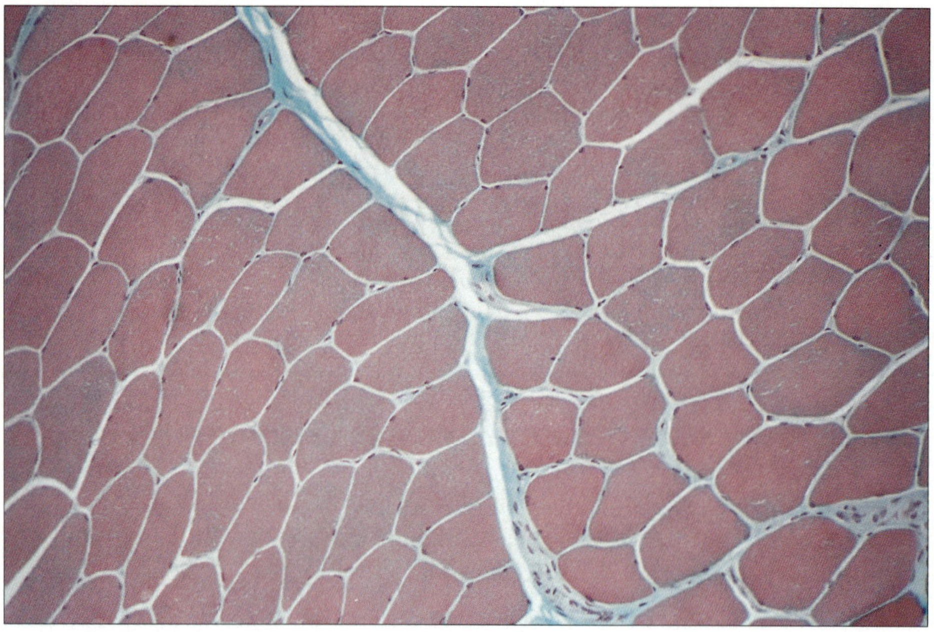

3.2 Scant endomysial connective tissue (stained green) containing some capillaries is present between muscle fibres.
Perimysial connective tissue between the fascicles is also sparse. All muscle fibres show similar dense staining.
(Masson's trichrome, ×140)

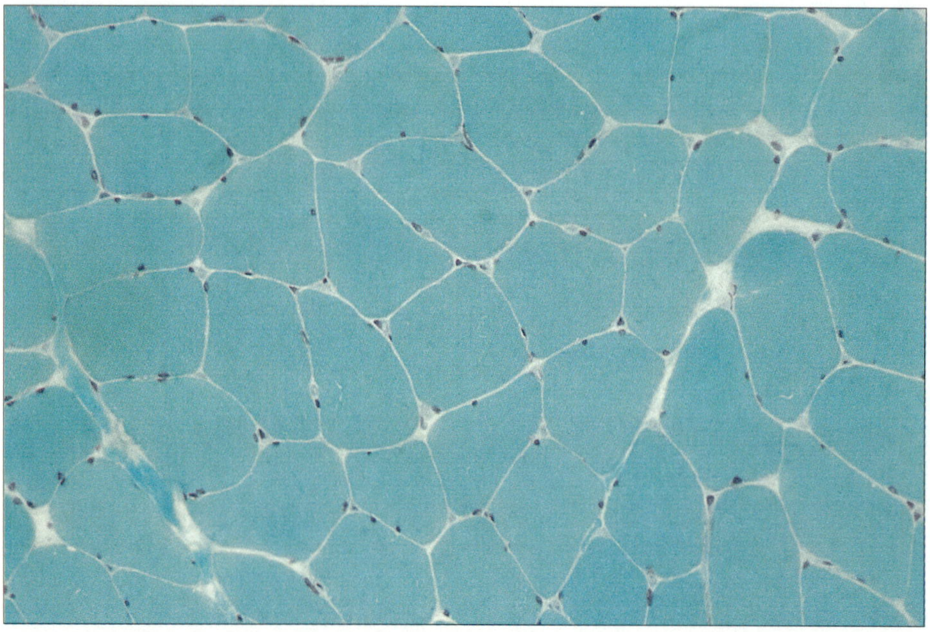

3.3 This reveals similar features to **3.1** and **3.2**.
Normal cytoplasmic staining is pale green, but nuclei and occasionally sites of high mitochondrial density stain red.
(Modified Gomori trichrome, ×217)

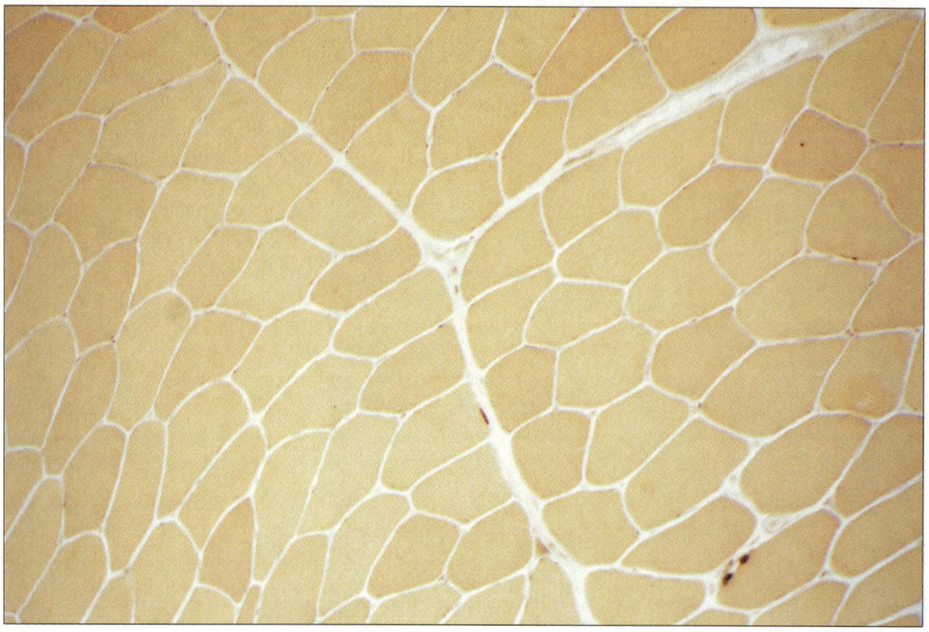

3.4 The background yellow-orange staining indicates practically no enzyme activity within the muscle fibres or amongst the connective tissues. Commonly, Type I fibres appear slightly darker than the Type II fibres. The dark brown deposits on the periphery of one muscle fibre indicate the site of a neuromuscular junction.
(Non-specific esterase, ×140)

and nerves) varied from 8–26%, with a median of 12%. Excess endomysial connective tissue, particularly between muscle fibres, would be considered pathological.

Skeletal muscle fibres are elongated multinucleated cells, with a regular arrangement of internal organelles. For light microscopical analysis they are best viewed in transverse section, where, under normal conditions of close packing, they appear polygonal and have smooth outlines (**3.1–3.4**) with a shape factor (*see below*) of around 0.9. Although the average muscle fibre size varies according to sex, age, site of biopsy and method of measurement (*see below*), within a typical biopsy sample the coefficient of variation for fibre diameter should not usually exceed 25%. Data for mean fibre sizes for the different fibre types is given in the next section.

Myonuclei appear as small, flattened, dark staining profiles, located directly beneath the sarcolemma (**3.1–3.2**). A typical number of nuclei per fibre in a cross-sectional profile is 2 or 3. In cryostat sections, myonuclei cannot usually be distinguished from similarly located satellite cell nuclei (see below). Variation in nuclear shape, density or position are important indicators of pathology. About 4% of fibres in normal muscle may possess internally located nuclei, and this value may be much higher near myotendinous junctions.

Cytoplasmic (eosin) staining should be homogeneous, without visible presence of vacuoles or inclusions. Areas of punctate basophilia may occur as interfibrillar granules, particularly in muscles from infants. However, cytoplasmic basophilia seen as diffuse, granular, or sub-sarcolemmal rims should be considered to be pathological.

The regular fascicular arrangement and smooth fibre outline may appear disrupted in places close to neuro-vascular hila or tendinous attachments. This should not be confused with pathological change. For this reason, non-specific esterase and acetylcholinesterase staining is particularly useful in demonstrating neuromuscular and myotendinous junctions (**3.4–3.5**).

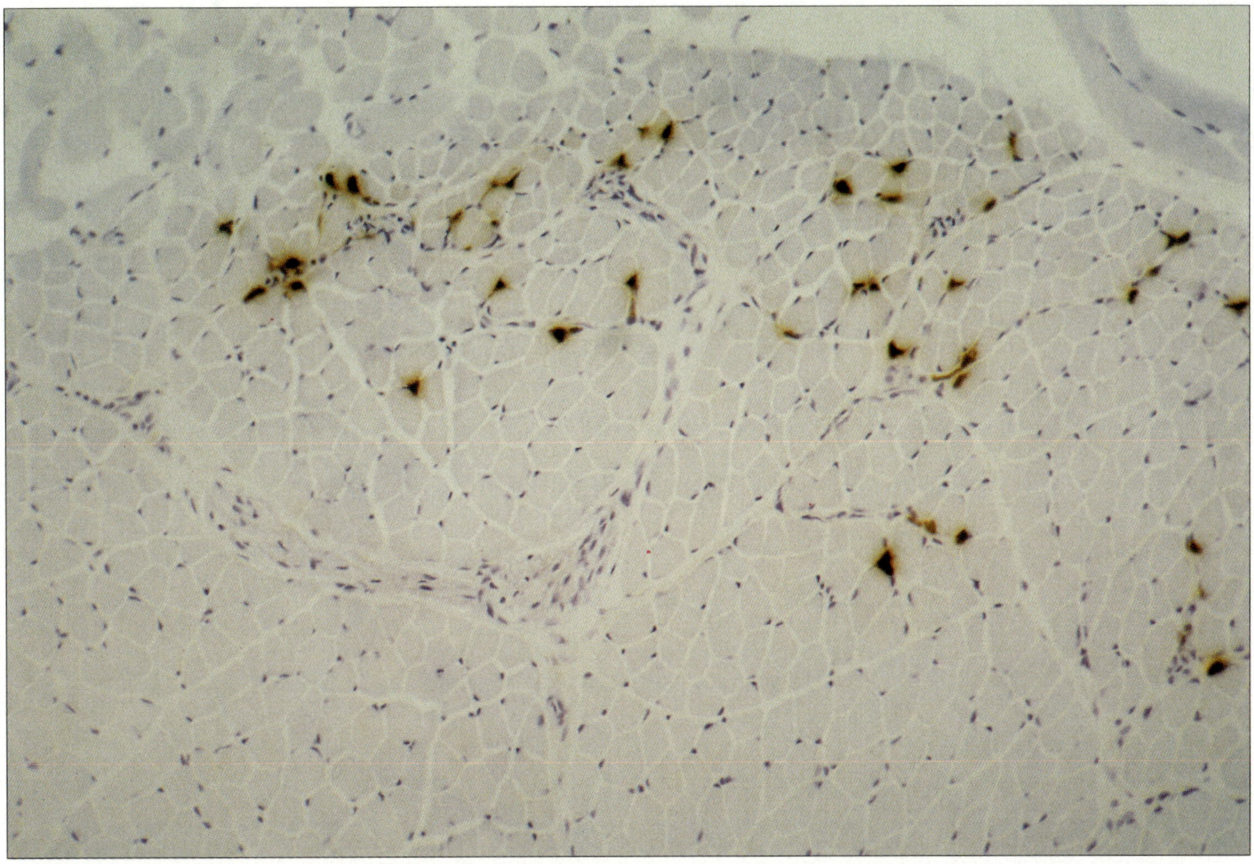

3.5 Enzyme activity (revealed by dense brown deposits), present only on the 'corners' of some muscle fibres in this biopsy indicates the sites of neuromuscular junctions. The staining is usually limited to a small region within the specimen, and marks the innervation band. (Acetylcholinesterase, ×189)

Normal muscle histochemistry

Muscle fibres innervated by a specific motor neurone are said to belong to the same motor unit. These fibres may be distributed fairly widely across a number of fascicles. The sizes of these motor units may vary from 10–2000 fibres between muscles, depending on the degree of neural control. However, the muscle fibres within a single motor unit function together and have similar metabolic and contractile properties that may differ markedly from those of fibres belonging to neighbouring motor units (**3.6**).

These differences in metabolic and contractile properties between muscle fibres may be revealed using histochemical techniques, and provide a spatial picture of the fibres' distribution within the muscle section. Although the use of numerous staining methods has enabled detailed histochemical profiles to be established, it has also lead to a confusing nomenclature for the different types of muscle fibre (*Table 3.1*).

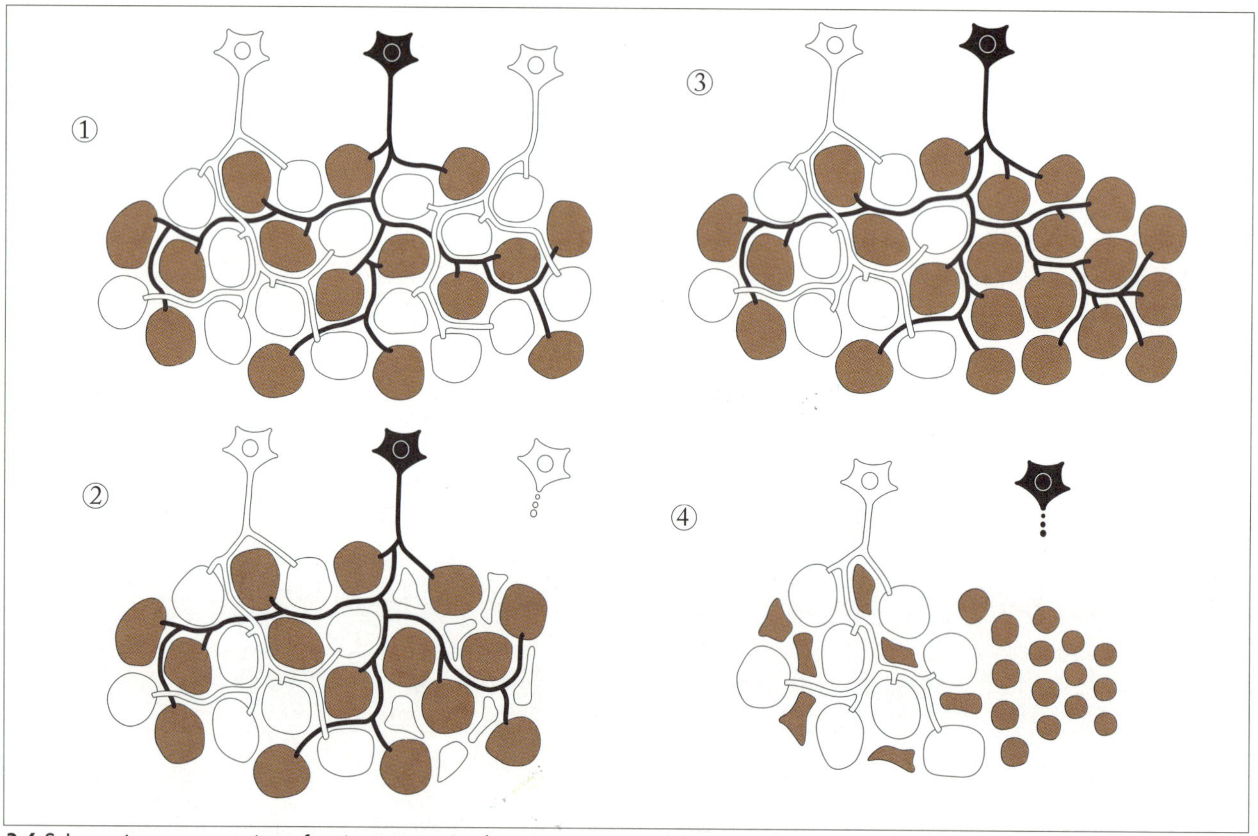

3.6 Schematic representation of various stages in chronic processes of denervation and reinnervation.
1) Three motor units: two Type I units (white) and one Type II (black).
2) Partial denervation leads to disseminated fibre atrophy.
3) Successful reinnervation by collateral sprouts from a nearby intramuscular axon. The mosaic pattern is replaced in part by a group of histochemically uniform Type II fibres.
4) The enlarged motor unit is denervated in its turn. There is a group of histochemically uniform atrophic fibres. Other atrophic fibres belonging to the same motor unit lie disseminated between nonatrophic fibres.

Table 3.1 The major fibre types in human skeletal limb muscles

	Type I		Type IIa		Type IIb		Type IIc	
Other names	Slow-twitch SO Red SR		Fast-twitch FOG Intermediate FR		Fast-twitch FG White FF		Transitional Undifferentiated	
Staining properties								
Oxidative enzymes (NADH-Tr, SDH)	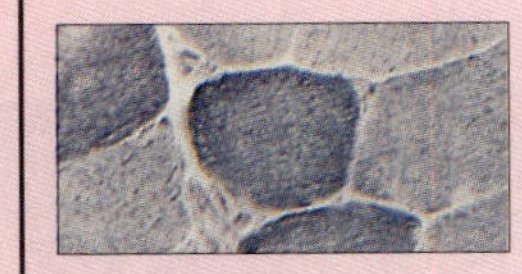	Dark		Medium		Pale		Medium/ dark
Glycolytic enzymes (phosphorylase)		Pale (yellow)		Medium/dark (brown/blue)	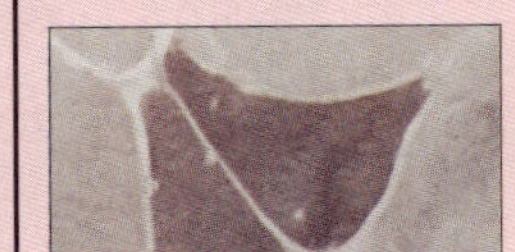	Dark (blue)		Medium
ATPase alkali **pH** 9.4, 10.2	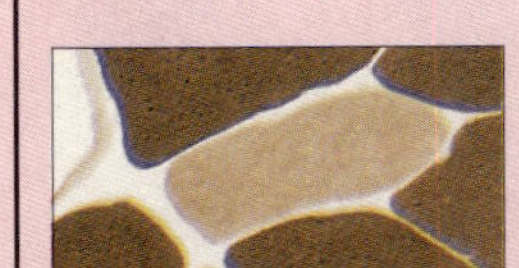	Pale		Dark	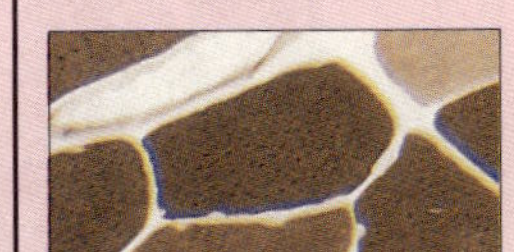	Dark	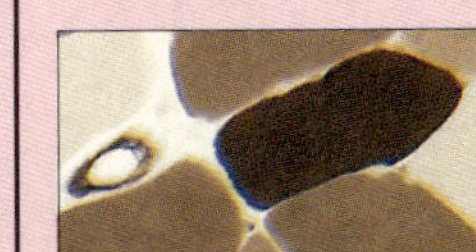	Dark
alkali fixed	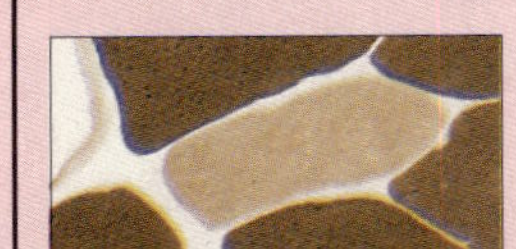	Pale		Darkest		Dark		Dark
acid **pH** 4.6	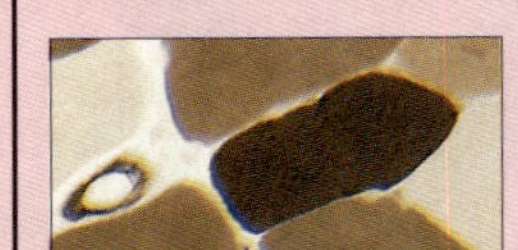	Dark		Pale		Medium	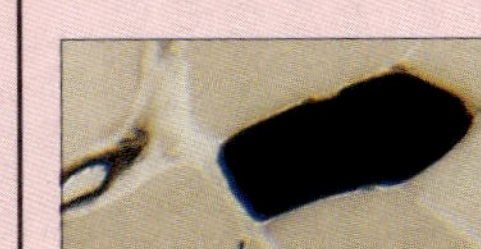	Dark
acid **pH** 4.3		Dark		Pale		Pale	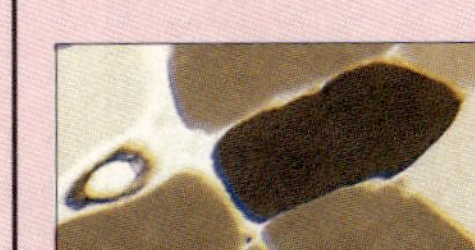	Medium/ dark

Abbreviations:

SO Slow Oxidative
FOG Fast Oxidative Glycolytic
FG Fast Glycolytic
SR Slow, Fatigue Resistant
FR Fast, Fatigue Resistant
FF Fast, Fast Fatiguing

Serial sections from the same specimen as **3.1–3.4** illustrate the various types of muscle fibre (**3.7–3.10**).

The two major types of fibre (slow-twitch Type I and fast-twitch Type II) are usually distinguished by the myofibrillar ATPase reaction at an alkaline pH (**3.7**). However, the results of this method may vary, or a pathological condition may render the fibre's type indistinct. It is therefore standard practice to take advantage of the property of the myofibrillar ATPase reaction to show different staining patterns at different pHs, and carry out the reaction on serial sections, following acid preincubation conditions, usually at pH 4.35 (**3.8**).

Further differentiation of a fibre's type, specifically into subclasses of Type II fibres, may be achieved by adding information from other pH preincubations (**3.9**) or by taking account of metabolic activity with NADH-Tr (**3.10**). By careful cross-correlation, the major fibre types may be more clearly differentiated (*Table 3.1*), and reliable estimates of their percentage frequency of occurrence evaluated (*Table 3.2*).

In general, postural or more continually active muscles, such as those of the calf, glutei and back have higher contents of Type I fibres than phasic muscles, such as the lateral portion of quadriceps. Variations also occur through the muscle depth, with a tendency to increased frequency of Type I fibres nearer the underlying bone (for detailed review, see Dubowitz, 1985; Loughlin, 1993).

A small percentage of fibres may not clearly fall into the main categories outlined above. Most of these fibres stain darkly with ATPase following both acid and alkaline preincubation. These have been designated Type IIc, and may represent fibres with intermediate or

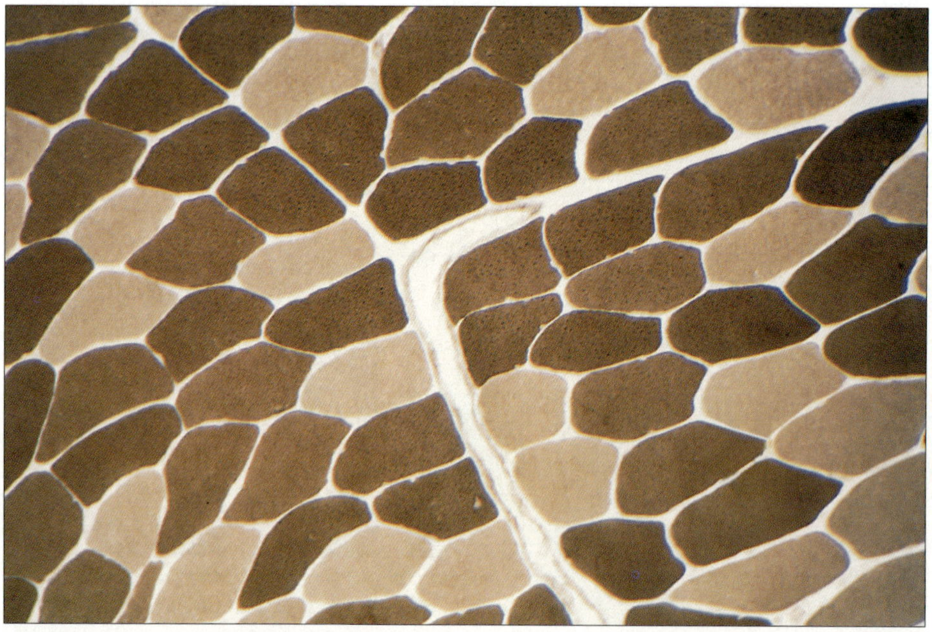

3.7 This shows the clear distinction between Type I ('histochemically slow') muscle fibres, which are palely stained, intermingled with Type II ('histochemically fast') muscle fibres, which have a dark brown stain.
Type II fibres predominate over Type I fibres in a ratio of 2:1. Staining is uniform, and intracytoplasmic organelles and connective tissue elements are not distinguishable.
(Myofibrillar ATPase, pH 10.2, ×140)

3.8 The 'reverse' ATPase reaction' stains Type I fibres very darkly and leaves Type II fibres unstained. Blood vessels in the connective tissue spaces also show positive staining.
(Myofibrillar ATPase, pH 4.35, ×140)

3.9 The 'additional' acid reaction enables differentiation of Type II fibres into Type IIa (unstained and pale brown) and Type IIb (brown). These occur in roughly equal proportions. Type I fibres remain as the darkest staining fibres. (Myofibrillar ATPase, pH 4.6, ×140)

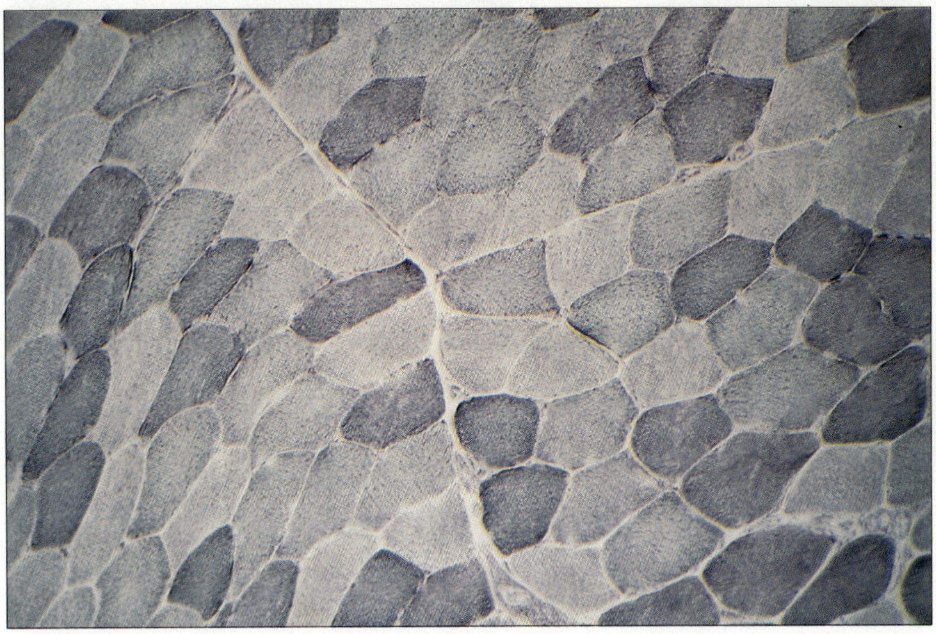

3.10 The presence of oxidative activity is shown as granular, blue-stained deposits at the sites of mitochondria.
The darkest staining fibres correspond to the Type I fibres in the ATPase sections. Fibres with intermediate activity are Type IIa fibres.
Note the relatively homogeneous distribution of stain deposits, save for occasional subsarcolemmal clumps in Type I and IIa fibres, both within and between fibres of the same histochemical type.
(NADH-Tr, ×140)

Table 3.2 Human vastus lateralis: morphometric data for young adult males

	All	Type I	Type IIa	Type IIb
Fibre diameter μm[1]	62	56	67	64
% number	(96)[2]	40	32	24
% total muscle area[3]	(97)	37	35	25

1 Coefficients of variation were approximately 20%
2 Approximately 4% of fibres did not correspond to the three main types.
3 Connective tissue accounts for 10% of the total sectional area (see Mahon *et al.*, 1984 for a review)

transitional properties (e.g. following reinnervation or regeneration) (**3.11**). They are commonly observed in infants, and may represent up to 10% of fibres.

Recently, immunocytochemical methods have been introduced which allow not only the accurate identification of different fibre types (**3.12**), but also the detection of pathological or developing fibres that possess more than one type of myosin. These methods also provide for visualisation of other cytoskeletal proteins (e.g. dystrophin) (**3.13**) and are now part of routine diagnostic investigations.

The distribution of fibre types in normal human skeletal muscle has undergone considerable study, because the detection of any deviation may be important in determining underlying pathological processes (Chapters 4 and 6). The normal arrangement has been variously described as checkerboard, random or hyperdispersed, and has a tendency for more Type II fibres at the periphery of fascicles in the commonly biopsied muscles (**3.7–3.10**, **3.12–3.15**).

An abnormality of this distribution, as seen in the tissue section, may be the result of perturbation occurring in the motor unit as a whole, or at any particular level. It cannot be overemphasised that consideration of the motor unit concept of muscle structure, as revealed by the histochemical methods described above, is vital in many

a

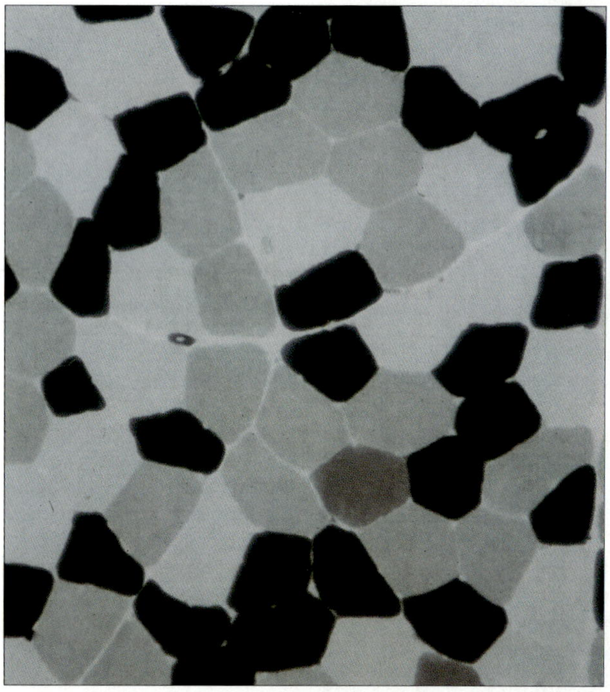

b

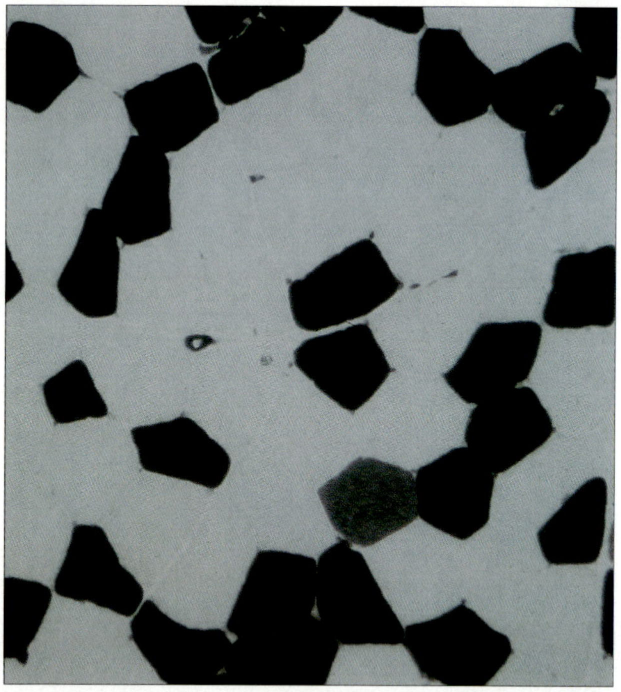

3.11 Type IIc fibres are darkly stained, and represent a few percent of normal muscle fibres. They are identified both by their lack of 'reversal' between alkaline and acid preincubations and by their prominence at intermediate pHs of the myofibrillar ATPase reaction.
a) Myofibrillar ATPase, pH 4.6, ×36;
b) Myofibrillar ATPase, pH 4.35, ×36.

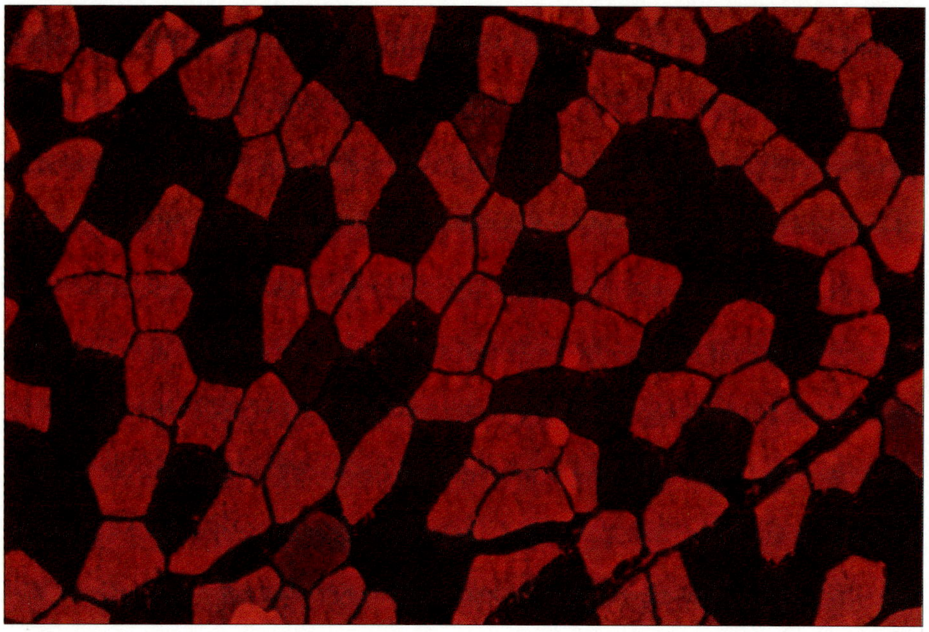

3.12 Immunocytochemical labelling for fast myosin reveals the clear differentiation between muscle fibre types. (Anti-fast-myosin, Texas red fluorescence ×140)

3.13 Immunocytochemical localisation of dystrophin, where labelling is continuous around the muscle fibre membrane.
(Antidystrophin, peroxidase ×140)

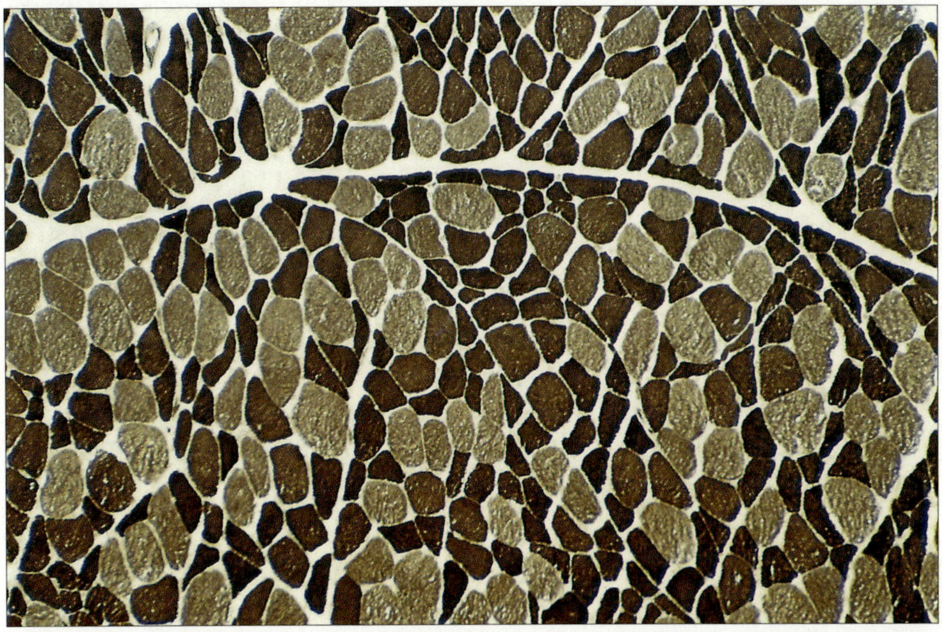

3.14 Muscle biopsy from a 65 year old showing atrophy of Type II fibres, which occur in small groups in some areas.
(Myofibrillar ATPase, pH 10.2, ×70)

3.15 Distribution of muscle fibre types in a muscle biopsy from a 15 year old boy showing the normal checkerboard appearance.
(Myofibrillar ATPase, pH 4.6, ×70)

instances to making the correct clinicopathological diagnosis (Chapter 4). A major advantage of using cryostat sections and histochemical staining in myopathology, is the ability to detect storage products and enzyme deficiencies or excesses within individual muscle fibres, and to be able to relate these to specific fibre types (**3.10**, **3.16–3.18**).

In normal muscle, most of these methods give homogeneous staining within the fibre cross-section, albeit with differences in staining intensity between the fibre

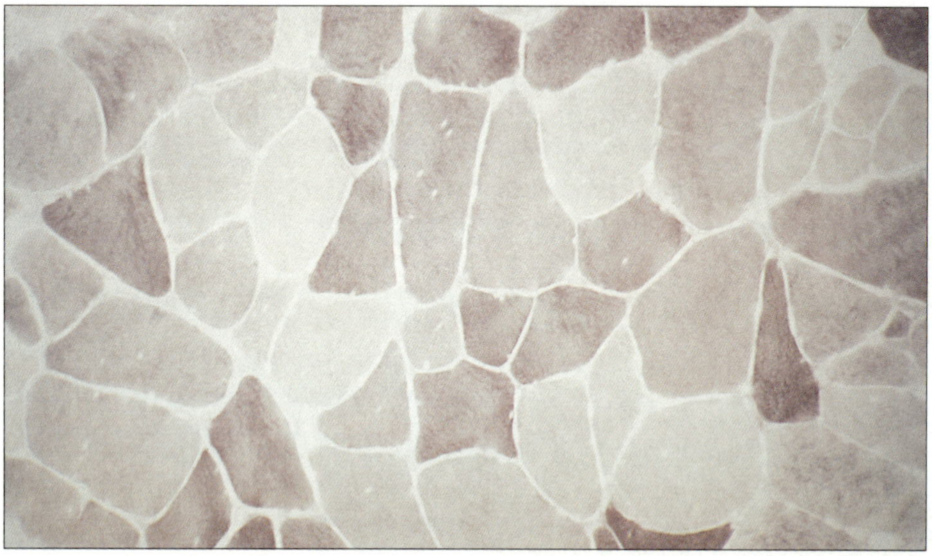

3.16 Blue to purple staining indicates Type II fibres, which have a high level of glycolytic enzyme activity. The background, pale-yellow staining of Type I fibres indicates a lack of glycolytic enzyme activity. Intermediate activity (frequently Type IIa fibres) is marked by various shades of brown staining.
Normally, staining should be homogeneous and nongranular. Occasionally however, thread-like networks, corresponding to the intermyofibrillar sarcoplasmic reticulum, are apparent at higher magnifications.
(Phosphorylase, ×140)

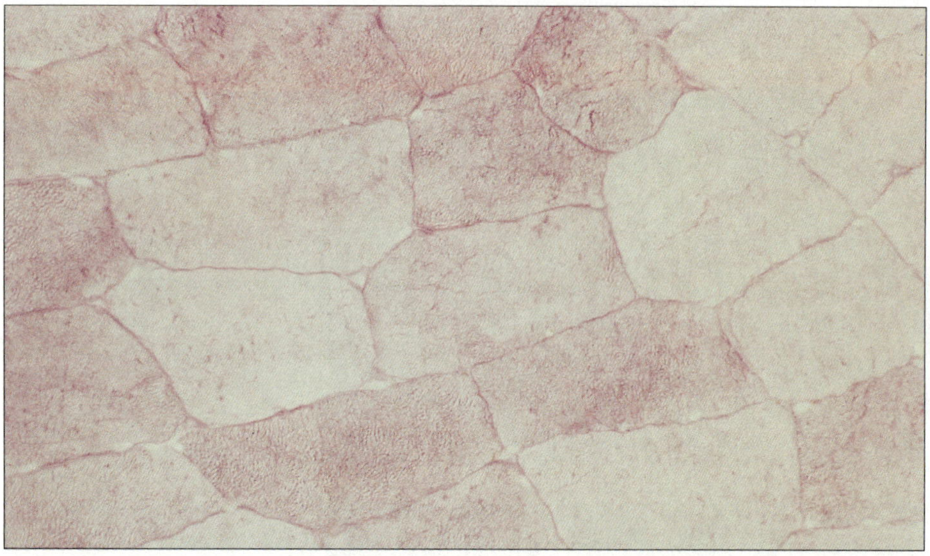

3.17 The storage of glycogen and associated carbohydrates within the muscle fibres is indicated by the density of the pink staining. Note that the fibres have variable staining, and that the intrafibre distribution is homogeneous and does not show subsarcolemmal accumulations.
At high magnifications, the network-like sarcoplasmic distribution of glycogen is visible.
(PAS, ×350)

3.18 Fine, evenly distributed black deposits indicate sites of stored lipid and dense collections of mitochondrial membranes. Some degree of differentiation into oxidative and non-oxidative fibres is apparent, although this is unreliable for fibre-typing purposes.
(Sudan black B, ×217)

types. Conversely, variations in staining may be indicative of a pathological condition (see **4.8**, **4.44**, **4.47**, **4.50–4.53**, **4.57–4.59**, **4.61** and Chapter 9). One of the oxidative enzyme reactions is employed as standard, and other reactions (e.g. cytochrome oxidase) are used if an abnormality is detected, or if the provisional clinical report is indicative of a mitochondrial disorder. Other methods, relating to glycolytic metabolism or the analysis of storage products and inclusion bodies, are dealt with in Chapter 9.

Anomalies in muscle structure

One must consider the potential misinterpretation of less commonly encountered features such as:

- Muscle spindles and myotendinous junctions.
- Artefacts (which are irreversible) that may have been produced during the sampling, transport or preparation of the biopsy.
- Improperly performed or controlled staining reactions, where the results may be considered as pathological by the inexperienced observer.

Emphasis has already been given to interpretation of the biopsy in relation to its site and the status of the subject from whom it was obtained. For instance, compare **3.14**, obtained from a healthy individual aged 65 years, to **3.15**, obtained from a 15 year old. Ageing muscle fibres (particularly of Type II) regularly show signs of atrophy, denervation, reinnervation and degeneration. Special consideration must be given to establishing the normal parameters when assessing biopsies from non-limb muscle.

The presence of neuromuscular junctions that disrupt the fibre outline has already been mentioned (**3.5**). Muscle spindles, consisting of small unequal-sized intrafusal fibres (up to 20μm in diameter) surrounded by a connective tissue capsule (**3.19**), must not be confused with atrophic fibres or fibrotic foci. When cut in the transverse section at their extreme polar ends, where they lack a capsule but sit within the perimysium, spindles are even more likely to confuse. At the equatorial level, the single or multiple internal nuclei may be seen.

Fortunately, muscle spindles are rarely encountered in the adult quadriceps, but are more common in postural and oxidative muscles. They are also more prominent in fetal muscles and are often preferentially preserved (i.e. appearing relatively large) in otherwise atrophic muscle. Myotendinous junctions are also a possible source of misinterpretation, since their invaginated sarcolemma and interdigitation of collagen fibres resembles fibre splitting (**3.20**).

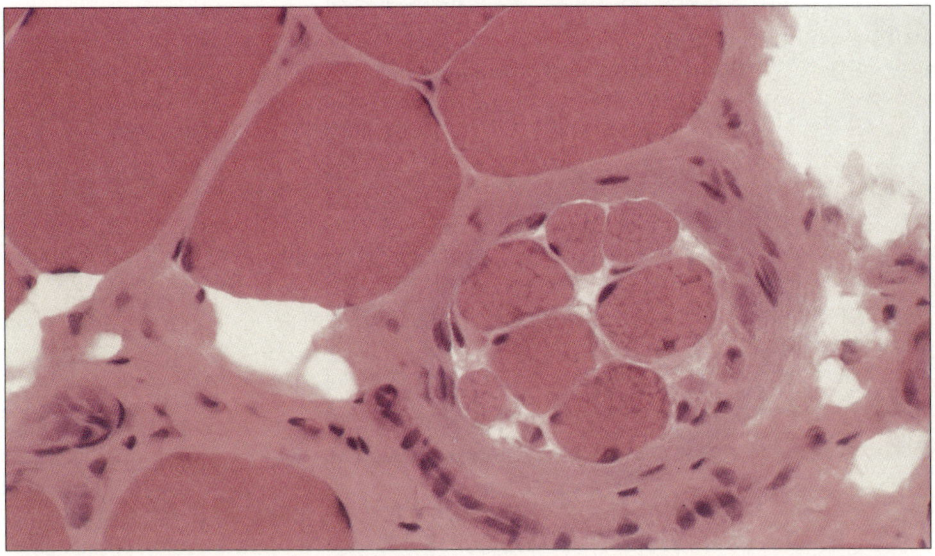

3.19 Muscle spindles are infrequently present in biopsies from large limb muscles, and may be confused for pathological features. They consist of 5–20, small diameter, intrafusal muscle fibres, enclosed for most of their length by a connective tissue capsule.
Normal features of these structures are great variability in intrafusal fibre size and histochemical type, centralised myonuclei, high esterase activity and prominent nervous tissue elements.
(H&E, ×438)

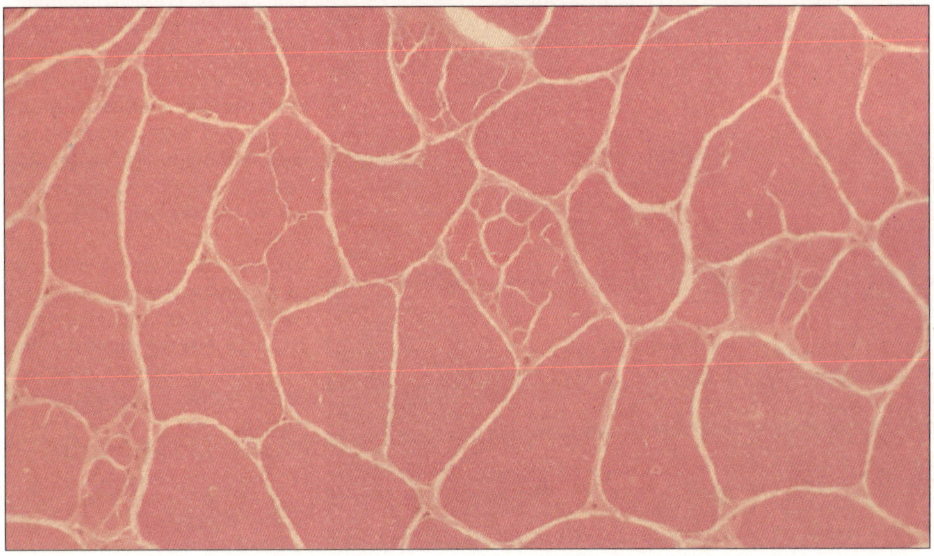

3.20 Many muscle fibres of relatively normal size show fragmentation and apparent features of fibre splitting.
These fibres are associated with excess, pale-pink staining endomysial connective tissue on their boundaries, and represent normal myotendinous junctional regions.
The splits within each fibre are irregular and multiple, and are not directly associated with myonuclei, although centralised nuclei can occur.
(H&E, ×350)

Preparation artefacts

Some of the common artefacts introduced by poor technique are illustrated in **3.21–3.25**. Their resemblance to a number of pathological features seen in later chapters emphasises the comments made in Chapters 1 and 2 concerning the organisation of 'muscle pathology laboratories'. In particular, the observer must be wary of apparently abnormal muscle fibres found only on the edges of the biopsy sample, which may arise from mechanical damage or drying. The manifestation of artefacts is further discussed in the next section.

It should be noted that **3.20–3.23** are not samples of normal, healthy muscle.

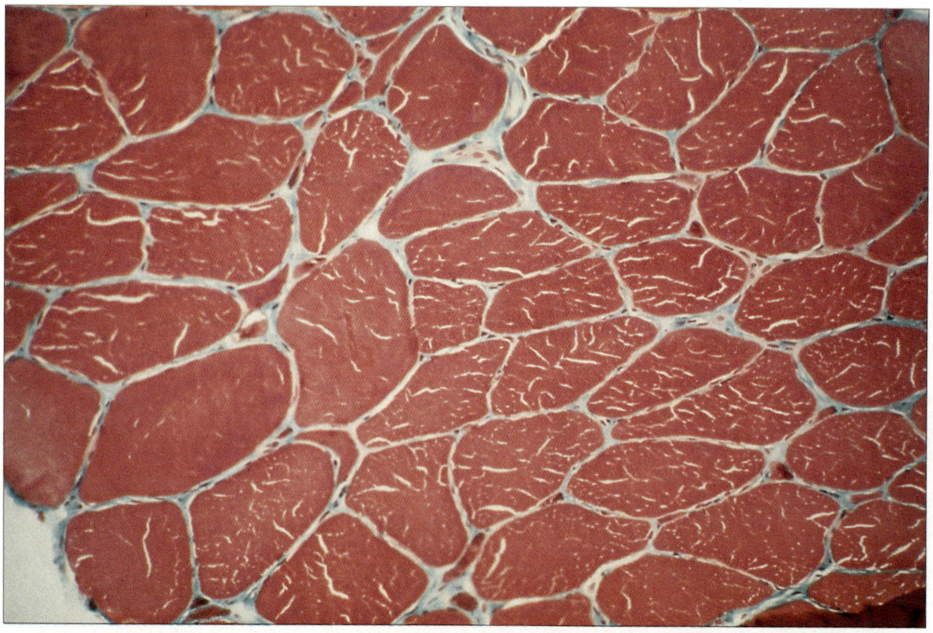

3.21 The splits, seen in nearly all muscle fibres of this section, are artefacts. These irregular tears are caused by sectioning with a blunt cryostat knife.
Some atrophic fibres are present in this abnormal muscle biopsy.
(Masson's trichrome, ×217)

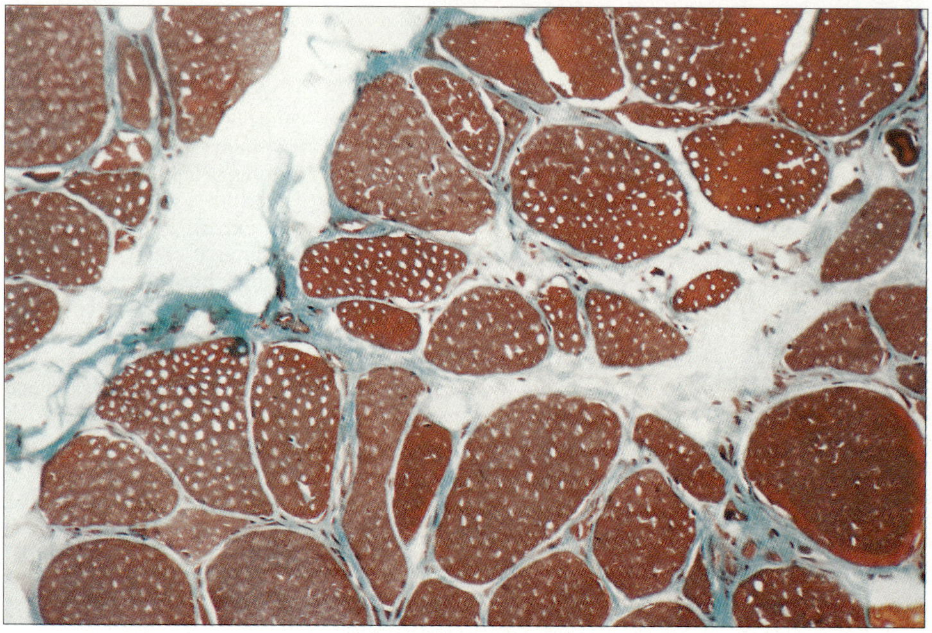

3.22 Ice crystal artefacts caused by slow freezing of the specimen appear as regular rounded holes within the muscle fibres. Ice crystal spaces caused by thawing and re-freezing of the specimen are usually much larger and often disrupt the fibre outline. In both cases, the artefact may be localised or occur throughout the tissue block.
Subsequent staining of serial sections for storage products confirms these 'apparent vacuoles' as empty spaces. The increased variability in the fibre size is indicative of abnormal muscle.
(Masson's trichrome, ×350)

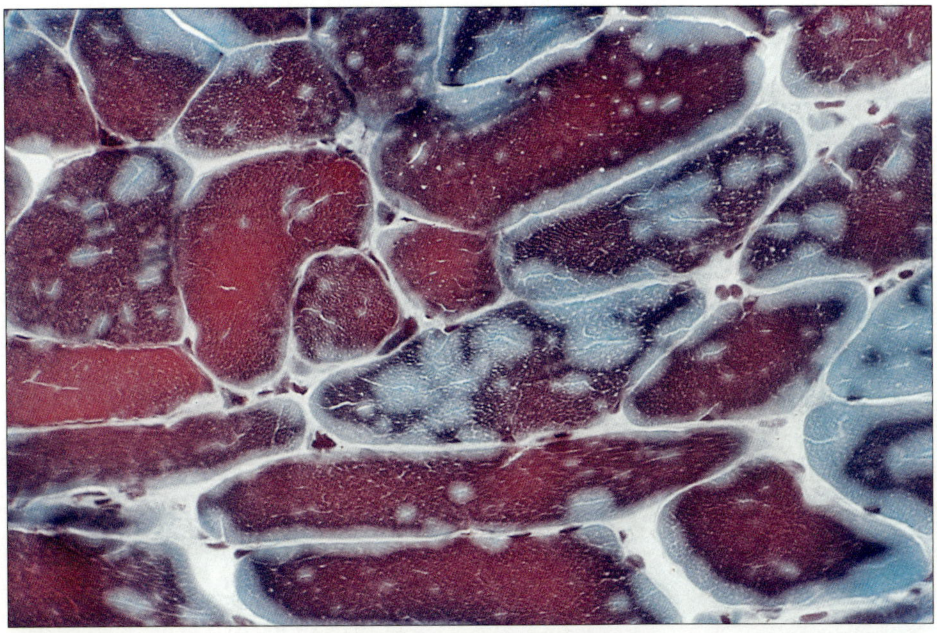

3.23 The uneven staining of these normal muscle fibres is an artefact, and may be caused by improper differentiation during staining, incomplete drying of the section, unclean microscope slides, or internal variations in section thickness caused by blunt knives.
Compare with **3.3**. Such appearances are often localised within the section, and are particularly problematic with this stain.
(Modified Gomori trichrome, ×350)

3.24 This crinkling or folding around the edges of muscle fibres is an artefact, and produces this typical 'pie crust' appearance. It may be caused by improper drying of the specimen before staining, cutting a block before it has warmed to cryostat temperature, mishandling during picking the specimen from the cryostat knife, or excessive contracture of the muscle fibres upon freezing.
Mild degrees of this artefact may be confused for abnormalities of the subsarcolemmal mitochondria.
(NADH-Tr, ×140)

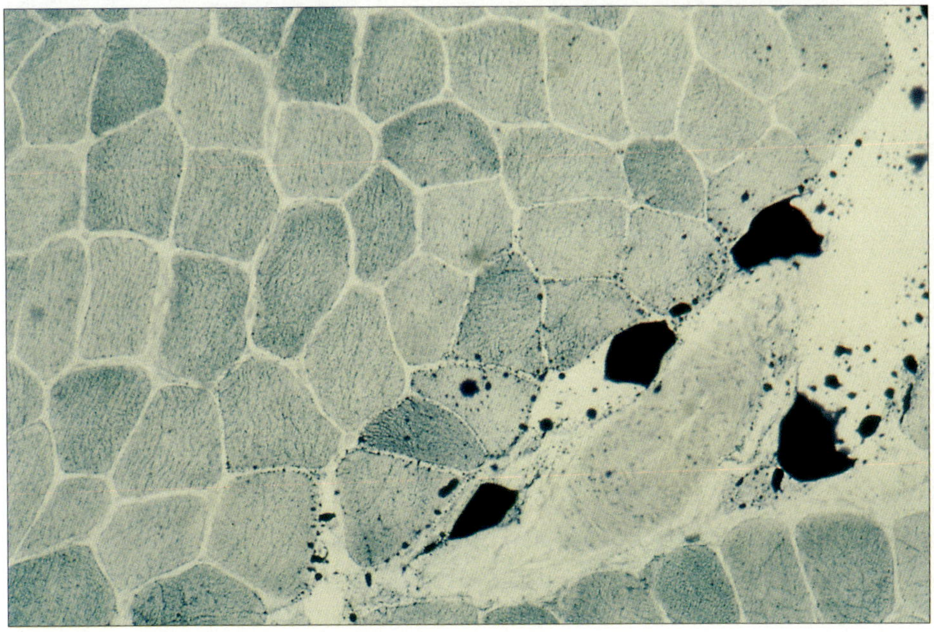

3.25 Artefactual deposits of large masses of Sudan black B, upon or between muscle fibres, are often due to spreading from interfascicular adipose tissue sites, incompletely washing away the excess stain, or to greasy microscope slides.
Abnormal storage of intramuscular lipid occurs as much smaller deposits. (See **4.47**)
(Sudan black, ×217)

Normal muscle ultrastructure

The electron microscopic appearance of normal skeletal muscle is illustrated in **3.26–3.39**. In longitudinal section, the regular arrangement of the thin actin and thicker myosin filaments that form the I-bands and A-bands of the sarcomere respectively, are visible (**3.26**). The appearance, in transverse section, of each of the bands of the sarcomere can be seen in **3.27**. Higher magnification of the electron dense Z-disc region (**3.28**) reveals a rectangular lattice structure into which the thin (actin) filaments are inserted. Actin and myosin filaments and the cross bridges that occur between the latter, in what is known as the 'M-line', can be seen in **3.29**. The interdigitation of the filaments in the region where they overlap can also be seen. Elements of the sarcoplasmic reticulum are situated between the myofibrils.

In a well-prepared biopsy, the arrangement of the filaments, the spacing of the myofibrils and the organisation of the membranous components is very regular. It is therefore important that processes which may cause alterations to the structure of the muscle fibre are minimised, in order that the appearance of artefacts is kept as low as possible.

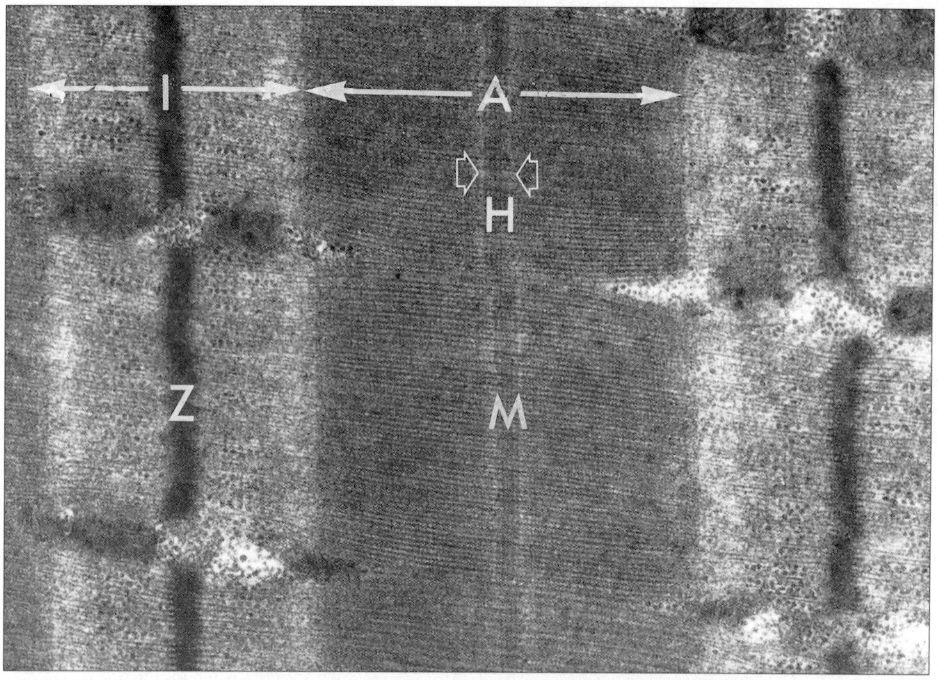

3.26 In longitudinal section, the components of the sarcomere are readily identified. The thicker myosin filaments form the A-band and the thinner actin filaments the I-band.
There is a lighter region in the middle of the A-band, delimited by the ends of the interdigitating actin filaments, which is called the H-zone. A series of cross-bridges traverse the centre of the A-band, linking the myosin filaments. This is known as the M-line. The I-band is bisected by the darkly staining Z-line or Z-disc.
(×36000)

3.27 Cut transversely, the A-bands and I-bands appear as irregular dark and light areas, depending upon the plane of the section. The darker staining Z-disc and lighter H-zone are also identifiable. (×25200)

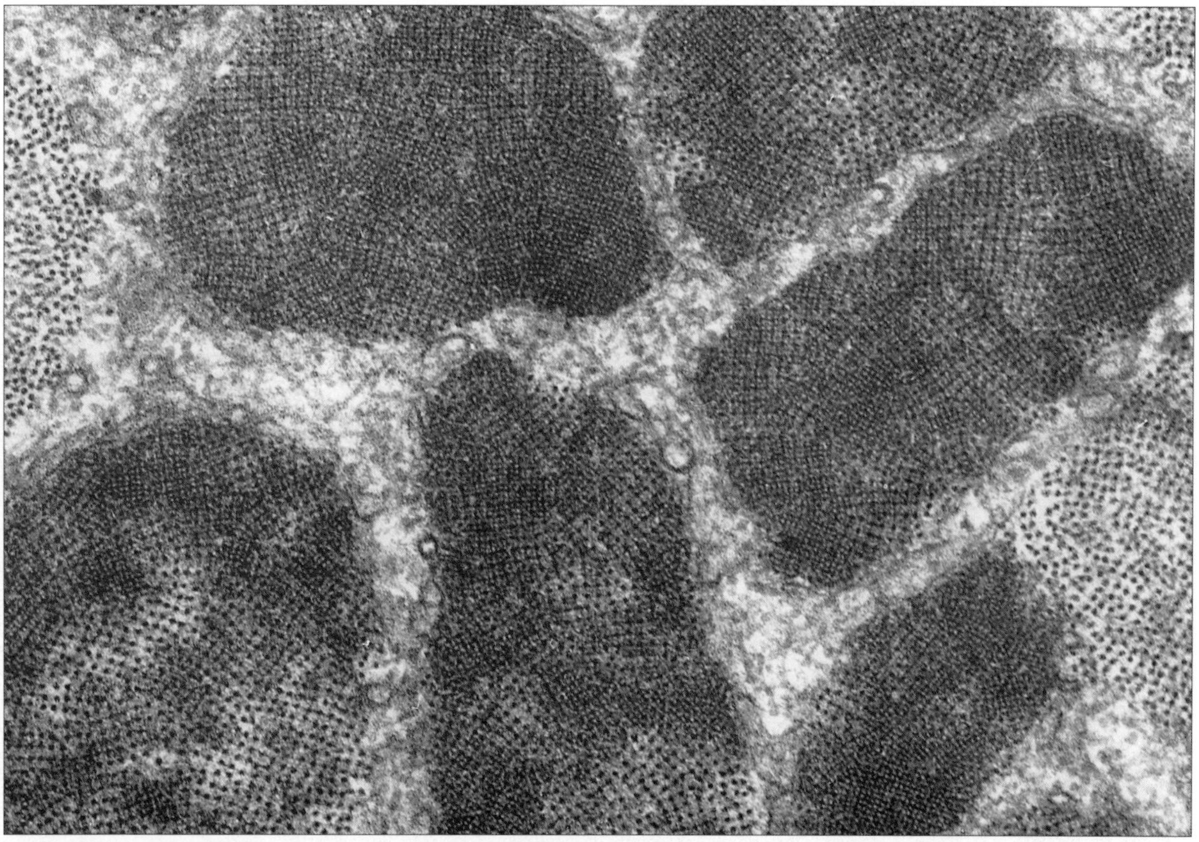

3.28 Viewed in transverse section, the Z-disc is densely stained and can be seen to be in the form of a rectangular lattice. The actin filaments are anchored to this structure. (x100000)

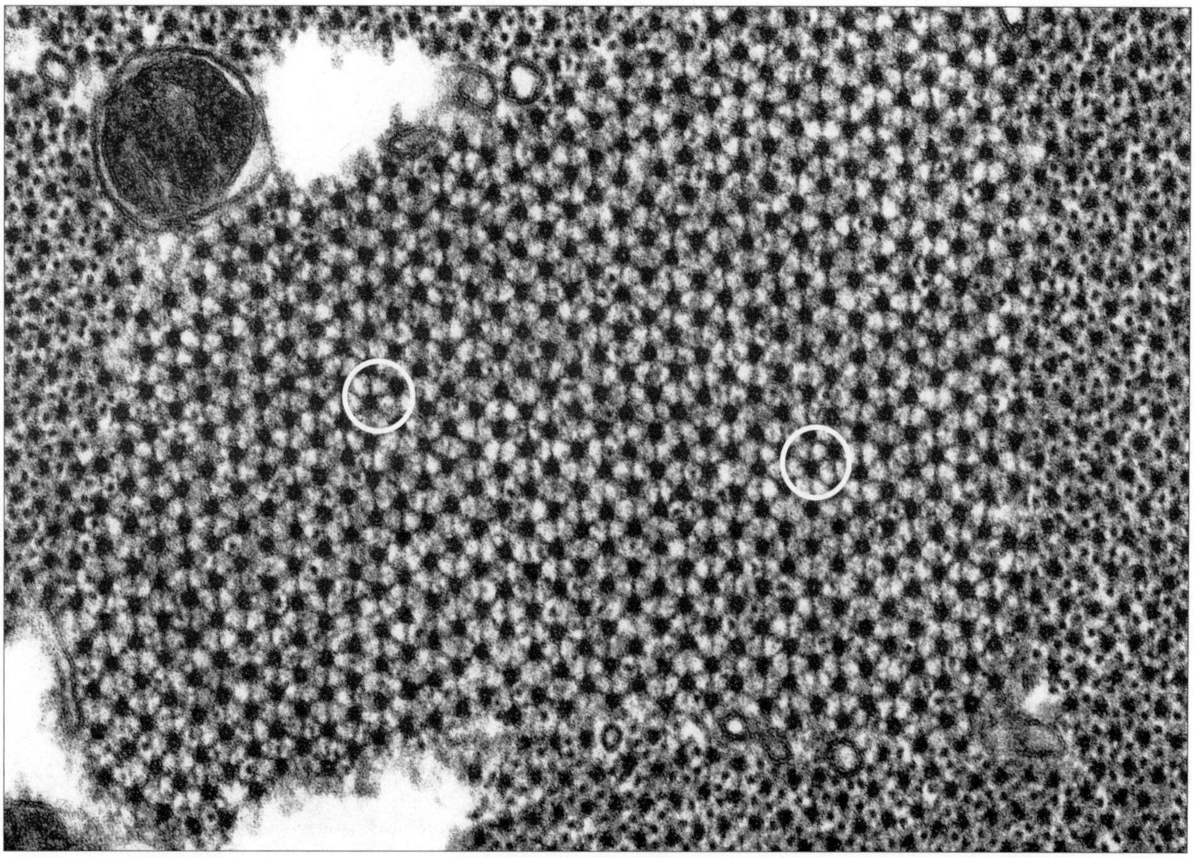

3.29 Cross-bridges can be observed linking the thicker myosin filaments to each other in a hexagonal array at the level of the M-line (ringed). The unlinked myosin filaments form the H-zone. (×160000)

The nuclei of the muscle fibre are normally situated subjacent to the plasma membrane (**3.30**). Satellite cells are also found on the edge of the fibre, lying beneath the basement membrane but external to the plasmalemma, and are enclosed within their own cell membrane (**3.31**). Displacement of the nucleus into the central portion of the muscle fibre should be regarded as abnormal (see **4.4**, **4.13**, **4.67**, **8.4–8.6**).

There are several other features that may be observed within the fibre. Tubules of the T-system run transversely across the fibre and connect with the sarcoplasmic reticulum at the junction of the A-bands and I-bands, to form structures called 'triads' (**3.32**). A neuromuscular junction is shown in longitudinal section in **3.33a**, and in transverse section at a higher magnification in **3.33b**. Intramuscular nerves are occasionally seen (**3.33c**).

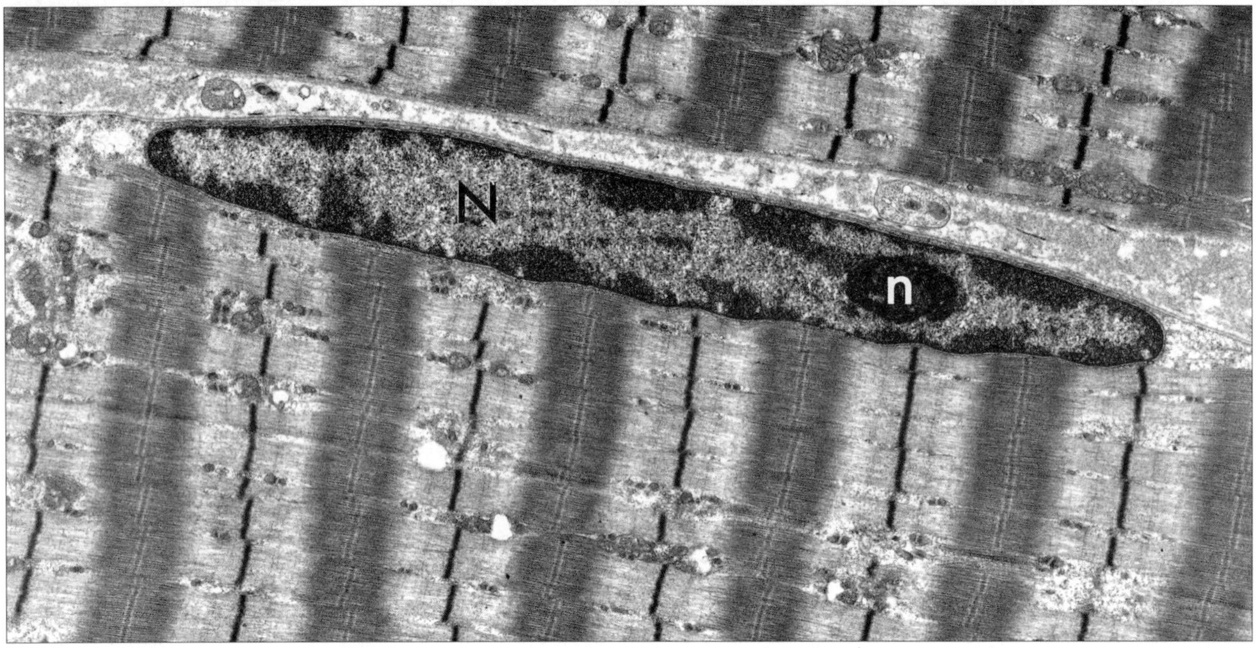

3.30 The nucleus is normally situated just beneath the plasma membrane. It is elongated and ovoid in shape, with the chromatin distributed towards the periphery. The nucleolus (n) is small and densely stained. (×12390)

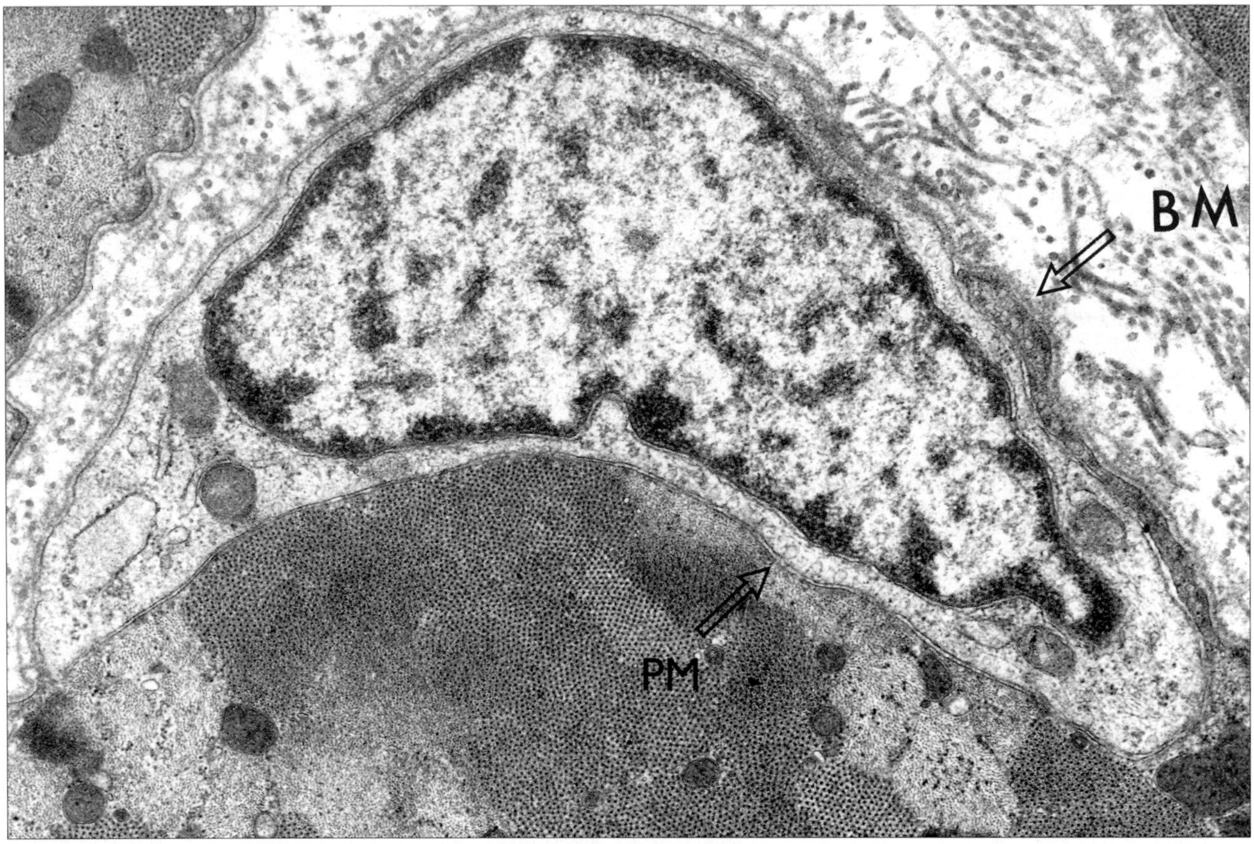

3.31 A satellite cell is a mononuclear cell, found closely adjacent to the muscle fibre beneath the basement membrane (BM). It lies external to the fibre, separated from it by the plasma membrane (PM) of each cell (arrowed). (×18900)

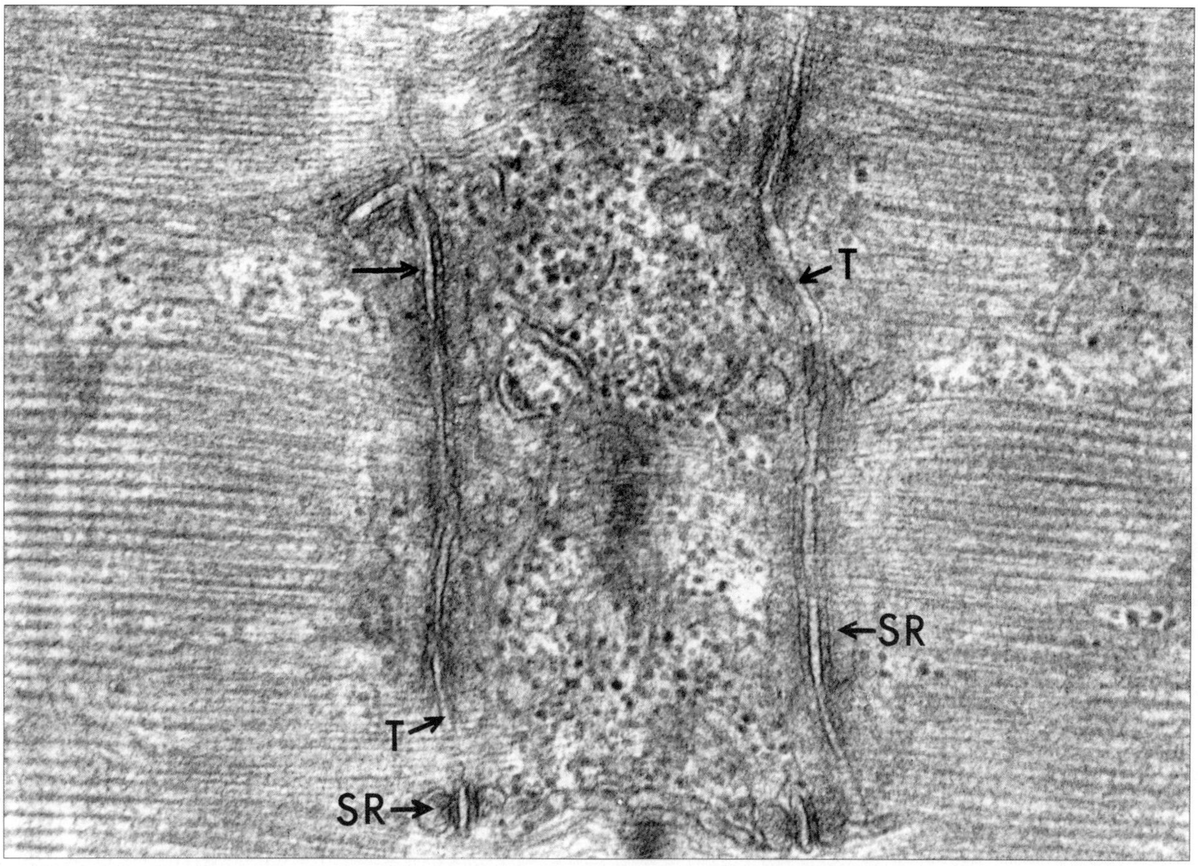

3.32 A system of fine tubular extensions of the plasma membrane traverse the fibre on either side of the Z-line to form the T-system (T). The junctions of the T-tubules with the terminal cisternae of the sarcoplasmic reticulum (SR) are called 'triads'; foot processes link the tubule to the wall of the terminal cisternae. (×75000)

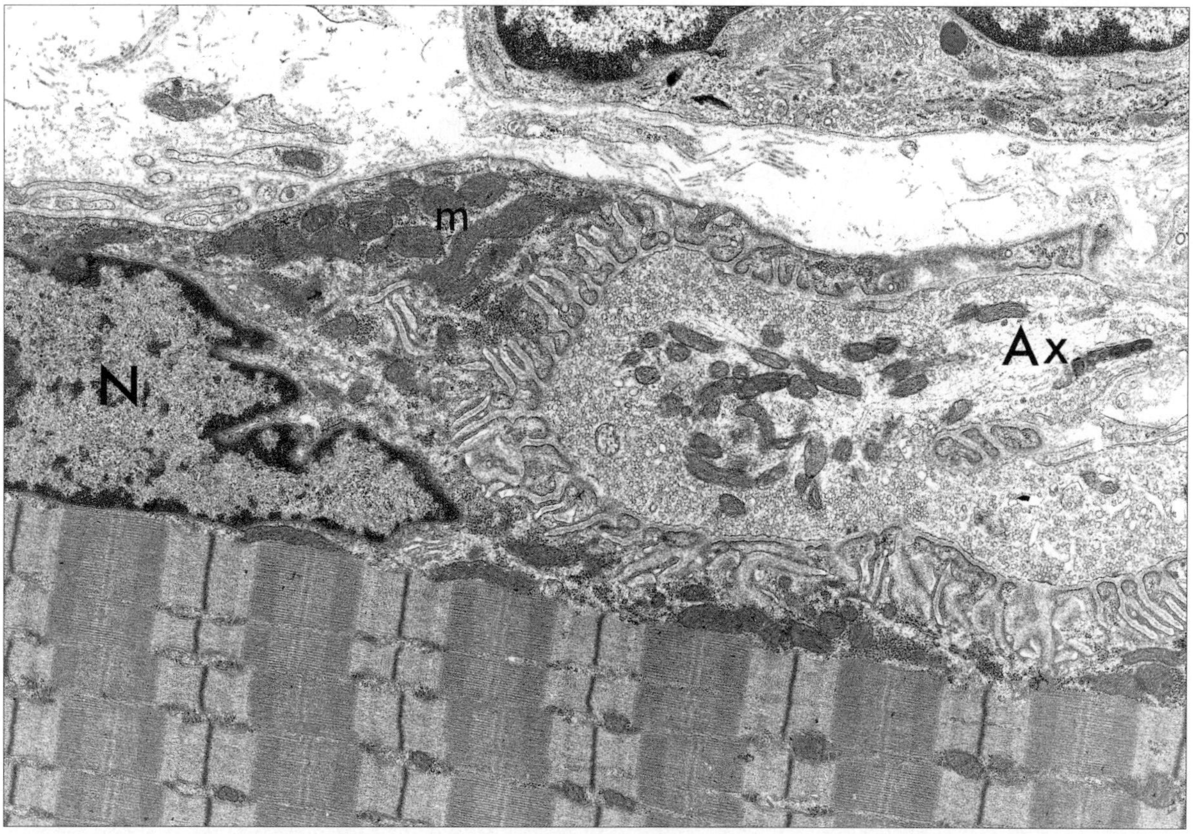

3.33a The neuromuscular synapse is a complex structure on the surface of the fibre. At the motor endplate, where the axon terminal (Ax) is apposed to the fibre, the plasma membrane is highly convoluted, forming numerous postsynaptic clefts. Aggregations of mitochondria (m) are seen in this region, and nuclei (N) may be adjacent (×11200)

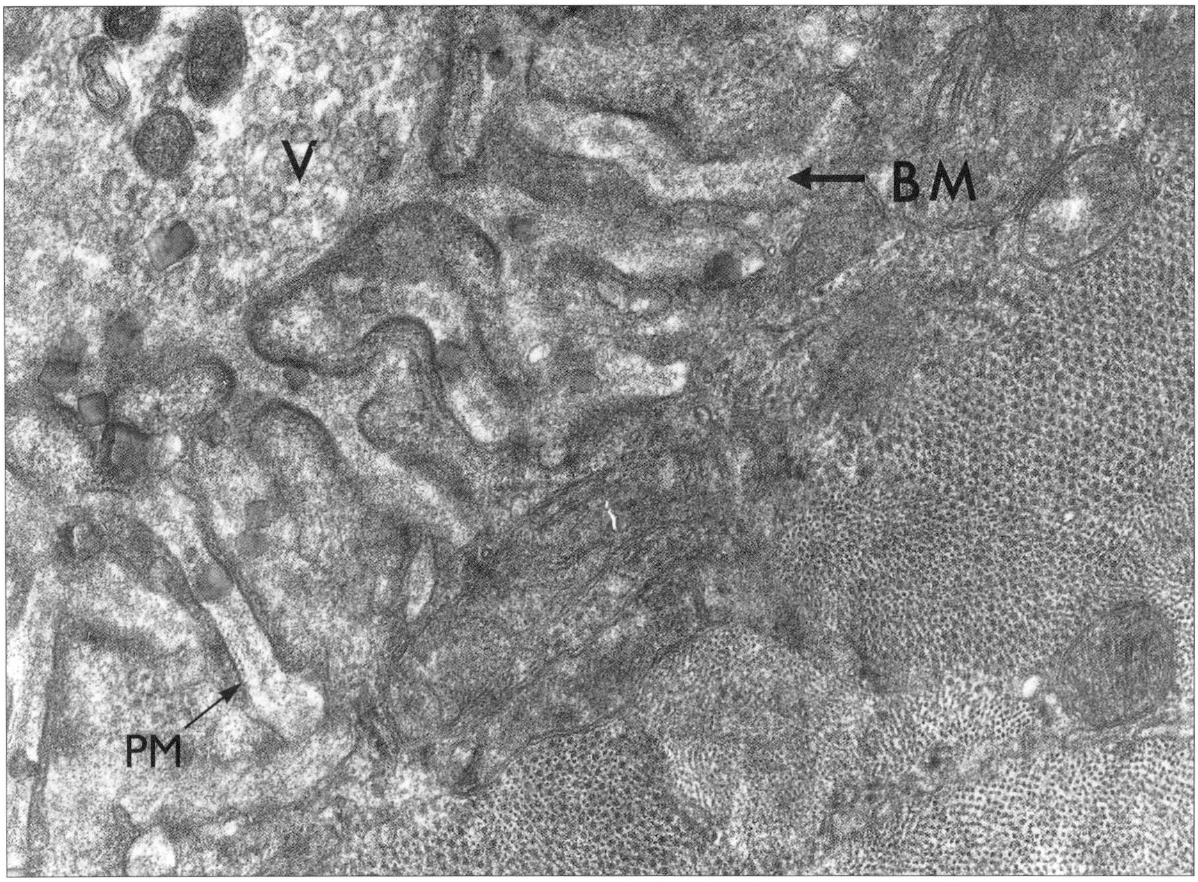

3.33b At high magnification, the basement membrane (BM) can be seen to extend down into the postsynaptic clefts. The plasma membrane (PM) appears thicker and more dense, and the presynaptic vesicles (V) can be seen within the axon terminal. (×70000)

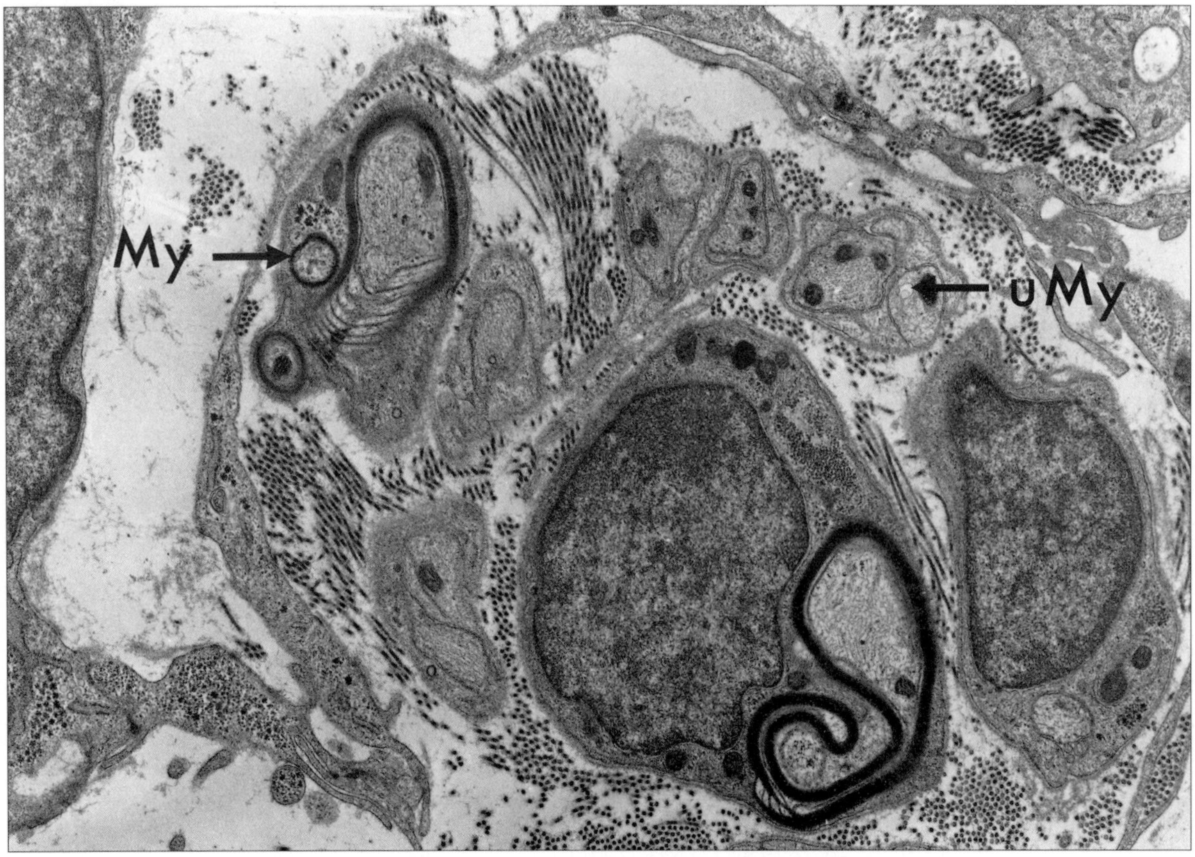

3.33c Both myelinated (My) and unmyelinated (uMy) nerves may be found in the interstitium. (×12000)

Mitochondria, which often have a pleomorphic appearance (**3.34**), are normally seen to lie between the myofibrils on either side of the Z-line, or as small subsarcolemmal aggregates. Glycogen appears as electron-dense granules lying between the filaments (**3.34**). The surface of the muscle fibre is covered by a moderately dense basement membrane, closely apposed to the sarcolemma; collagen fibrils lie in the interstitium (**3.35**).

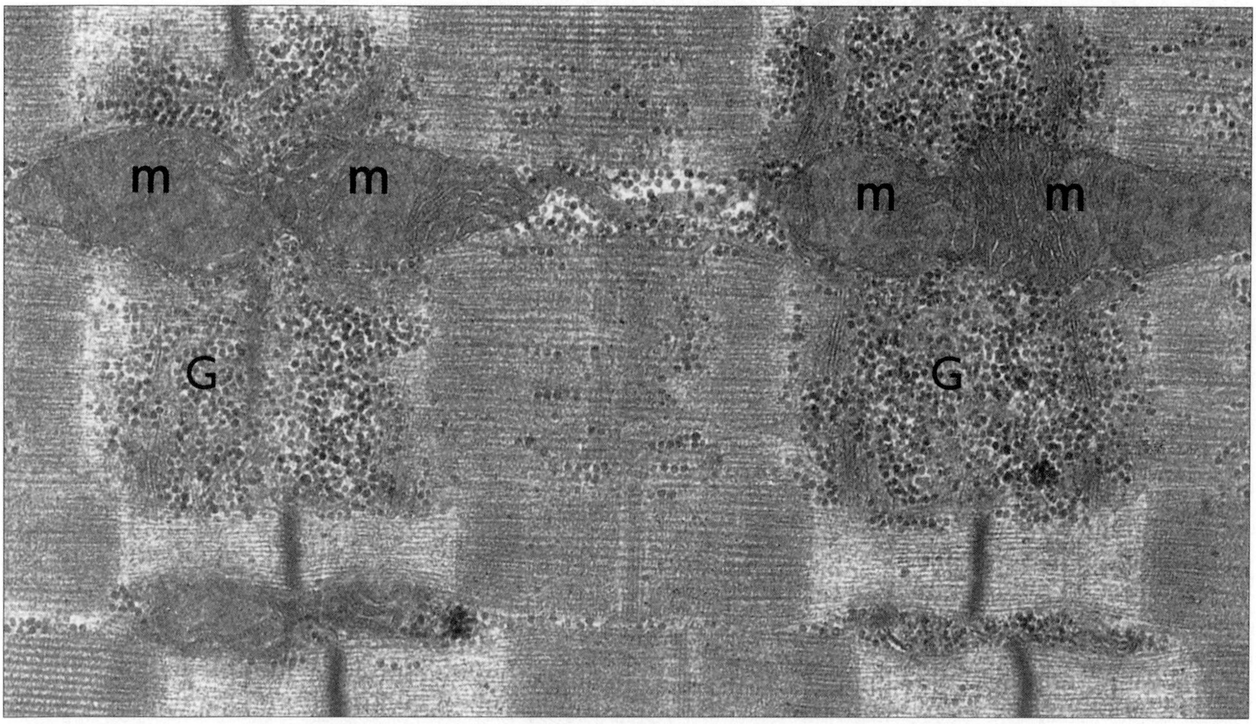

3.34 The mitochondria (m) vary in size and shape, and are normally situated between the myofibrils. Glycogen (G) is seen as small, electron-dense granules, lying between the filaments, mostly within the I-band region. (×36750)

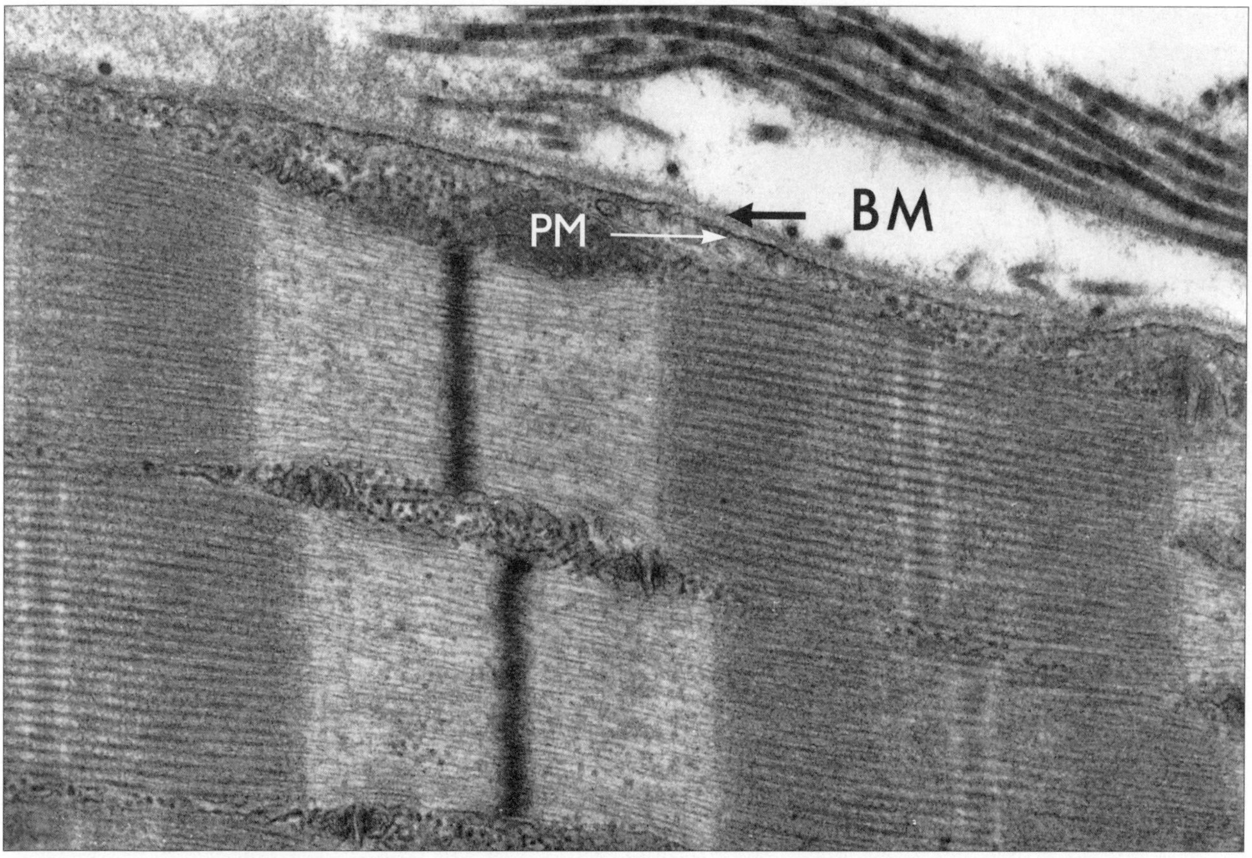

3.35 The basement membrane (BM) is 30–50nm thick, and is closely apposed to the external surface of the plasma membrane (PM), compared to which it appears less dense. (×49613)

At the ends of the muscle, where it forms a junction with the tendon, its outline becomes irregular and extensions of the fibre interdigitate with collagen fibrils (**3.36–3.37**); in transverse section, these extensions can give the appearance of pathological fibre splitting (Chapter 4). Muscle spindles may be encountered by chance in biopsy specimens from large muscles such as quadriceps. The spindles may be misinterpreted as artefacts (**3.19**), but their ultrastructural configuration is unmistakable, if only because of innervation (**3.38)**

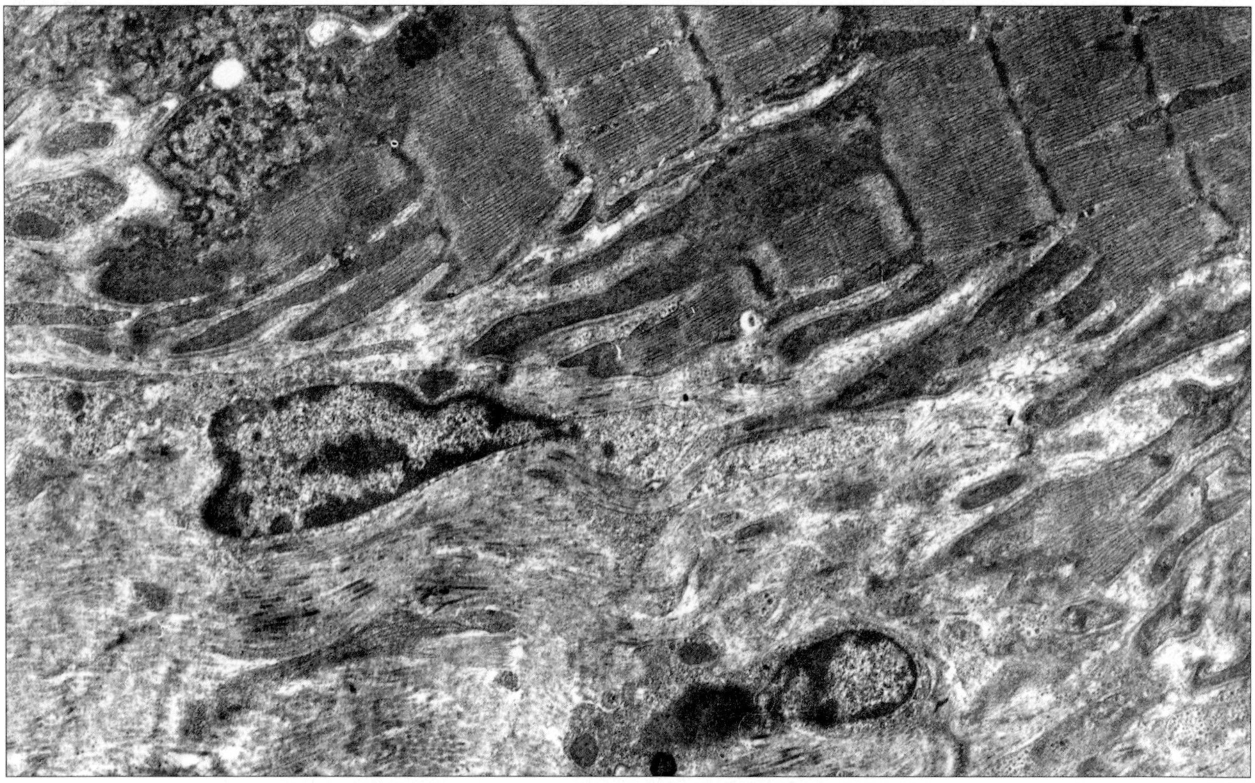

3.36 The myotendinous junction is characterised by the interdigitation of elongations of the muscle fibre with collagen fibrils. There are darker staining regions within the sarcoplasm adjacent to the plasma membrane. (×11025)

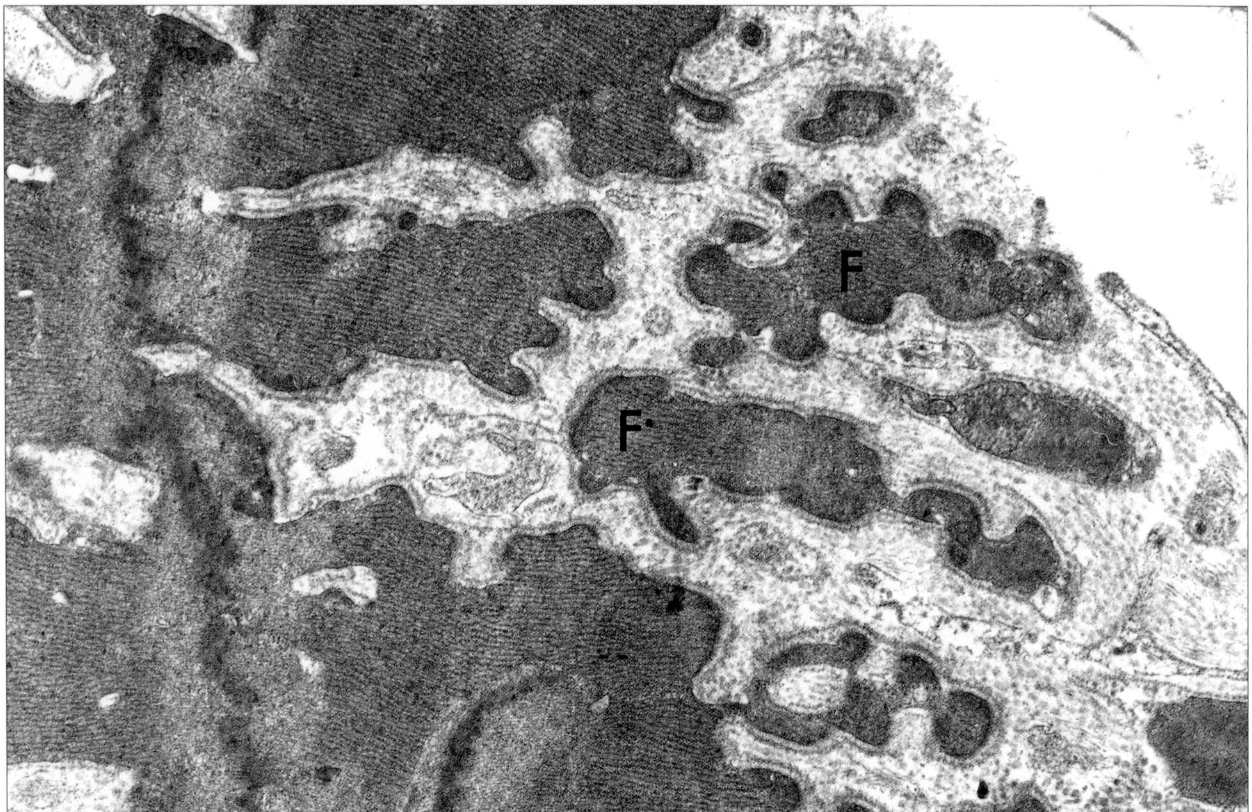

3.37 In this transverse section of the myotendinous junction, the extensions of the muscle fibre (F) surrounded by collagen fibrils can be seen. (×21000)

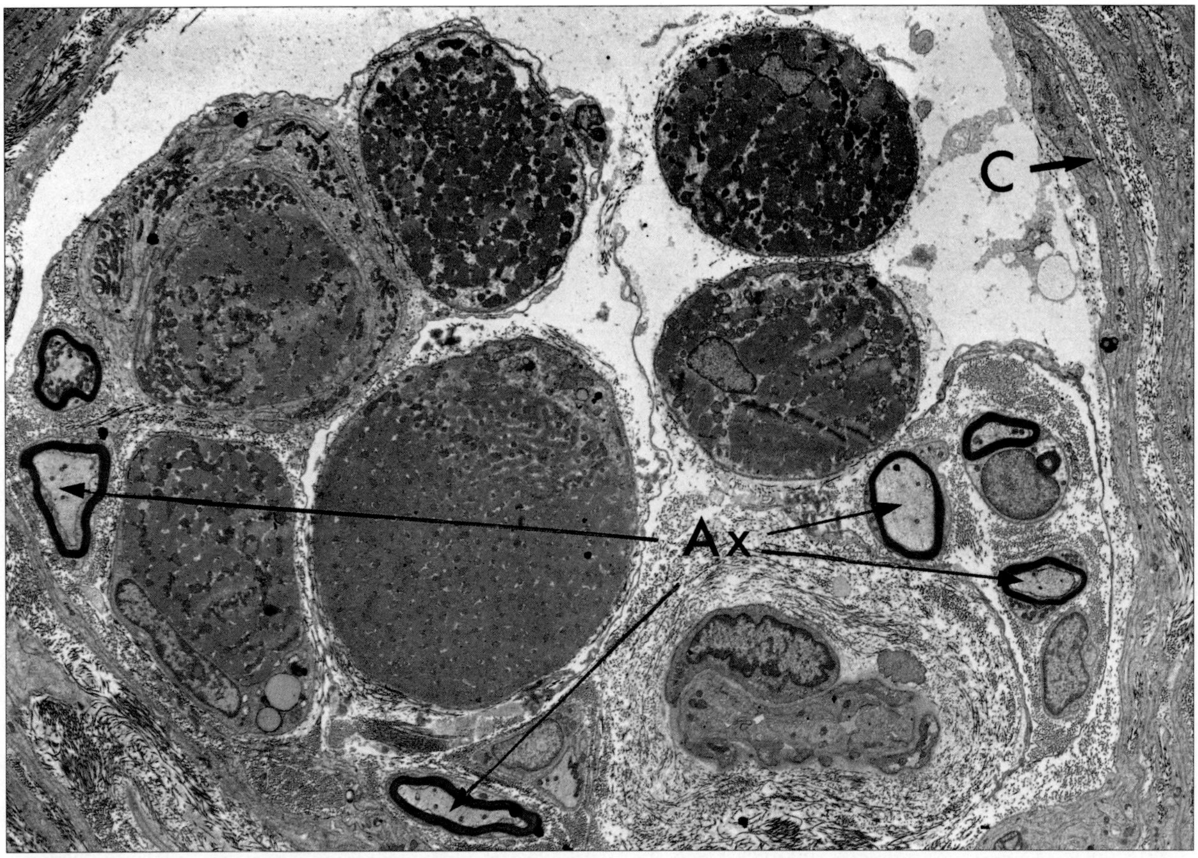

3.38a Muscle spindles are rarely seen in normal muscle, but are more commonly observed in denervated muscle where more fibres are sampled. Their structure is complex, and different components are observable in different sectional planes. There are two types of intrafusal fibres: nuclear bag fibres that contain a central cluster of nuclei, and the narrower nuclear chain fibres that contain a central chain of single nuclei. The spindle illustrated is comprised of a few small intrafusal fibres surrounded by a connective tissue capsule (C). Myelinated axons (Ax) are present. (×3000)

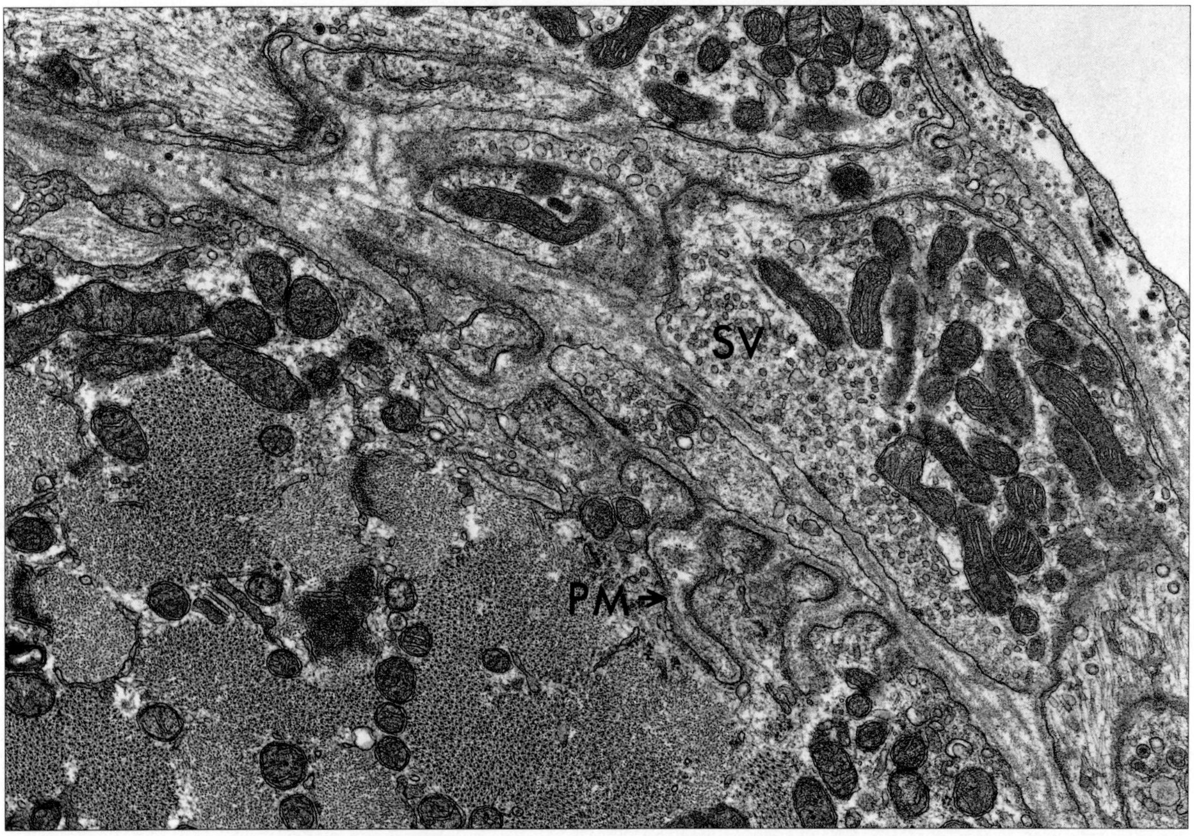

3.38b Both sensory and motor innervation is present in the spindle. This section, taken near to one pole, shows the postsynaptic folding of the sarcolemmal membrane (PM) of the motor endplate, and a sensory nerve ending rich in mitochondria and synaptic vesicles (SV). (×26250)

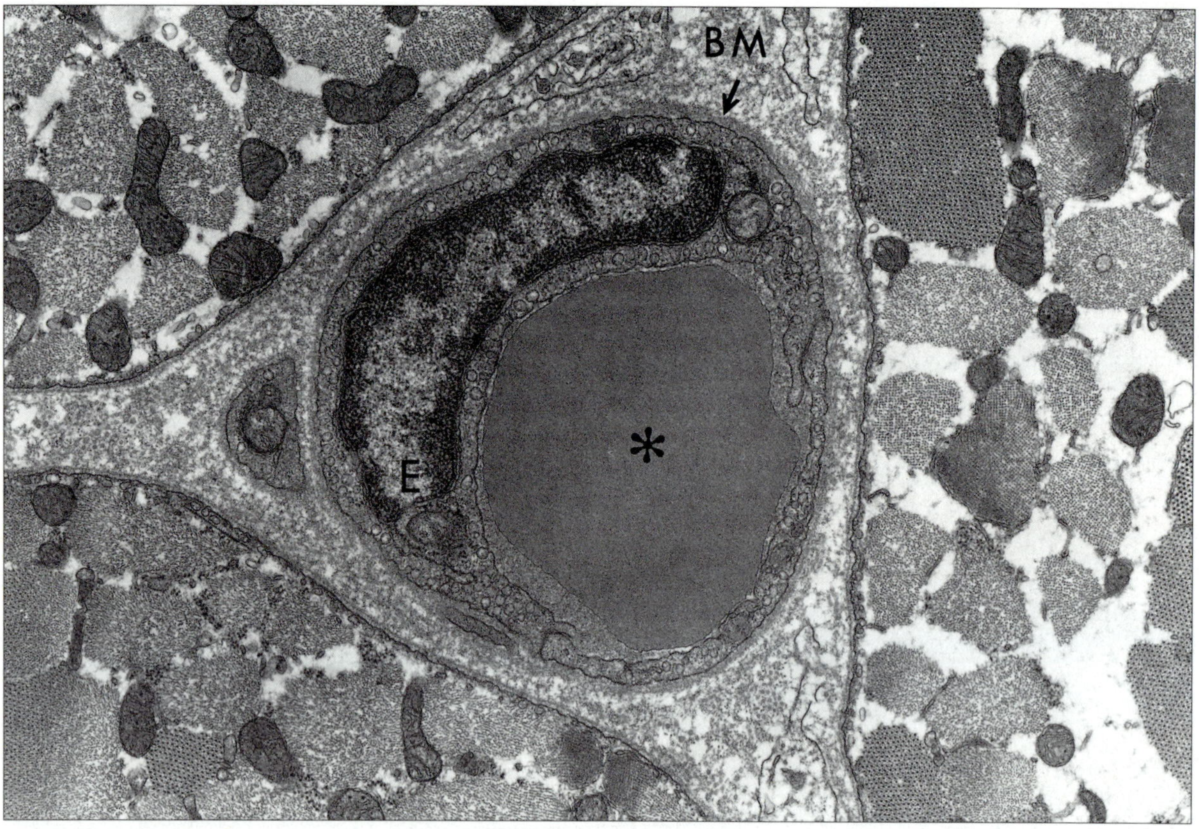

3.39 A normal capillary, showing the endothelial cells (E) surrounded by a basement membrane (BM) and an erythrocyte within the lumen (*). (×22500)

3.40 Fibres on the surface of a biopsy may show contraction and intense staining (similar to hyaline degeneration) if allowed to dry before fixation. (1 µm section, toluidine blue, ×600)

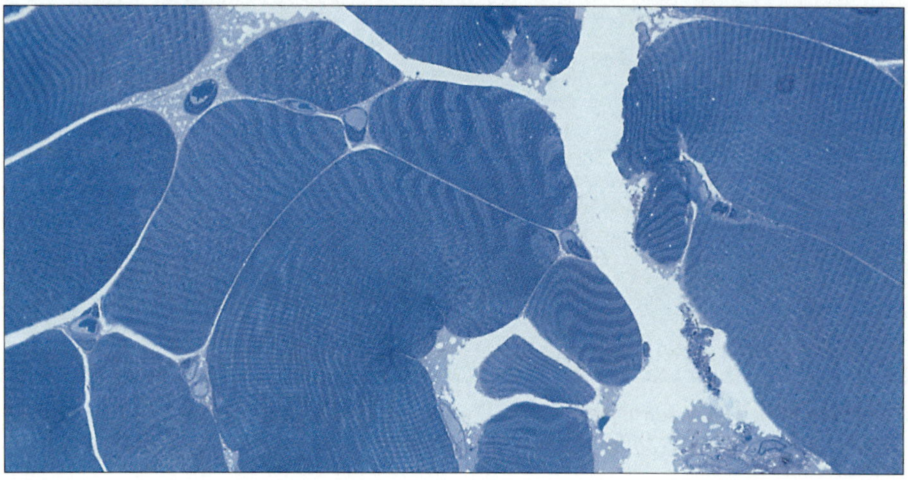

3.41 Fibres that are not held under tension during fixation will contract and become disorientated, making the interpretation of genuine pathological changes difficult. (1 µm section, toluidine blue ×600)

Muscle capillaries are seen in endomysium in close juxtaposition to their 'client' muscle fibres (**3.39**).

The importance of a correctly removed biopsy, handled in such a way as to minimise the production of artefacts and processed using a reliable fixation and embedding protocol, cannot be emphasised too strongly. The subsequent sectioning and staining of the tissue prior to microscopic examination must also be performed correctly. Some of the artefacts that can arise during the preparative stages are illustrated in **3.40–3.45**.

3.42 Focal contraction of the fibres can result from mishandling of the sample. The sarcomeres on either side of this type of contracture appear normal, unlike those seen in pathological lesions (see **4.29a**, **5.3a**). (×7875)

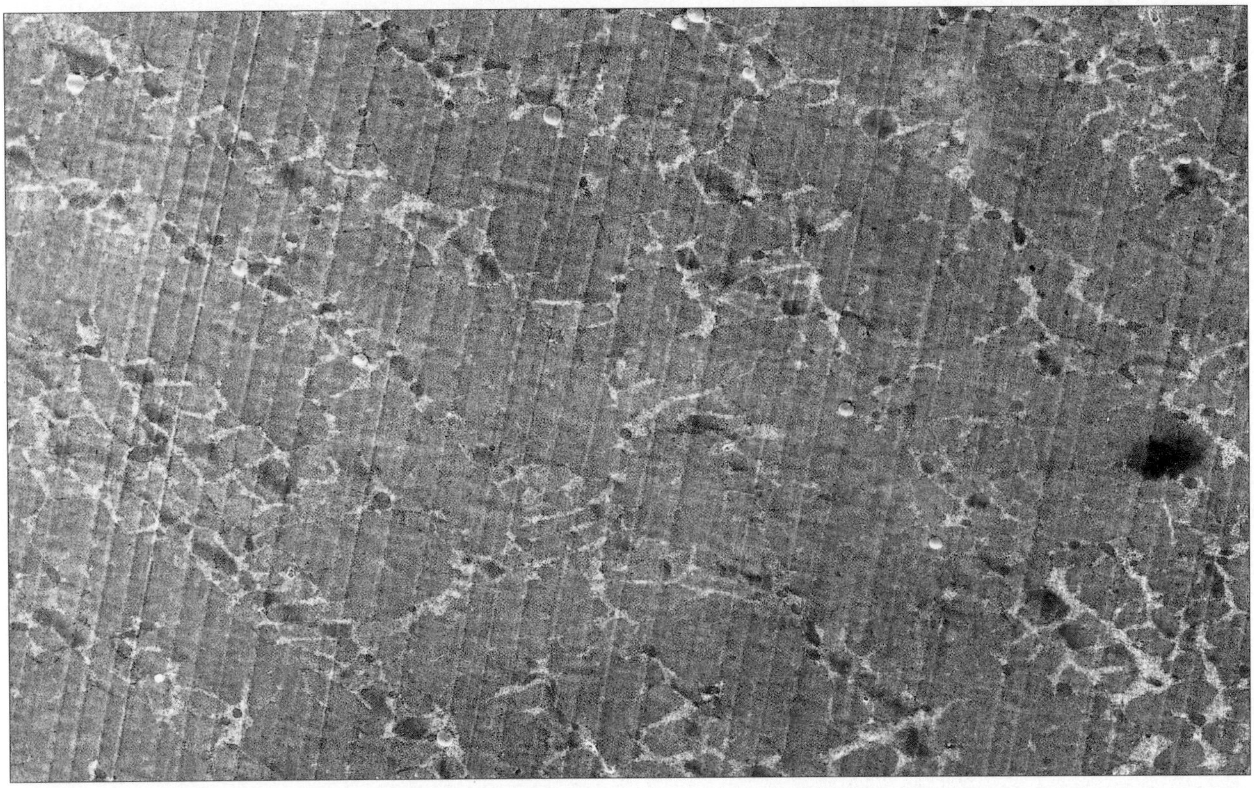

3.43 This illustrates a badly orientated specimen that was sectioned with a damaged knife, causing severe scoring, such that almost all detail is obscured. (×5775)

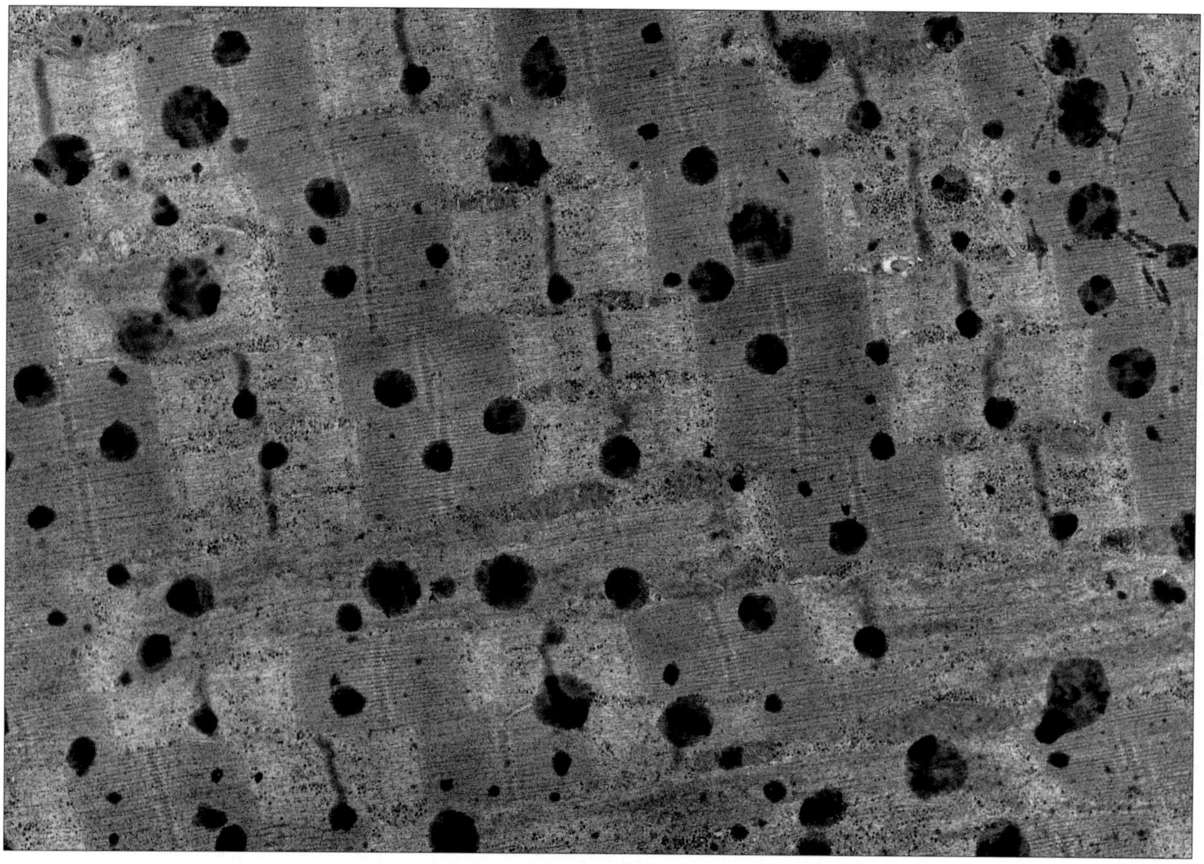

3.44 Stain precipitate on the surface of the section can obscure detail within the specimen. (×16000)

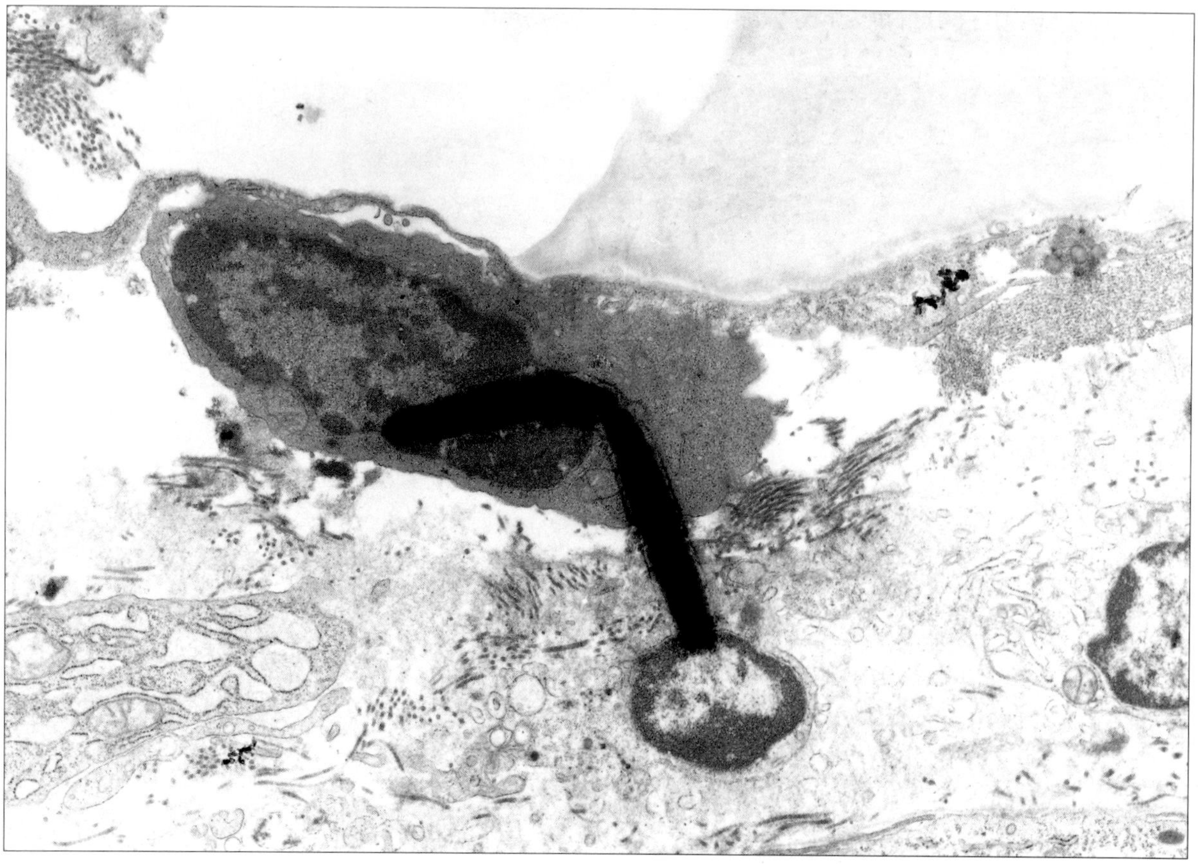

3.45 Micro-organisms may contaminate solutions used during the cutting and staining stages of section preparation. (×13750)

Analysis of the biopsy

Evaluation of the biopsy preparations should be carried out in an organised manner. Alterations in muscle morphology from the histological preparations must be taken into account before a basic histochemical screening is made. The observations may then suggest the need for:

- Further sections.
- Additional cytochemical stains.
- Immunocytochemistry.
- Examination of the ultrastructure, especially when unusual vacuoles or inclusions are present.

It is useful to employ a semiquantitative scoring schema (e.g. + to ++++) when recording the results to ensure consistency and easy communication of information (**3.46**). A checklist of features which must be assessed is given below. Some staining reactions fade quickly and need to be examined immediately, and, if necessary, the sections should be photographed. In some circumstances, morphometric analysis may also be required in order to identify deviations from the normal.

The observations should not be viewed in isolation, and need to be interpreted in the light of an understanding of the basic pathological responses of the muscle. These are listed at the end of this chapter and discussed in relation to their aetiology and pathological processes in Chapter 4.

Since the pathological responses of muscle tissue are limited and similar features may occur in widely different neuromuscular diseases, it is imperative, as stated in Chapters 1 and 2, that a final diagnosis is only reached by interpreting the pathological findings in the context of clinical data (Chapter 11).

3.46 A muscle biopsy pathology report form.

Neuromuscular Unit
Muscle Biopsy Pathology Report

#HD:	Patient:	Sex:	Age:
Clinical:			
Investigations:	C K	EMG	
Biopsied Muscle:	Needle / Open		Quality:
Provisional Diagnosis:			
BIOPSY	Normal	+/++/+++/++++	Comment
Architecture			
Connective tissue			
Inflammatory			
Necrosis/phagocytosis			
Regeneration			
Fibre size variation			Scattered/grouped
Fibre shape			Angular/rounded
Nuclei: Subsarcolemmal			Size
Internal			Single/multi/central
Hyaline/splits/whorl			
Basophilia			Subsarcolemmal/granular
Vacuoles			
Esterase			Fibres/Conn. tissue/Lipofuscin
Modified Gomori			Ragged-red/nemaline
NADH pattern			Central core/Target/'Moth eaten'/Subsarcolemmal/granular
grouping			
ATP %			
size			
grouping			
Phosphorylase			
Other: PAS/SB/			
SDH/Cytox			
Other: Dystrophin			
Morphometry:	%	Size	SPAM
Comments: Neurogenic / myopathic			
Further investigations			
Signed:	Date:		

Histological assessment

When examining the muscle biopsy, the following questions should be addressed:

- Is the biopsy normal or abnormal?
- Is there an excess of connective tissue or fatty replacement?
- Is there an inflammatory response?
- Are there changes in muscle fibre size? If so, are these changes general, focal, scattered, grouped, or perifascicular?
- Are rounded or angulated fibres present?
- Are hyaline fibres present?
- Are internal nuclei present within the muscle fibres? If so, are they multiple?
- Are phagocytes present within the muscle fibres?
- Are necrotic and/or regenerating fibres present?
- Are there structural changes within the fibre?
- Are split or whorled fibres present?
- Are vacuoles or inclusions visible within muscle fibres? If so, where are they located within the fibre?
- Is there uneven cytoplasmic staining? If so, where is it located in the fibre?
- Are rod bodies present?
- Are there any abnormalities of other structures, e.g. muscle spindles, nerves, or blood vessels?

Histochemical assessment

The following questions should be addressed:

- Are fibre types clearly discernible?
- Are the fibre type proportions and spatial distributions correct for this muscle?
- Is there a fibre type predominance?
- Is there grouping of a particular fibre type?
- Is there atrophy or hypertrophy of a particular fibre type?
- Is degeneration or regeneration occurring within the fibres?
- Are there alterations in the staining pattern within fibres, e.g. cores, target, moth-eaten, ringbinden, lobulation, clumps, tubular aggregates, etc.?
- Are there reductions or excesses of storage products, e.g. lipid, glycogen? If so, where are they located and are they limited to a particular fibre type?
- Are vacuoles present? If so, where are they located and what is in them?
- Are any specific enzymes lacking, e.g. phosphorylase?
- What is the nature of any inclusion bodies?
- If inflammation is present, what cell types are involved?

Immunocytochemical assessment

The following questions should be addressed:

- Is there an abnormality of dystrophin antibody labelling (in terms of distribution and intensity), compared to C′-terminal domain, rod domain and N-terminus?
- Are there deficiencies or accumulations of other cytoskeletal and membrane associated proteins, e.g. desmin, nebulin, spectrin, dystrophin-associated glycoprotein (DAG), laminin and vimentin?
- Is there any abnormality of myosin isoform distribution?

Ultrastructural screening

It is important that several blocks of both longitudinally and transversely embedded tissue are examined by light microscopy to select appropriate areas for examination using the electron microscope. Careful examination of the toluidine blue stained sections can provide useful information, bearing in mind that only a very small sample is available and only a few fibres will be observable. This can serve as a screening process to eliminate both the time-consuming preparation of ultrathin sections and the long periods of searching with the electron microscope for pathological changes that may not be present in the sample.

Where the disease process is uneven in distribution, or where there is patchy involvement of the muscle fibres, it is possible that many of the fibres will appear normal, or even that abnormalities may not be found in the sample examined. Examination of transverse sections will answer the following questions:

- Are the fibres normal, polygonal, angular or rounded in outline, and are they split?
- Are the fibres larger or smaller than those of a representative, non-pathological muscle (prepared in the same manner) from a subject of the same age and sex?
- Are there densely or palely stained fibres, or is there variation within individual fibres not related to the normal sarcomeric pattern?
- Do the fibres show a normal, close relationship to each other, or are they abnormally separated from one another?
- Is the interstitial space occupied by an excess of connective tissue, fat cells or inflammatory cells?
- Are there changes in the blood vessels, nerve fibres and spindles that are normal constituents of muscle tissue?
- Are the nuclei situated beneath the sarcolemma, or are they displaced towards the centre of the fibre?
- Are there increased numbers of intramuscular nuclei?
- Are the nuclei normal in outline, size and appearance?
- Is the nucleolus visible?
- Do the nuclei contain inclusions?
- Are the myofilaments closely packed, or loosely distributed throughout the sarcoplasm?
- Are ring fibres present, or any other abnormal arrays of the myofilaments visible?
- Is there evidence of myofilament loss involving the fibre as a whole, or within the centre to form a core?
- Are there multiple cores?
- Have the filaments undergone necrosis and been replaced by palely staining amorphous debris?
- Are phagocytic cells present?
- Is there vacuolation of the fibres, and is there anything contained within the vacuoles? If so, does this substance exhibit metachromasia?
- Is there excess glycogen present? If so, what is its distribution and appearance?
- Is there excess lipid within the fibres?
- Are there abnormal numbers of mitochondria present, either as subsarcolemmal aggregates or elsewhere within the fibres?
- Are inclusions such as nemaline rods, cytoplasmic bodies, tubular aggregates or myelin figures present?

Examination of longitudinal sections will answer the following questions:

- Is there hypercontraction of the fibres?
- Is there loss of myofilaments?
- Is there loss of the A-band or I-band?
- Is there loss of the Z-line?
- Does the Z-line show streaming?
- Is there thickening of the Z-lines to form nemaline rods?
- Are there chains of centrally placed nuclei?
- Is there reduplication or thickening of the basement membrane?
- Are there defects in the sarcolemma?

Quantitative analysis

A morphometric assessment should be carried out as part of the routine evaluation of muscle biopsies. This is particularly useful:

- As an objective adjunct to the clinicopathological report.
- For comparison with normal data.
- For use in follow up studies, for example in assessing disease progression or the effects of treatment.
- In research studies.

It must be stressed, however, that in order to provide valid and reproducible morphometric data, rigid adherence to the sampling and preparation protocols discussed in Chapter 2 is essential. Small or poorly orientated samples may only provide data on a limited number of variables.

The measurements discussed in the following (bar that of staining density) may be performed readily in most laboratories and require a minimum of inexpensive equipment.

The measurement may be performed directly by microscope, using an eyepiece graticule or test lattice projected by drawing tube. Alternatively, the same information may be obtained from photomicrographs or camera lucida drawings, using planimetry and sizing lattices.

Connective tissue content

The connective tissue content (i.e. the proportion of tissue cross-sectional area occupied by non-muscle fibre components) is easily estimated by simple 'point counting' procedures (**3.47**). About 400 test points intersecting the measured component is sufficient for statistical analysis (Williams, 1977). Samples of any orientation provide valid data.

Some typical results are given above in the section on normal muscle histochemistry.

Relative frequencies of fibre types

The relative frequency of occurrence of the different muscle fibre types is usually recorded by counting contiguous fibres in a number of representative microscopic fields, to account for interfascicular variations. About 100–200 fibres is usually an adequate sample size. Some normal values are given in *Table 3.2*, although much inter- and intra-individual variation has been noted (Johnson *et al.*, 1973; Mahon *et al.*, 1984; Loughlin, 1993).

Spatial distributions of fibre types

The spatial distributions of the fibres may also be estimated by simple counting methods (e.g. enclosed fibre, SPAM) (**3.48**), or by measuring nearest-neighbour distances.

In healthy muscle, fibres of the same type have a tendency to spread out (hyperdispersion), but in disease their distributions may become random or grouped (Mahon, Kristmundsdottir and Cumming, 1987; Loughlin, 1993; Chapters 4 and 6).

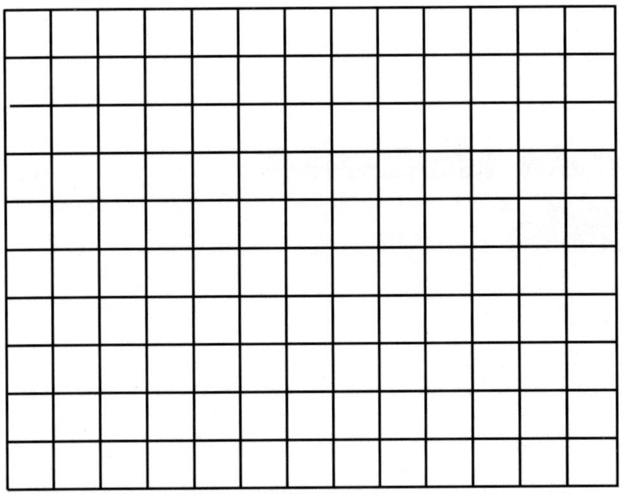

3.47 Point count lattice, for use in volume fraction estimations of tissue composition.

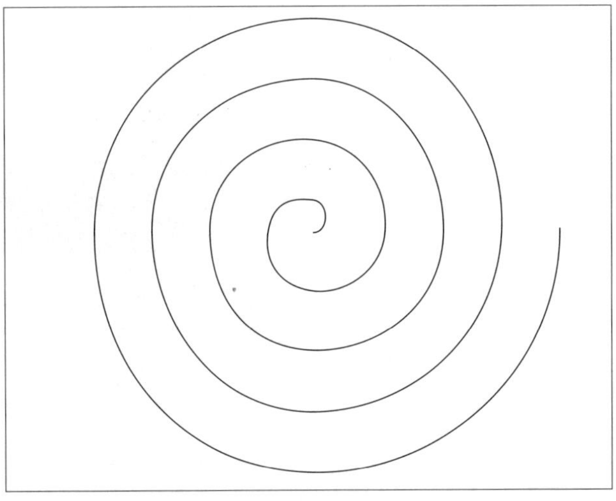

3.48 Spiral test line for estimation of muscle fibre distribution by the SPAM technique.

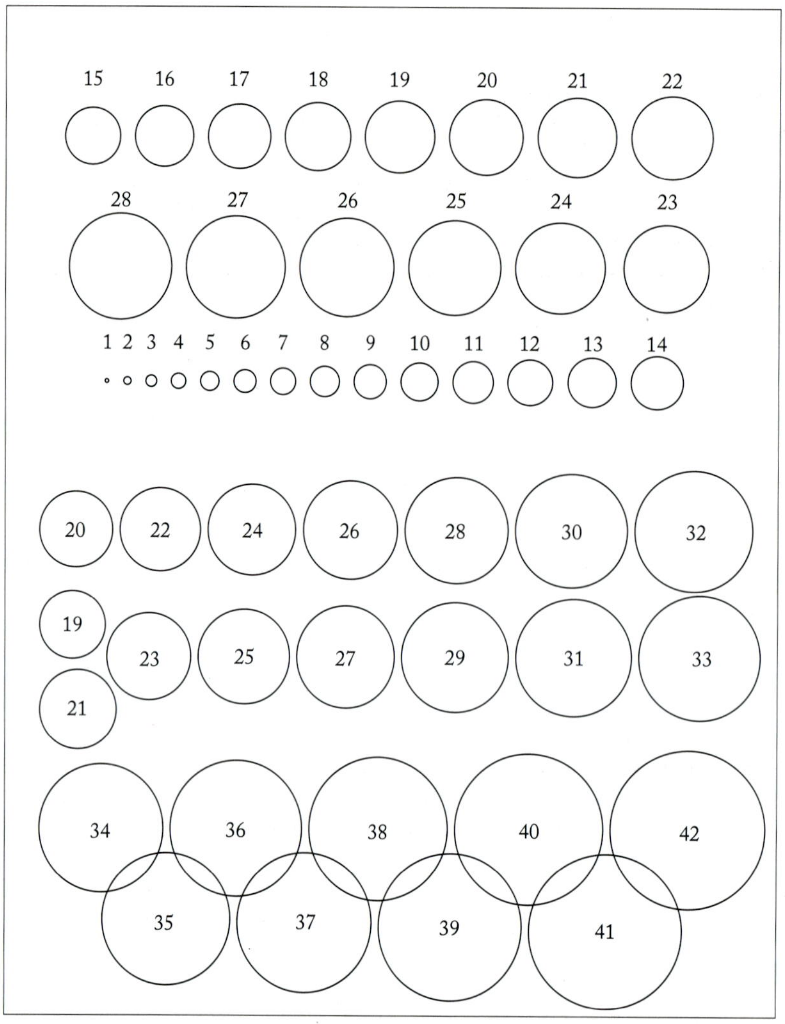

3.49 Sizing lattices, for estimation of 'narrow' muscle fibre diameter.

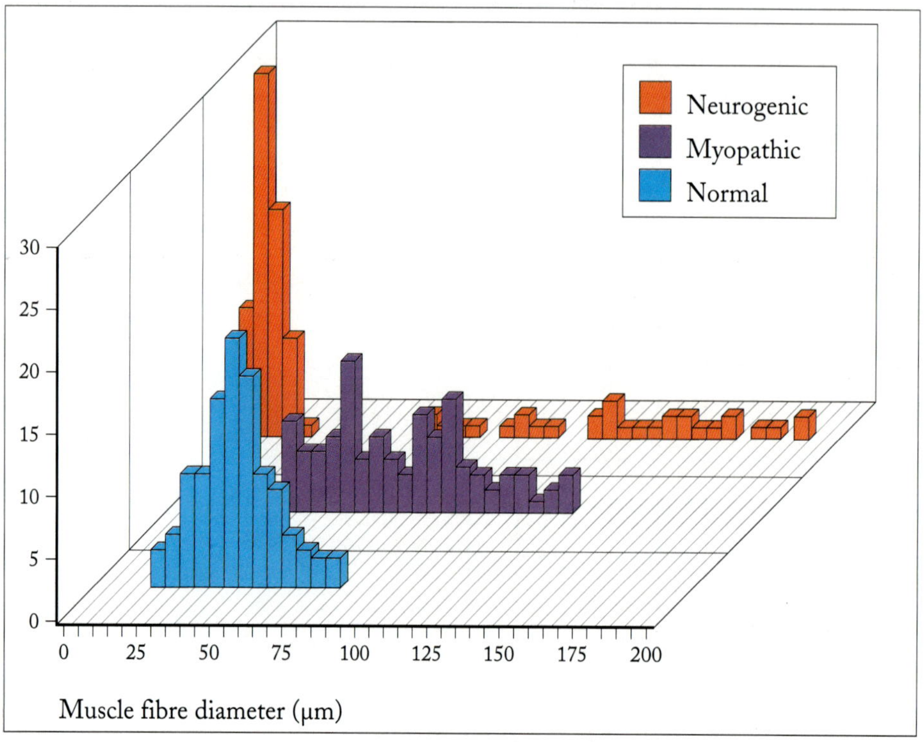

3.50 Typical histogram of 'narrow' fibre diameter distribution for normal and abnormal muscle (adult, vastus lateralis).

Muscle fibre size

Muscle fibre size is best estimated by measuring the cross-sectional area or diameter of about 100–200 fibres. The narrowest diameter is least prone to influences of section obliquity, is the least time consuming, and requires no more than a ruler or sizing screen (**3.49**). Data should be recorded separately for each fibre type. Although mean values are often used to express the data from an individual, a histogram showing atrophic or hypertrophic peaks and a coefficient of variation are more informative (**3.50**).

Fibre shape

If more sophisticated equipment is available (**3.51**), recording of fibres' cross-sectional areas and profile perimeter allows for estimation of shape factors (e.g. $4\pi \times$ area over perimeter^2 is a measure of 'roundedness', and gives a value of 1.00 for a perfect circle).

Relative occurrence of pathological features

The number of fibres of each type with specific pathological features (e.g. internal nuclei, vacuoles, necrosis, basophilia, etc.) is recorded by simple counting in a few representative microscopical fields. If the total number of fibres per field is also recorded, this data may be subjected to statistical analysis.

Staining density

Staining intensity may be used to estimate both enzyme activity and the amounts of storage products within skeletal muscle fibres by the technique of microdensitometry. However, this requires extreme care and attention to preparation techniques, section thicknesses, uses of controls, dedicated equipment and some understanding of stoichiometric histochemical reactions in order to provide useful data. Such methods may be especially useful for the 'metabolic' stains (e.g. SDH, NADH, cytochrome oxidase, PAS).

Ultrastructural morphometry

Ultrastructural morphometry is usually only employed in research studies. Direct measurement of filament length, width or banding thickness may be applied to exact longitudinal or transverse sections. Standard stereological procedures may be used to estimate organelle volumes, surface areas and numbers in obliquely oriented sections, to take account of anisotropic organisation.

Having achieved valid, objective, histological, histochemical and ultrastructural data concerning the muscle biopsy, it is useful to perform statistical analyses. Details concerning the type and extent of the data distribution can be very informative. The recent use of digitising tablets, microcomputers, automatic image analyzers and disc storage facilities in myomorphometry (**3.51**), is now replacing the more laborious manual methods, and is making data storage and retrieval, statistical analysis, cross-correlation with clinical data, and graphical presentation of results more efficient.

Interpretation of the quantitative data should be made in comparison to control data from similarly processed material from the same site in normal individuals of the same sex and age. In some cases, further information on a patient's height, weight, or biophysique may be required, particularly with respect to studies of muscle from children with systemic growth disturbances.

If collected in a reliable and systematic manner, data from both qualitative and quantitative analyses of the muscle biopsy, clinical data (such as muscle strength, disease severity, disease duration) and perhaps information from other investigations (e.g. CAT or MRI scans, EMG, NCV and biochemical tests), may be usefully compared and given to proper statistical evaluation. This facilitates more objective and accurate diagnoses, and hopefully contributes to our understanding of neuromuscular disease processes and treatment regimes.

3.51 Semi-automatic digitiser system (Leitz-Kompira Video Imagan) with microscope TV attachment, used in the estimation of fibre size and shape.

Pathological reactions of skeletal muscle

As mentioned earlier in this chapter, the basic pathological responses of muscle tissue to internal or external stimuli are limited. They are summarised as follows.

Alterations in histological composition:
- Fibrosis.
- Inflammation.

Alterations of the muscle fibre:
- Atrophy, hypotrophy or hypertrophy.
- Degeneration or regeneration.
- Intracellular or sarcolemmal alteration.
- Hypoplasia or hyperplasia (rare).

Combinations of these processes often occur and may change with the natural history of the disease. Their aetiology is best considered as of either **myopathic** or **neurogenic** origin, as will be discussed in Chapter 4. Later chapters will then consider in more detail the X-Linked Recessive Myopathies (Chapter 5), Spinal Muscular Atrophies (Chapter 6), Inflammatory Disorders (Chapter 7), Congenital Myopathies (Chapter 8), Metabolic Disorders (Chapter 9), and other miscellaneous, less commonly biopsied, and toxin-induced disorders (Chapter 10).

The interpretation of the pathological changes in skeletal muscle must be made in the light of the clinical information outlined in the following chapters. Only then can a differential diagnosis be attempted (Chapter 11).

4

Pathological reactions of the muscle fibre

*All colour figures, except **4.6**, are of unfixed frozen specimens from the lateral portion of quadriceps femoris muscle.*

Although we have stressed in the previous chapter that the muscle cell has a limited response to insult, the pathological changes of the muscle cell may be the result of any one, or more, of the following:

- Interruption or alteration of its nerve supply.
- Disturbances, and other defects of the cell membrane.
- Disturbances and defects of muscle cell metabolism, which may be primary or secondary by nature.
- Abnormalities of growth and development.
- A response to systemic disease.
- Interruption or alteration of the blood supply to the muscle.

The overall pattern of the pathological response may, however, be indicative of the major groups of clinical neuromuscular disorders, i.e. neurogenic, myopathic and metabolic.

Neurogenic disorders

Beginning with the motor unit (**3.6**), we will first consider the neurogenic disorders, that is, the diseases affecting the anterior horn cell, its axon and the myoneural junction. In practice, the commonest disorders that will require a muscle biopsy are those derived from abnormalities of the anterior horn cell. The principal effect of denervation is to cause the muscle fibre to atrophy (**4.1**, see also **3.6**). This atrophy can be reversed if effective reinnervation takes place within a short period of time.

In some denervating processes of central origin, the atrophied fibres are arranged in groups of varying sizes (**4.2–4.3**). This observation can be of considerable diagnostic help, although its physiological basis remains obscure. Atrophy may progress until the bulk of the sarcoplasm is lost, and all that remains are clumps of nuclei within a redundant basement membrane (**4.2**, **4.4–4.7**). The nuclear membranes are preserved intact (**4.4**).

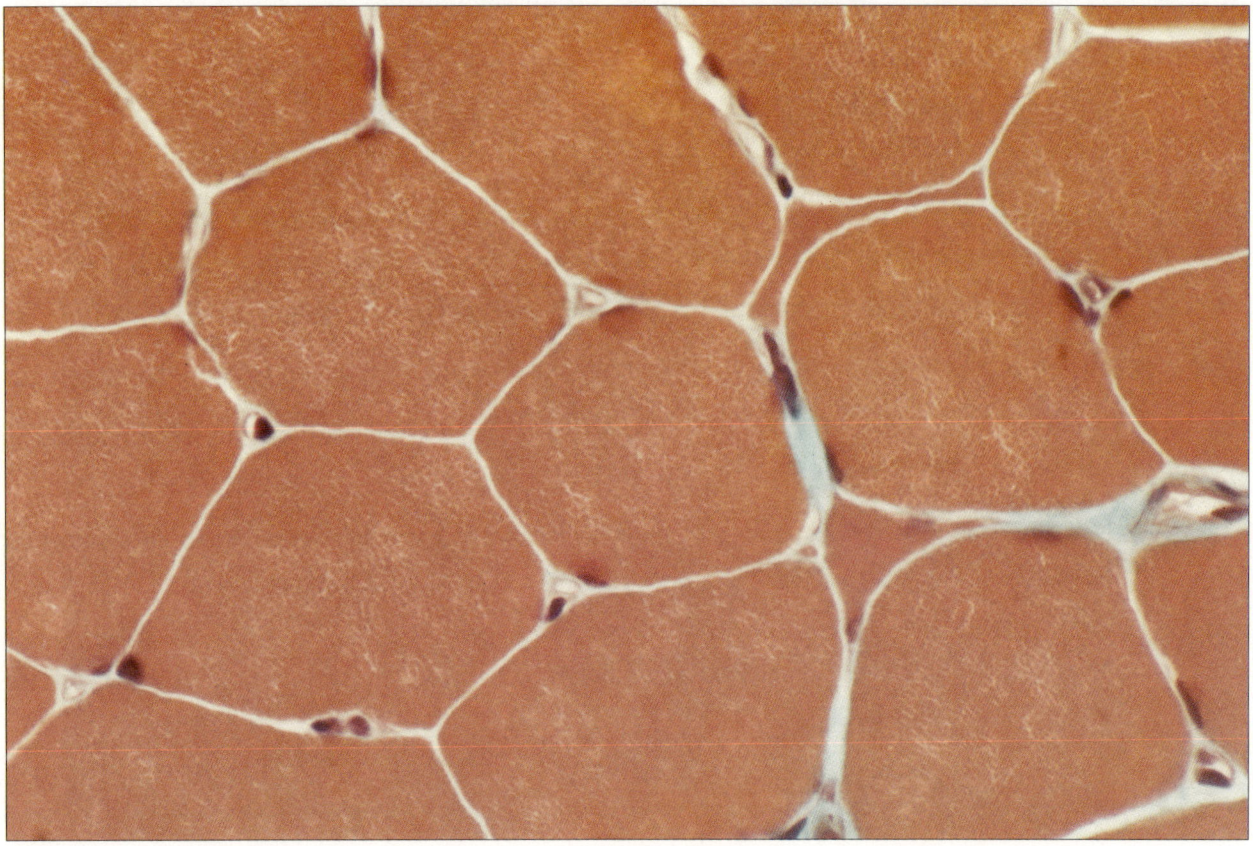

4.1 Occasional angular and atrophic muscle fibres are scattered amongst fibres of normal size and shape. Both histochemical fibre types may be affected. (Masson's trichrome, ×733)

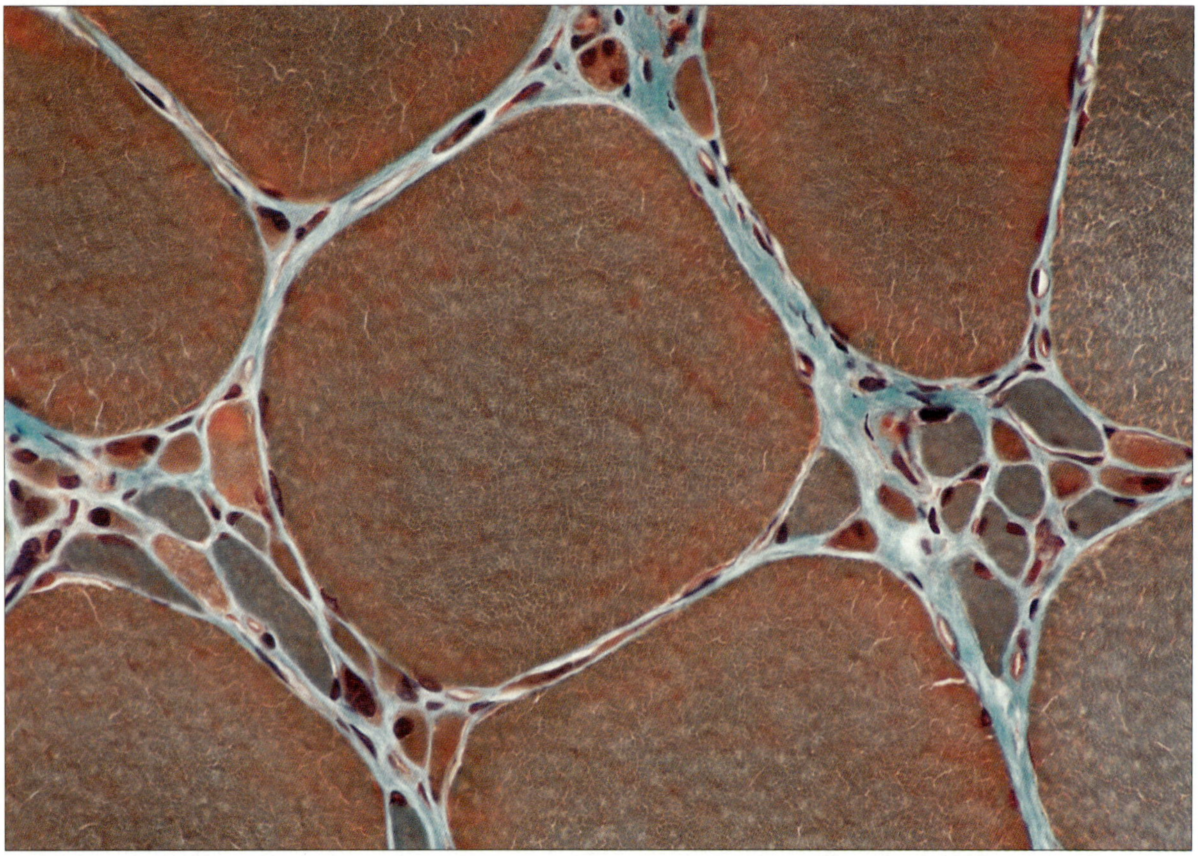

4.2 Groups of atrophic muscle fibres can be seen amongst normal fibres. Many atrophic fibres are angular in outline, and some contain clumps of pyknotic heterochromatic nuclear material. These fibres stain darkly with NADH-Tr and non-specific esterase.(Masson's trichrome, ×473)

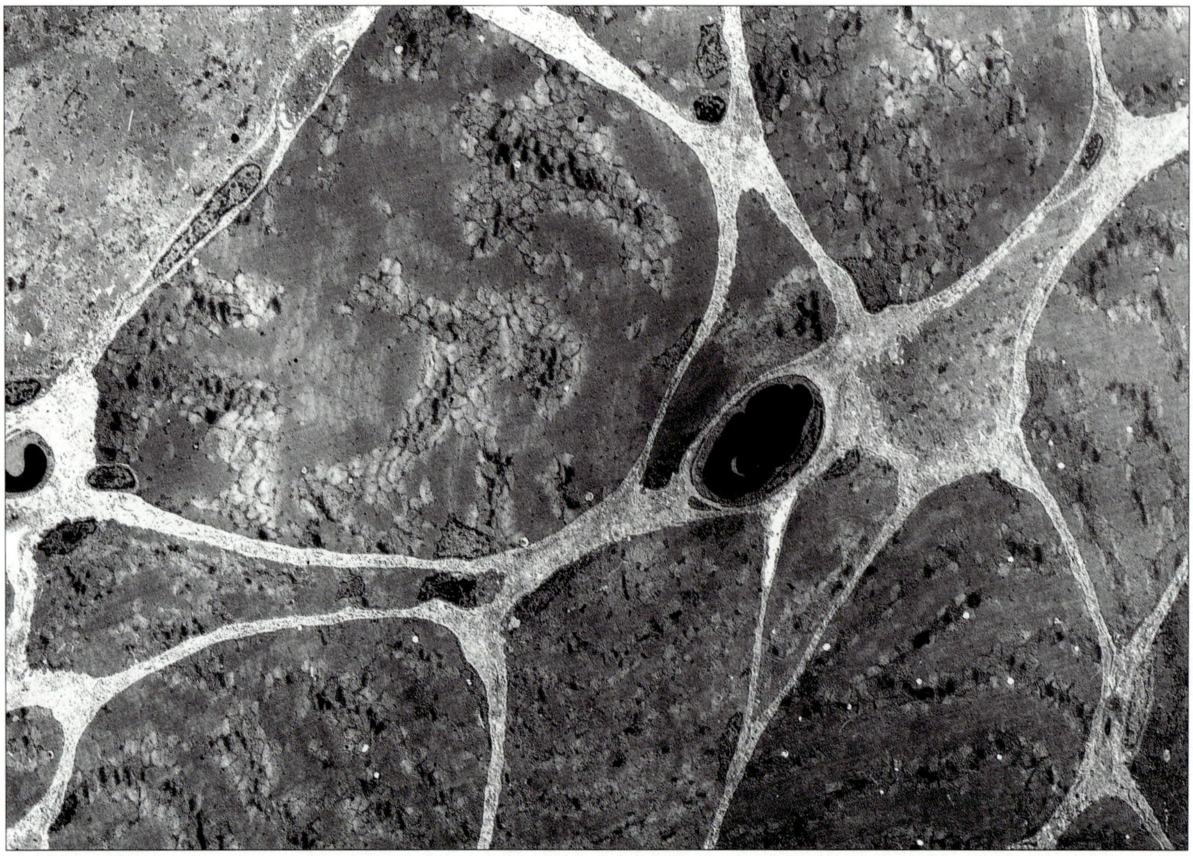

4.3 The selective atrophy of muscle fibres due to anterior horn cell disease is illustrated. The small fibres are very angular in outline. (×2750)

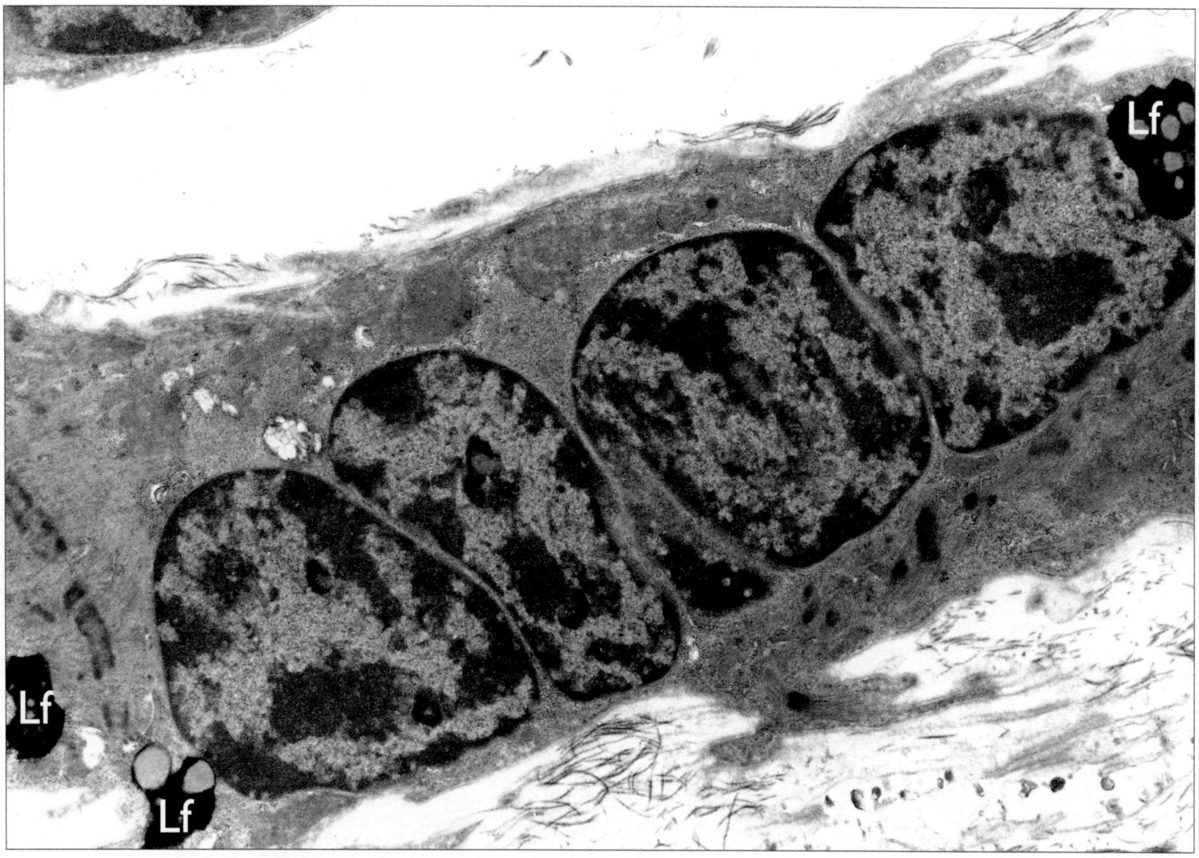

4.4a Rows of nuclei may be seen in longitudinal sections of atrophic fibres. There are gross alterations to the myofibrillar structure, with destruction of the sarcomeres, and, in chronic disease, the perinuclear accumulation of electron-dense lipofuscin droplets (Lf). (×7500)

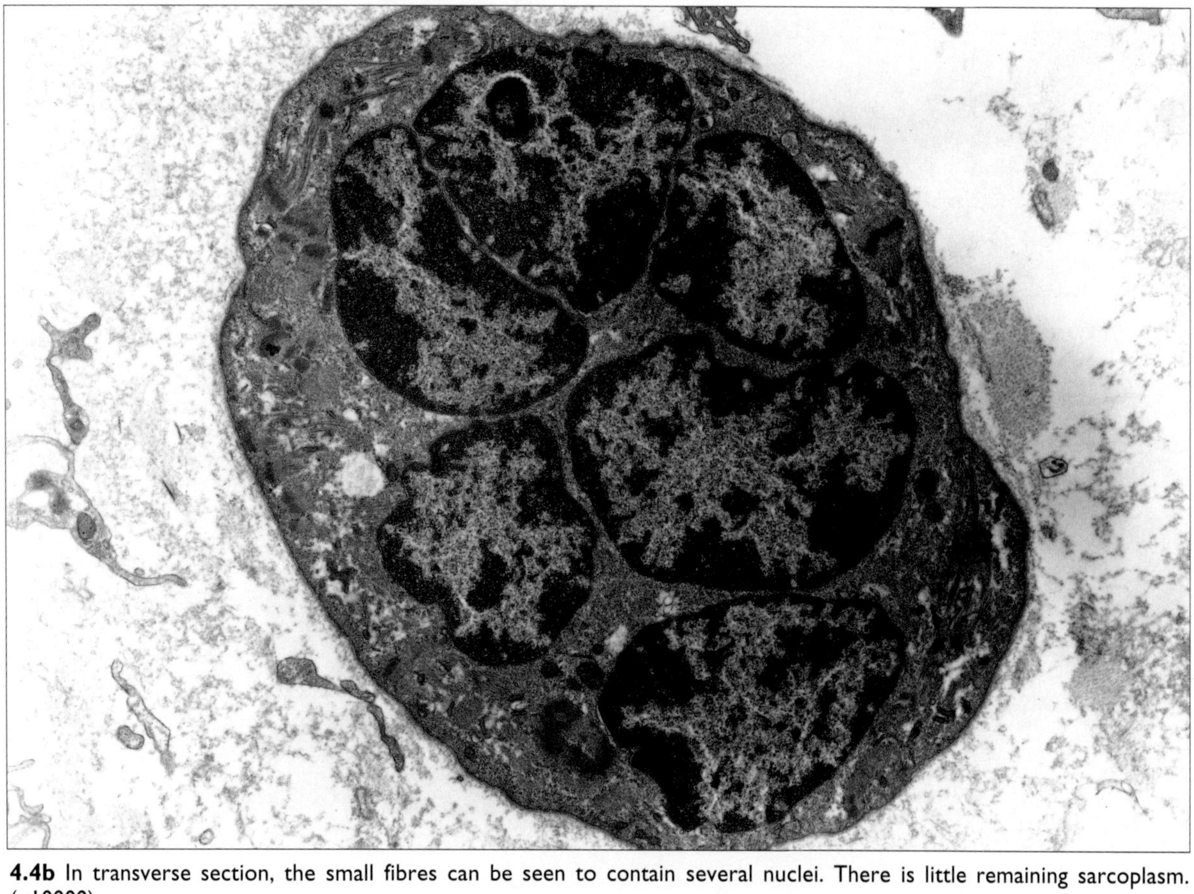

4.4b In transverse section, the small fibres can be seen to contain several nuclei. There is little remaining sarcoplasm. (×10000)

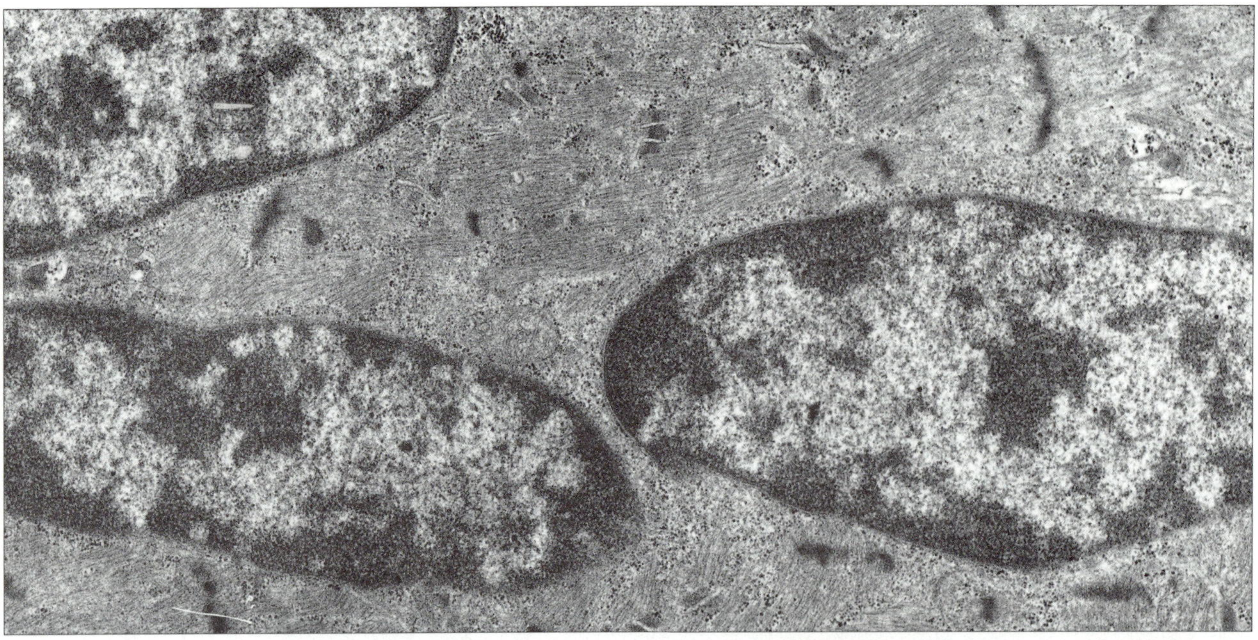

4.4c The denervated fibre shows disruption of the normal sarcomeric structure. There is loss of both myosin and actin filaments, and the disorientated fibrils are reduced in width. Few mitochondria are present (see **4.62**). (×18375)

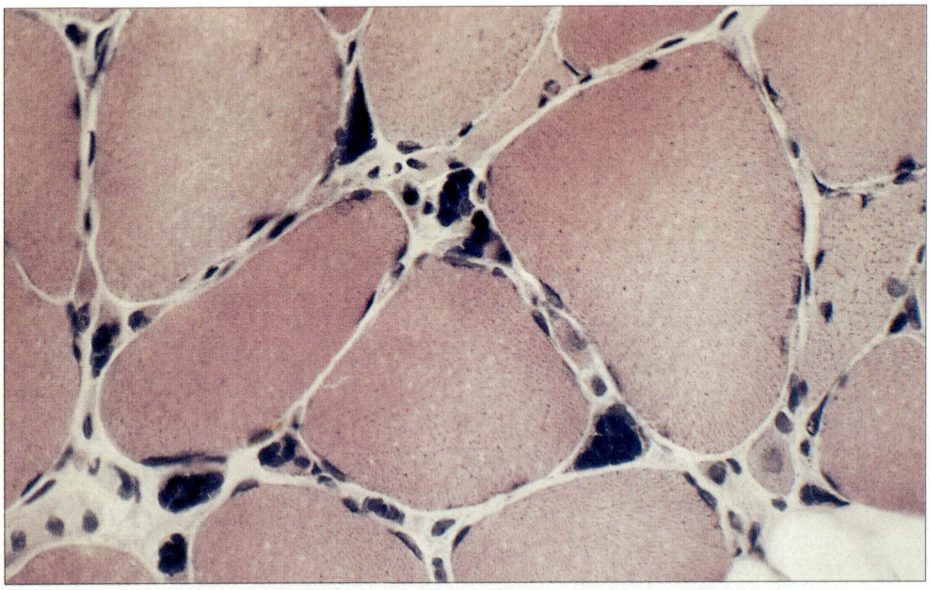

4.5 Scattered atrophic fibres containing dense nuclear clumps.
(H&E, ×350)

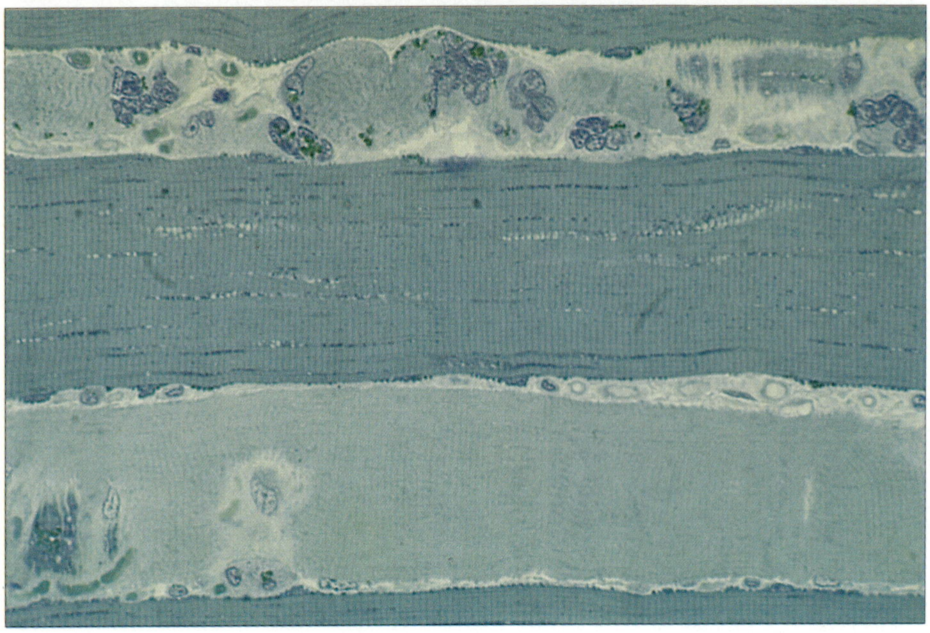

4.6 Longitudinal, semithin, epoxy resin section from the same specimen depicted in **4.5**. Note the presence of nuclear clumps at intervals along the pale-stained atrophic muscle fibre. The nuclear membranes can be seen clearly in these clumps.
(Toluidine blue, ×438)

Simultaneous denervation and reinnervation are seen frequently in chronic denervating processes of any kind, but are particularly common in Spinal Muscular Atrophy (SMA), where fibres which have lost their normal innervation are reinnervated by collateral sprouting from surviving motor nerve axons. The major consequence of this is the acquisition, by the reinnervated fibres, of the histochemical profile of the reinnervating axon. This results in large numbers of adjacent fibres being either uniformly Type I or II in profile, a phenomenon known as 'grouped reinnervation'; target/targetoid changes may be seen in both cases (**4.8**, **6.7**–**6.9**).

Fibres that have undergone chronic denervation or reinnervation for long periods of time often exhibit features of what is known as 'secondary' myopathic degeneration (**4.9**).

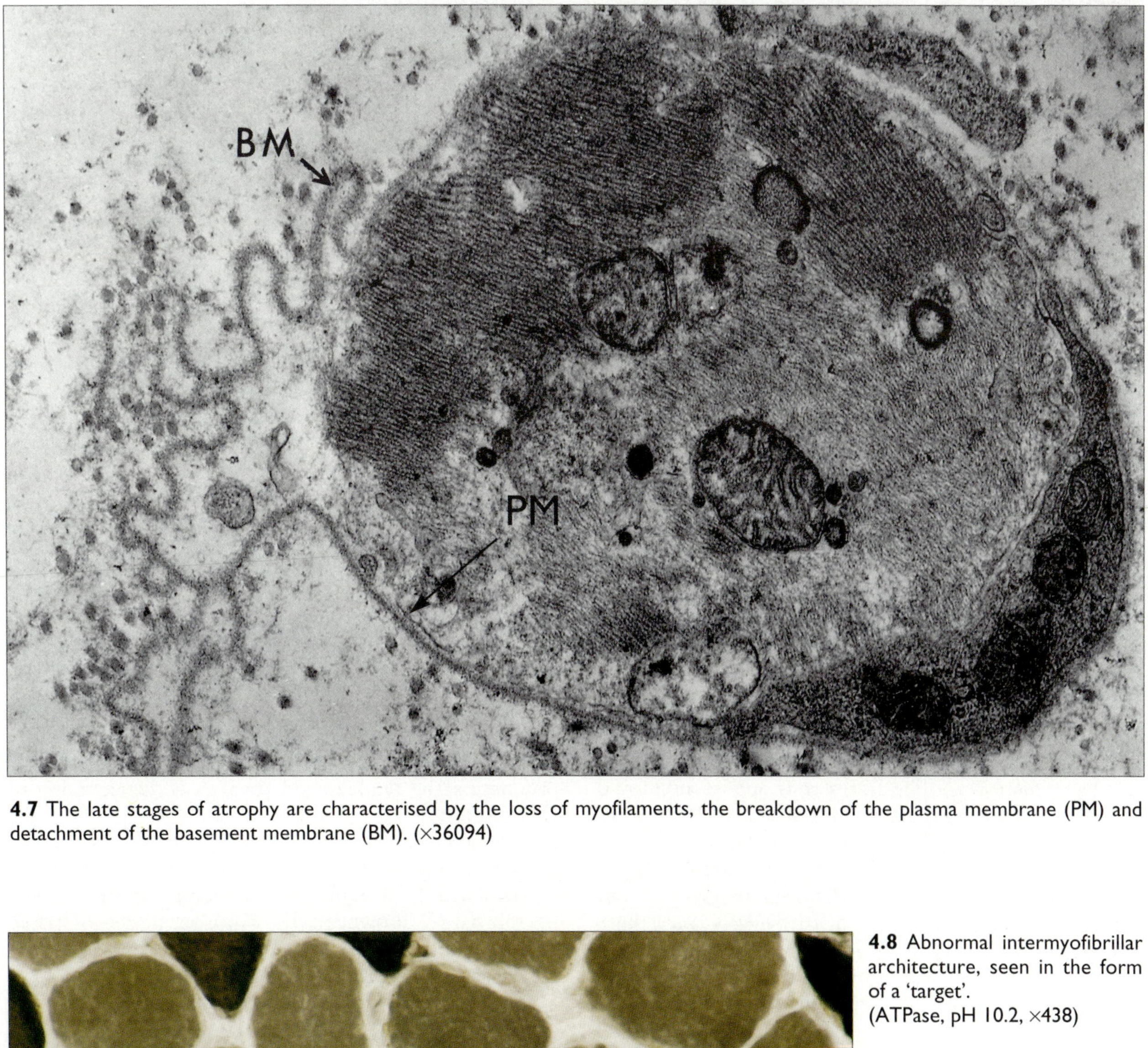

4.7 The late stages of atrophy are characterised by the loss of myofilaments, the breakdown of the plasma membrane (PM) and detachment of the basement membrane (BM). (×36094)

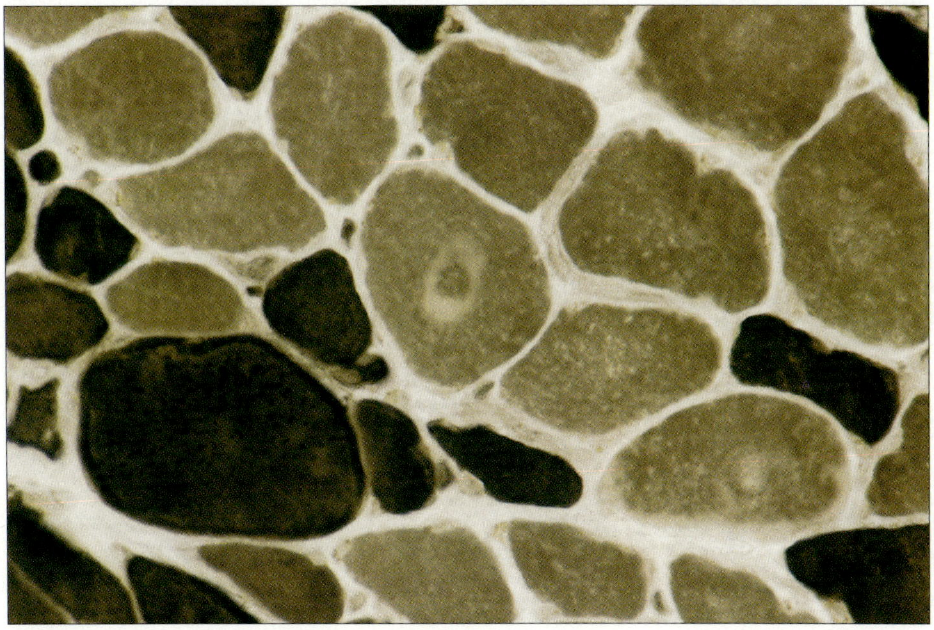

4.8 Abnormal intermyofibrillar architecture, seen in the form of a 'target'. (ATPase, pH 10.2, ×438)

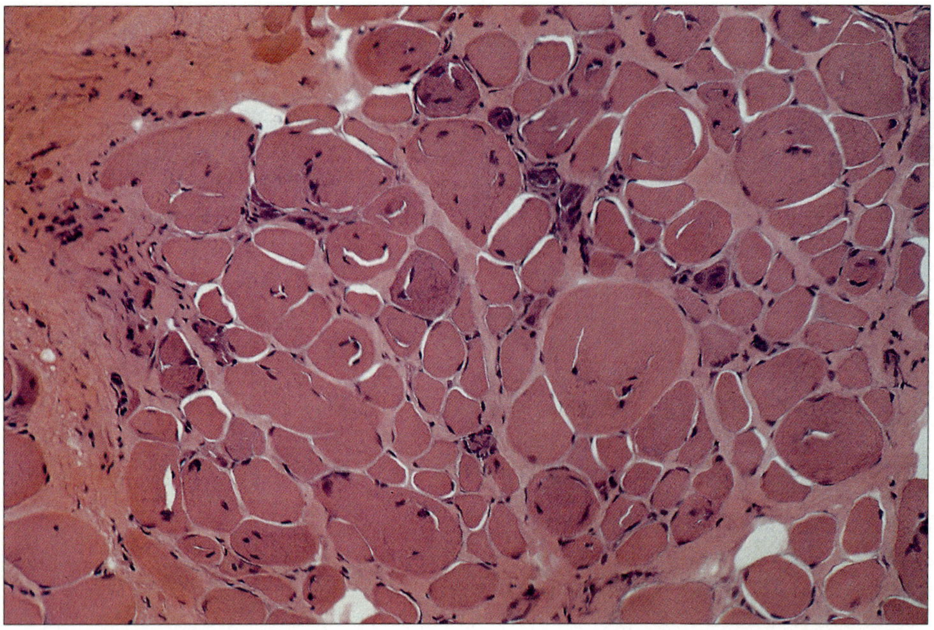

4.9 Excessive endomysial and perimysial connective tissue, rounding of many muscle fibres, great variation in fibre size, splitting associated with internal nuclei; whorled fibres, basophilia, and phagocytosis. (H&E, ×175)

Growth and development disorders

In diseases thought to be due to the failure of growth and development, the muscle fibres have never attained their normal dimensions. The fibres thus appear small and rounded, and are described as hypotrophic (**4.10**, **8.3**). In this situation, the hyoptrophic fibres tend to show Type I staining characteristics. As far as we are aware, the hypothesis that this is due to a disorder of growth and development remains unproven, particularly since these changes may be seen in Myotonic Dystrophy (MyD) (Chapter 10). The small fibres seen in the myopathic conditions may have a different pathogenesis (below).

Systemic disorders

In systemic disorders, muscle fibre atrophy also occurs as a non-specific reaction, notably in disuse, drugs, cachexia, ageing and carcinoma. In many of these it tends to be selective, affecting Type II fibres more severely than Type I (**4.11**).

Vascular disorders

In the vascular disorders, attention is directed to the vessel wall (e.g. the vasculitides **5.20**, **7.6**) and to changes in the surrounding myofibres as a consequence of ischaemia (see **7.19**).

Myopathic disorders

In the primary myopathic disorders, considered to be due to intrinsic defects of the muscle cell membrane and its related structures (e.g. Xp21.2 deletional myopathies) or to external insults (e.g. immunological, toxic or traumatic), the fundamental reaction is degeneration of the myofibre.

The sequence of events leading to cell death, and subsequent regeneration, are broadly similar, irrespective of the natural or experimental insult offered to the muscle fibre. However, the time sequence of these changes varies considerably, and may depend upon the nature and degree of the insult or defect.

The older accounts of myopathology tended to concentrate on the more chronic 'myopathic' changes, which include:

- Atrophy and hypertrophy of the muscle fibres (**4.12**, **5.15**).
- Internalisation of the muscle nuclei (**4.13**, **5.12**).
- Necrosis.
- Regeneration and an increase in the amount of interstitial fibrous connective tissue (**4.10**, **5.15**).
- Splitting of the larger fibres (**4.14–4.16**).

Such splits can be demonstrated to be of the same histochemical type as the parent fibre. Serial sections reveal that the appearance of the splits varies along the length of the fibre. Hypertrophy of muscle fibres is also commonly seen in this situation (**4.17**). Endomysial fibrosis may be gross, and in some instances (**4.17–4.18**) causes severe deformational changes in the surviving muscle fibres (**4.18**, **5.17**).

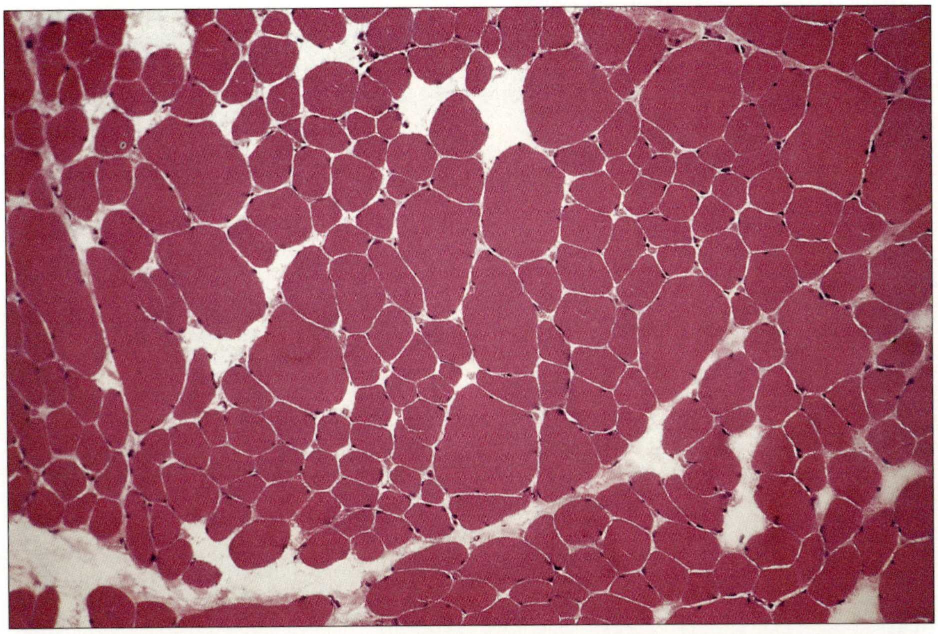

4.10 This abnormal degree of variation in fibre size is due to hypotrophy, together with some compensatory hypertrophy, in a case of congenital fibre type disproportion.
Unlike denervation atrophy, the muscle fibre shape is normal.
(H&E, ×140)

4.11 Type II fibre atrophy is predominant in this case of non-specific atrophy due to disuse. Note that the distribution of fibre types remains relatively normal.
(Myofibrillar ATPase, pH 10.2, ×175)

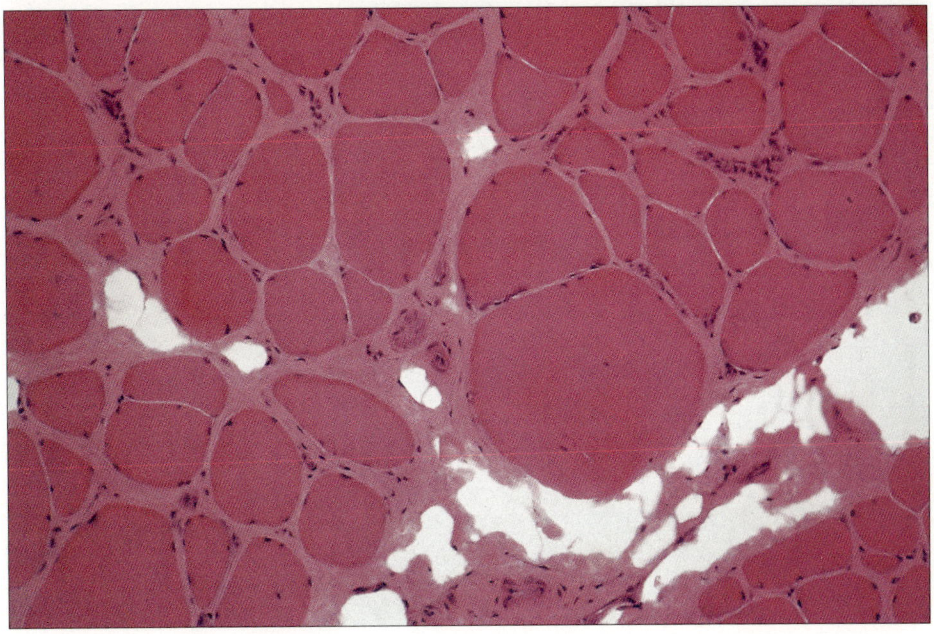

4.12 Increased variation in fibre size and shape and excess endomysial connective tissue throughout the biopsy are typical of a chronic myopathic condition.
(H&E, ×175)

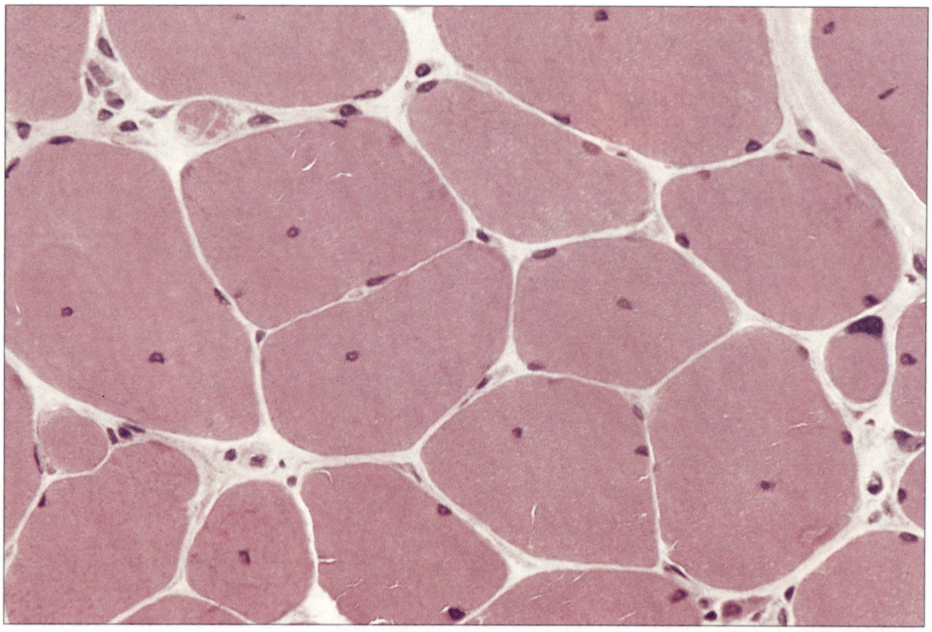

4.13 Internal myonuclei are present in many muscle fibres that are otherwise normal in appearance.
(H&E, ×350)

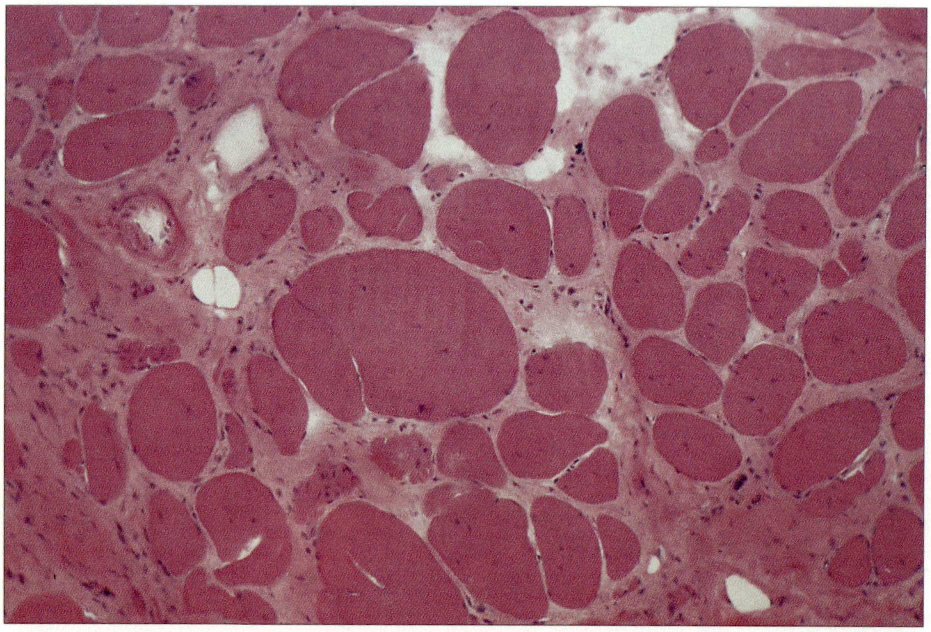

4.14 A chronic myopathy showing increased variation in fibre size. There is gross compensatory hypertrophy of some fibres, which exceed 150 µm in diameter.
Excess connective tissue, fatty infiltration, internal myonuclei and split fibres are also present.
(H&E, ×140)

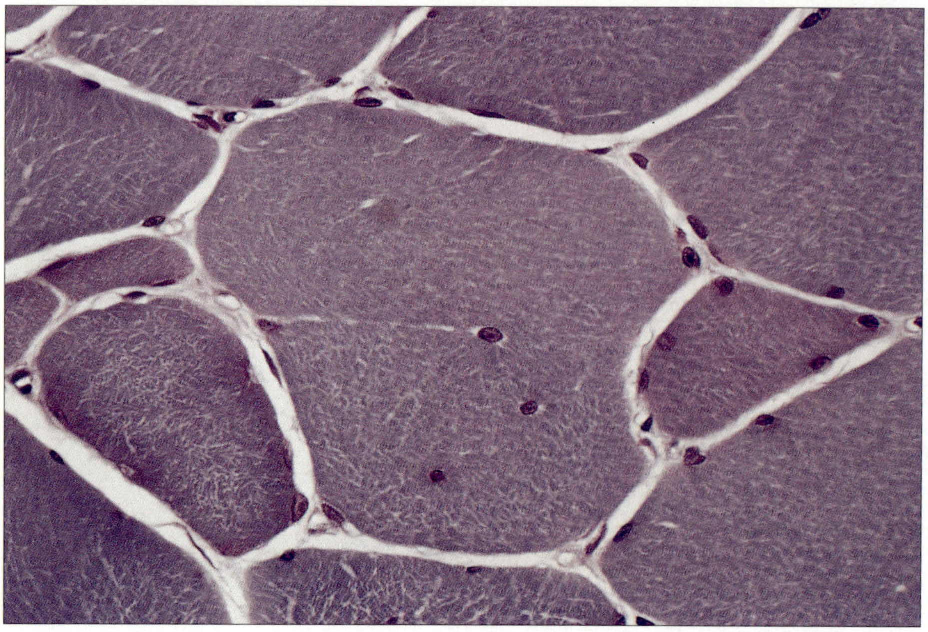

4.15 The large fibre in the centre of the field shows a split 'guarded' by nuclei at each end. Other internal nuclei, large 'activated' nuclei, and evidence of fibre splitting are also present.
(Toluidine blue, ×350)

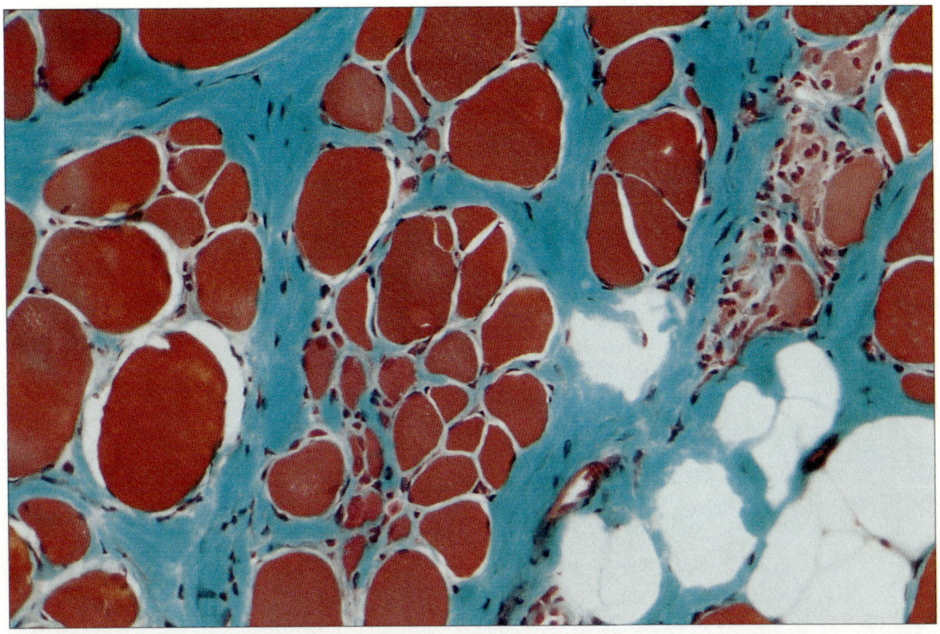

4.16 The marked variation in muscle fibre size, fibre splitting, necrosis, and the extensive proliferation of endomysial and perimysial connective tissue with fatty replacement throughout this specimen is typical of a long standing myopathy.
(Masson's trichrome, ×217)

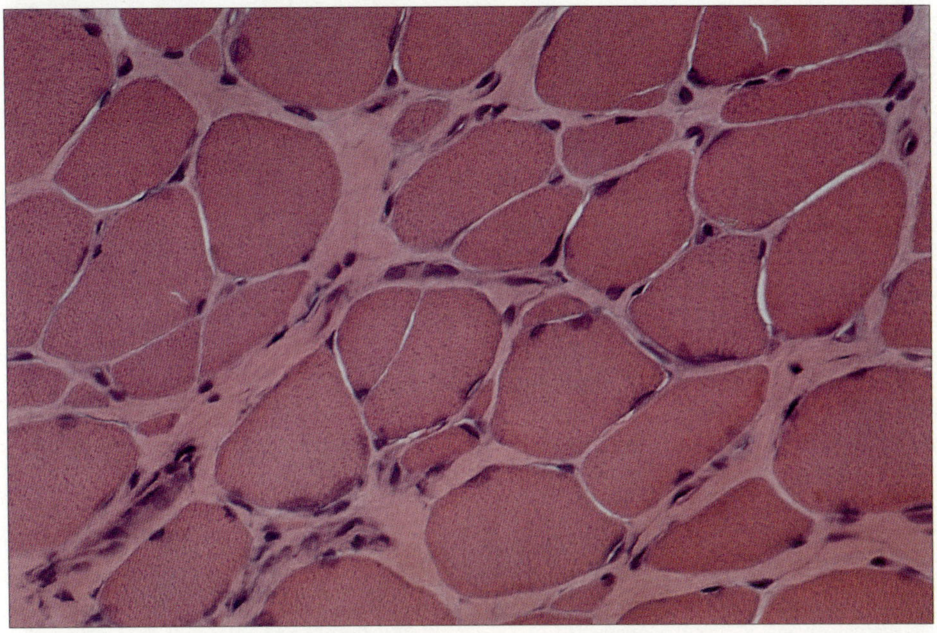

4.17 Fibre splitting, gross variation in fibre size and shape, and excess connective tissue.
(H&E, ×350)

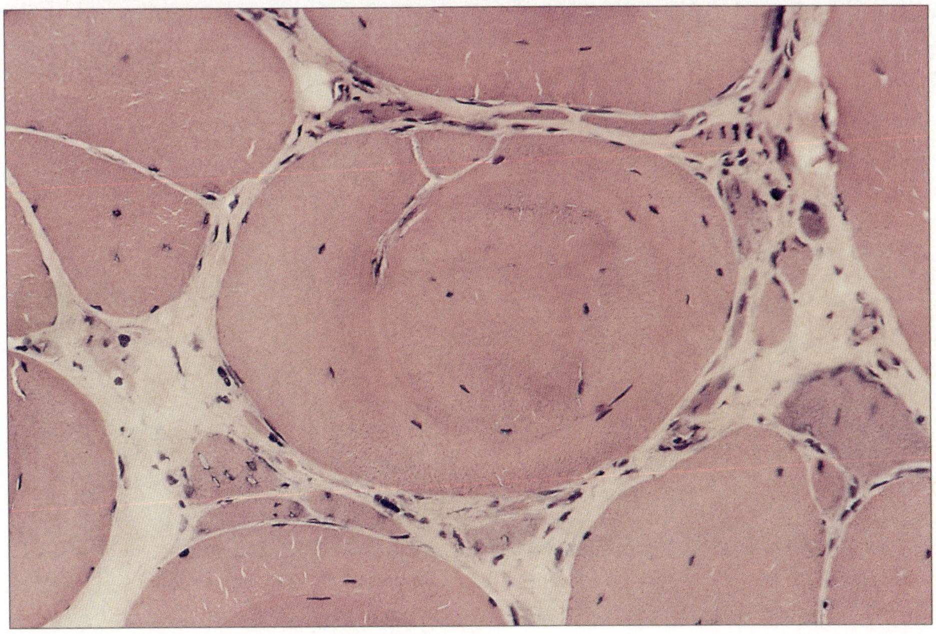

4.18 Excessive splitting and whorl-like features, associated with internal nuclei in a large muscle fibre.
(H&E, ×350)

These chronic changes are typical of the secondary myopathic changes seen in the chronic denervating states (**4.9**). One histological abnormality seen in both chronic myopathic and denervating states is the so-called 'ring' fibre, which is produced by myofibrils wrapping themselves circumferentially around the surface of the fibre (**4.19-4.20**). The pathophysiological basis for the development of the ring fibres is unknown. The end-stage of myopathic degeneration is illustrated in **4.21**, which shows only a few scattered atrophic muscle fibres surrounded by swathes of coarse connective tissue and islands of fatty infiltration. Biopsy of this material is of no value diagnostically.

The changes seen in muscle fibre in more rapidly progressive myopathies are necrosis, with simultaneous phagocytosis (**4.22–4.23, 7.23–7.24**) and sequential

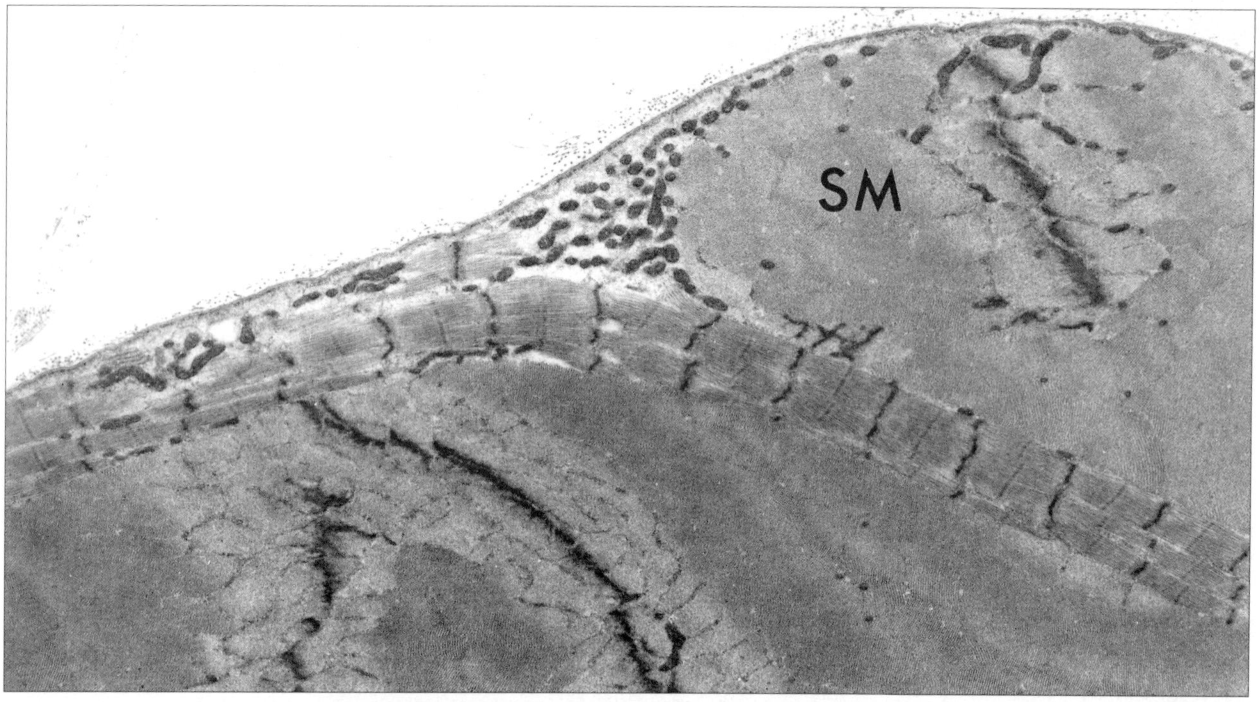

4.19a Myofibrils wrapped circumferentially around the fibre are seen in 'ringbinden' or ring fibres. They are a nonspecific abnormality and may be found in several disorders. The misaligned fibres have an isolated peripheral sarcoplasmic mass (SM), by turning inward from the sarcolemma. (×6300)

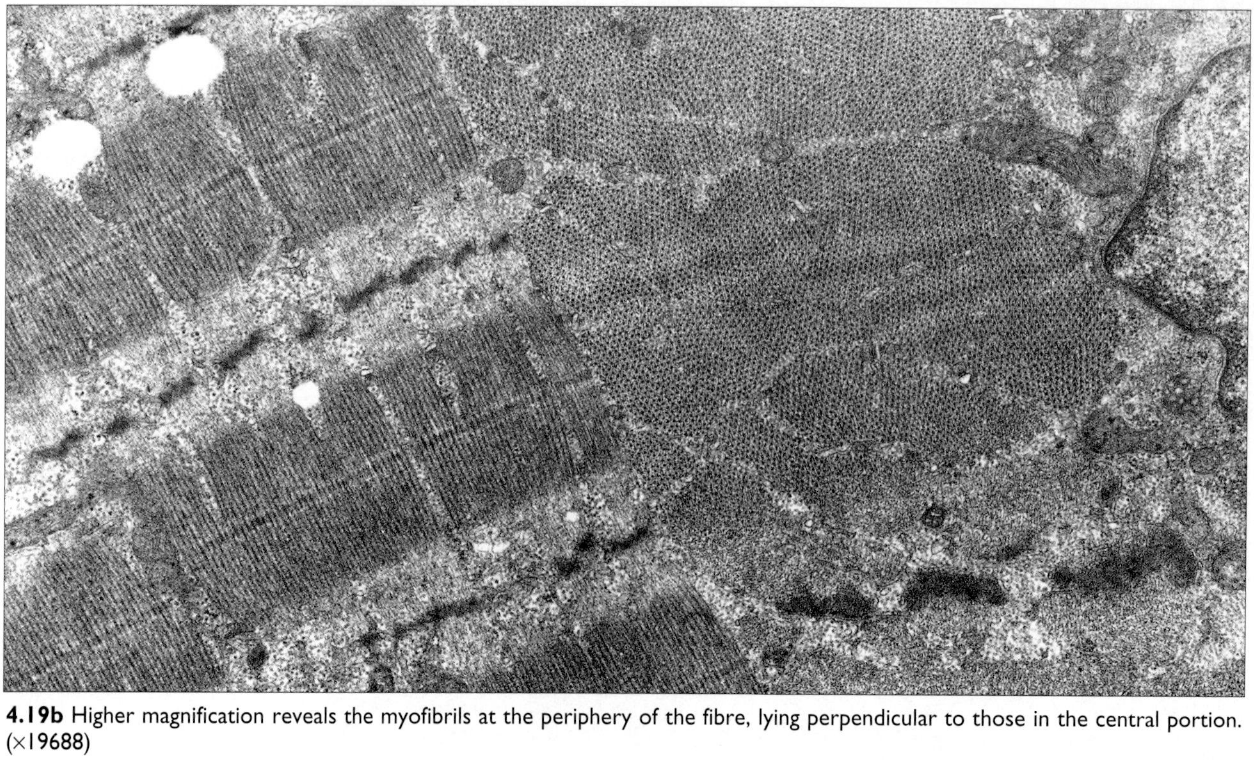

4.19b Higher magnification reveals the myofibrils at the periphery of the fibre, lying perpendicular to those in the central portion. (×19688)

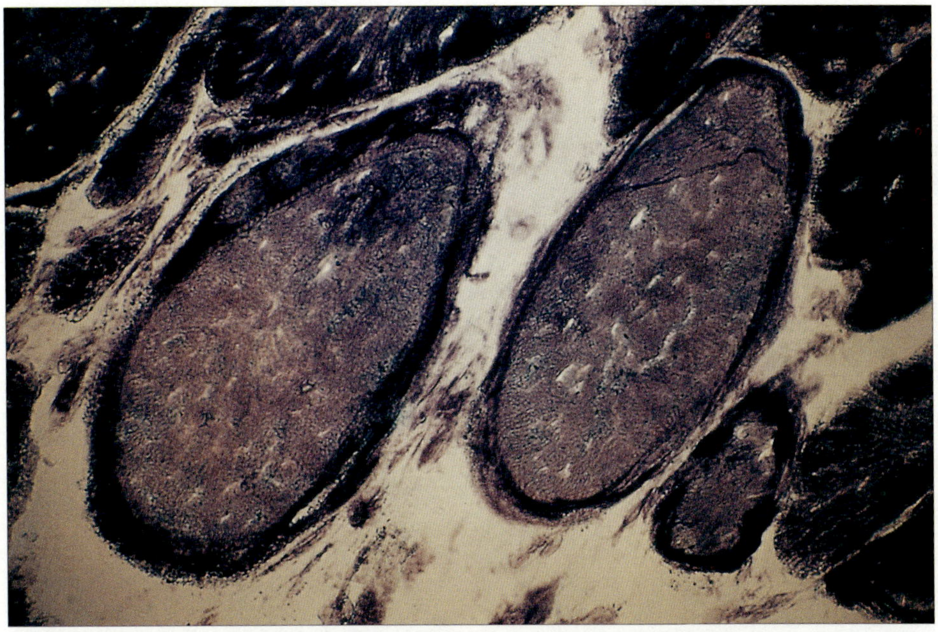

4.20 Ring fibres are visible as striations beneath the densely stained subsarcolemmal rim. They are due to myofibrils lying circumferentially to the main axis of the muscle fibre.
(NADH-Tr, ×350)

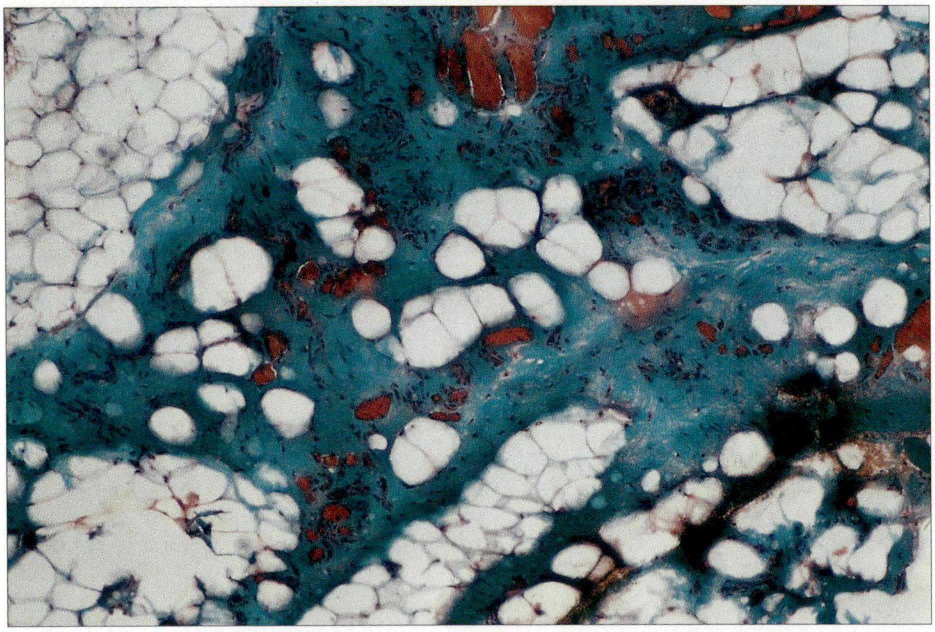

4.21 Typical end-stage appearance of a muscle disease, in which fat and fibrous connective tissue have almost completely replaced the muscle fibres.
(Masson's trichrome, ×70)

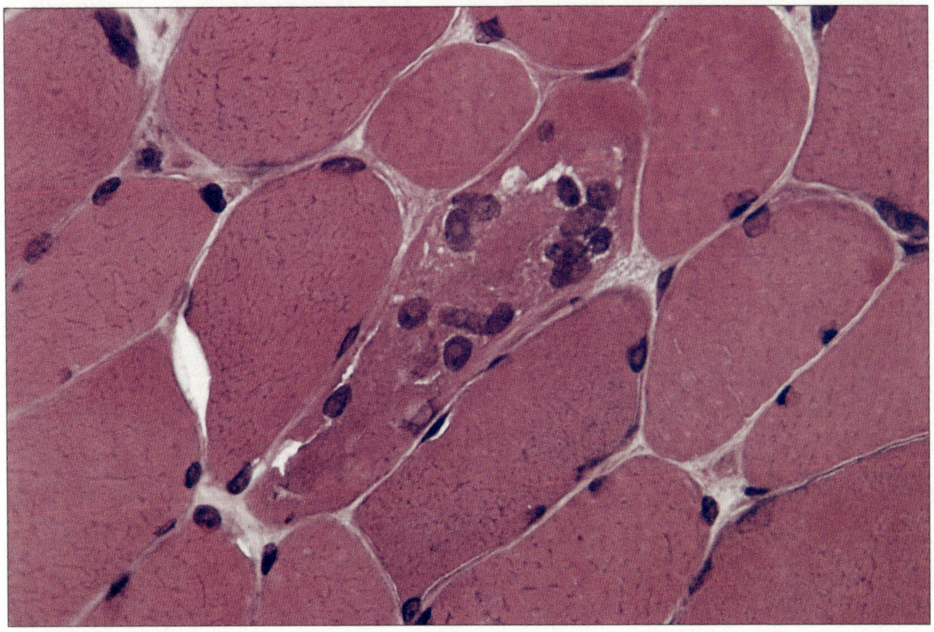

4.22 Numerous macrophages are visible inside a recently necrosed muscle fibre.
(H&E, ×473)

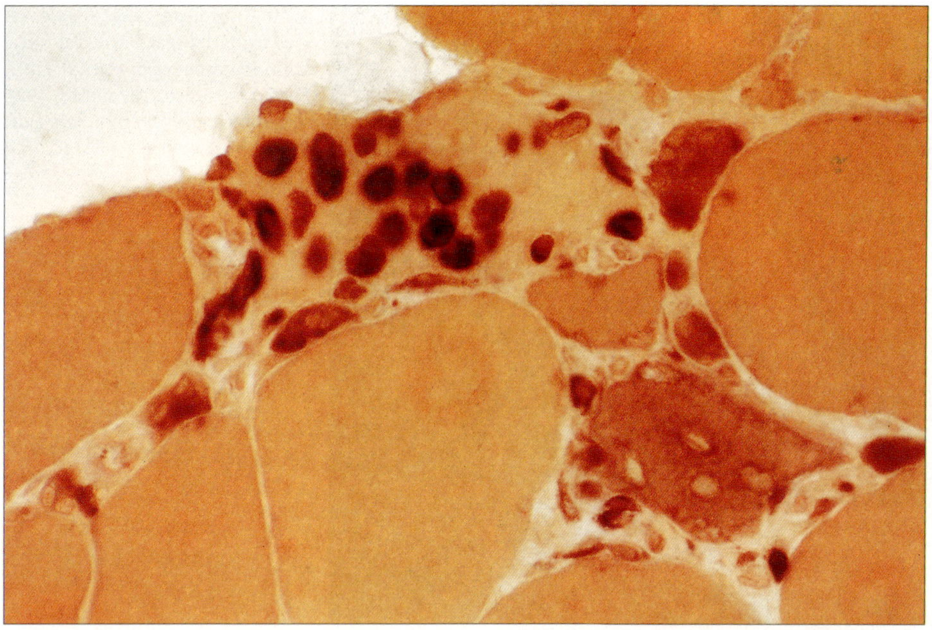

4.23 Both the muscle fibres undergoing phagocytosis and the phagocytes themselves are heavily stained because of their high lysosomal enzyme content. (Non-specific esterase, ×438)

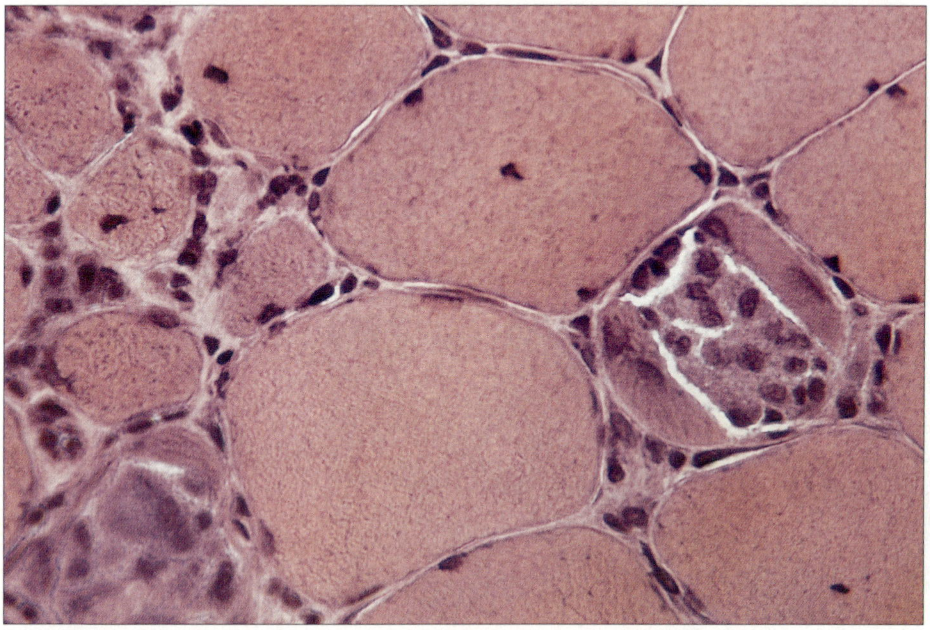

4.24 Necrotic fibres show early regeneration immediately beneath the sarcolemma, whilst necrosis and phagocytosis are evident in their centres. Central 'myonuclei' are present in small and normal sized fibres. (H&E, ×473)

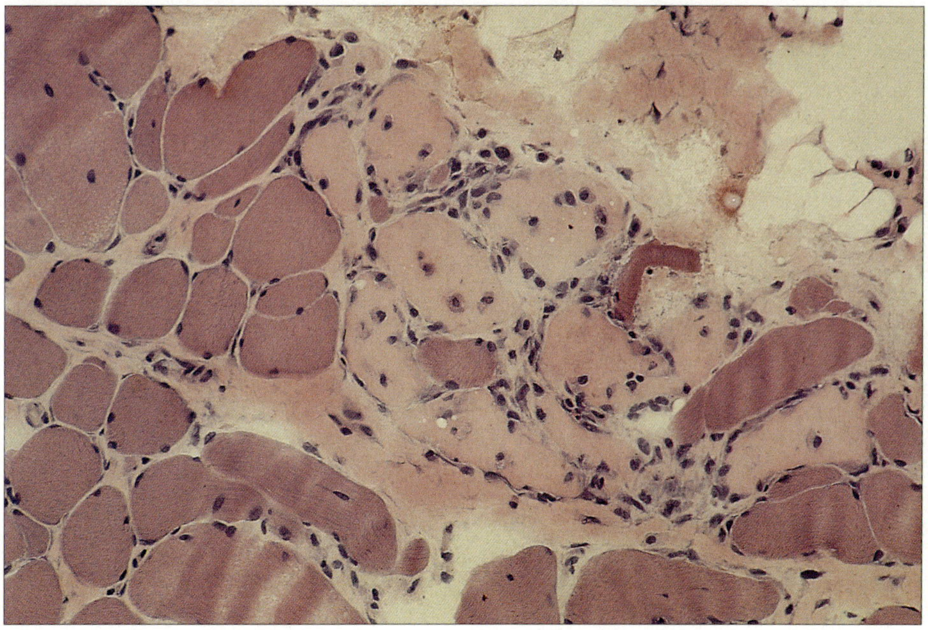

4.25 A focus of necrosis, containing groups of pale-staining muscle fibres invaded by macrophages can be seen. Fibre splitting and excessive connective tissue are also present. (H&E, ×175)

regeneration (**4.24**). Irrespective of the insult to the muscle cell membrane, this permits an excess influx of calcium which leads to disruption of homeostasis within the fibre. When the concentration of Ca^{2+} exceeds a critical level, activation of the process of degeneration is inevitable (**4.25**).

In many active necrotising myopathies, the primary event is thought to commence with the formation of 'delta' lesions (**4.26–4.27**). These lesions appear as focal subsarcolemmal dissolution of the myofilaments, and the loss of the plasma membrane.

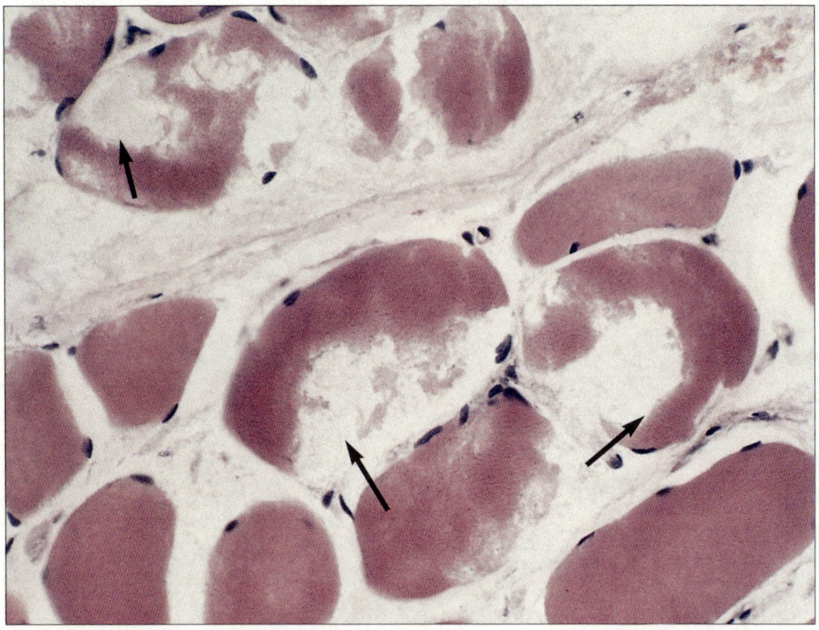

4.26 Delta lesions are visible in several muscle fibres as large pale areas lacking cytoplasmic detail (arrows). (H&E, ×350)

4.27 A pathological change that involves the myofibrils is hypercontraction of the fibre. This may be focal, and cause the clear distinction between A-bands and I-bands to disappear. The Z-lines are obscured by filaments compacted against them. (×5000)

It is thought that the excess intracellular Ca^{2+} produces a state of hypercontraction of the myofibrils from which they are unable to relax, resulting in the formation of hypercontraction bands (**4.28–4.30**, **5.3**).

It is possible that the stretching of the plasmalemma that is consequent upon this hypercontraction, may be an alternative explanation for the gaps in the plasmalemma that are seen in the muscle fibres (see **5.5**).

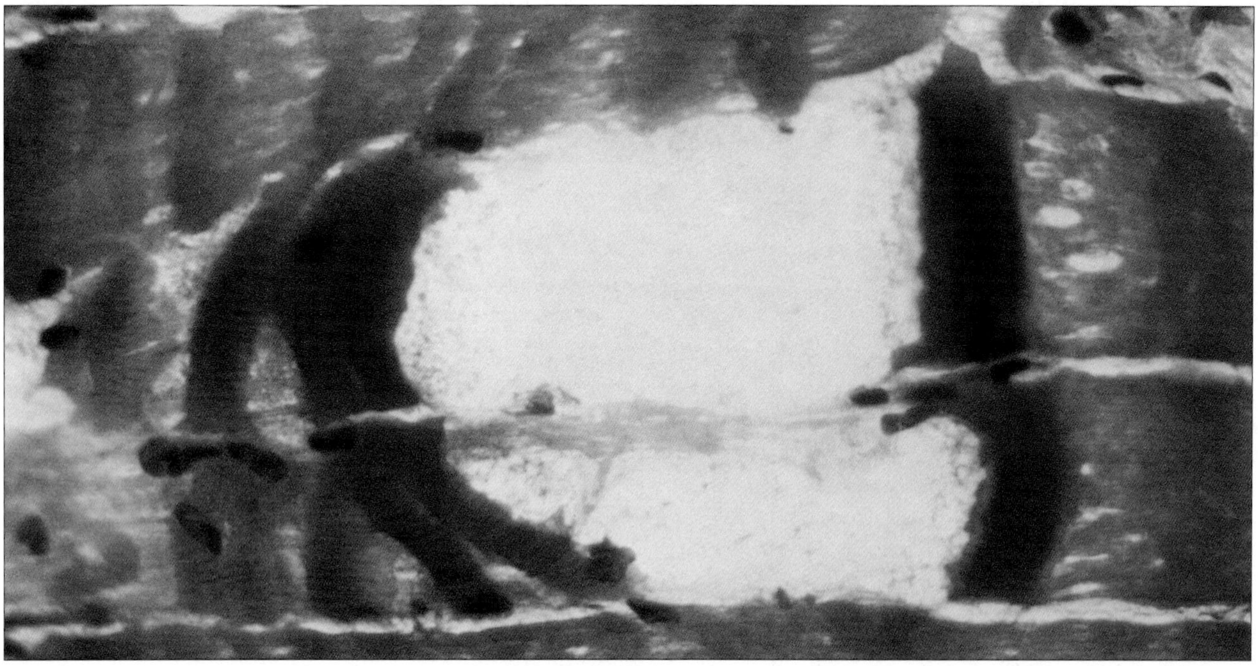

4.28 Longitudinal section showing hypercontraction bands separated by empty spaces. (H&E, ×638)

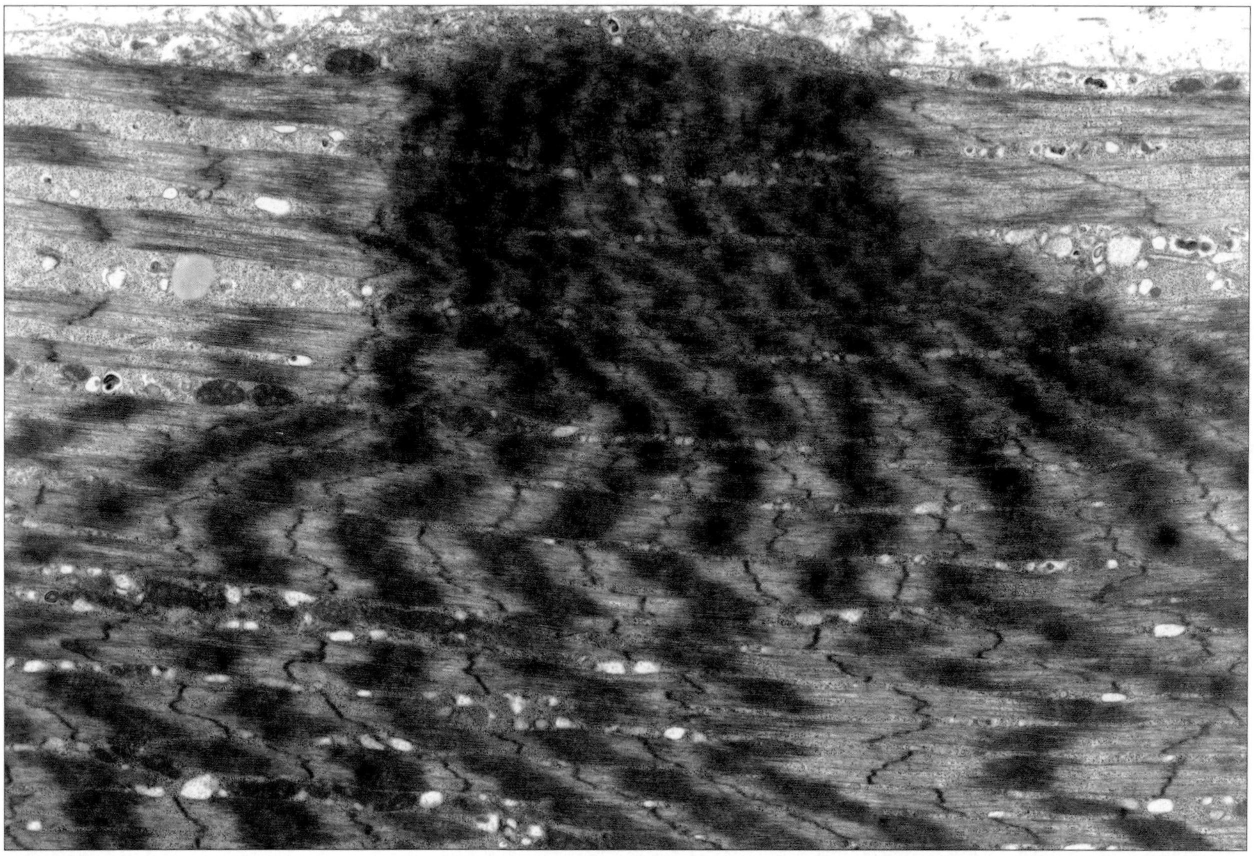

4.29a Hypercontraction, which may involve few or many sarcomeres, leads to hyperextension of those adjacent. Some filaments are still seen linking the Z-lines, and there is accumulation of glycogen-like granular material throughout the sarcoplasm. (×10500)

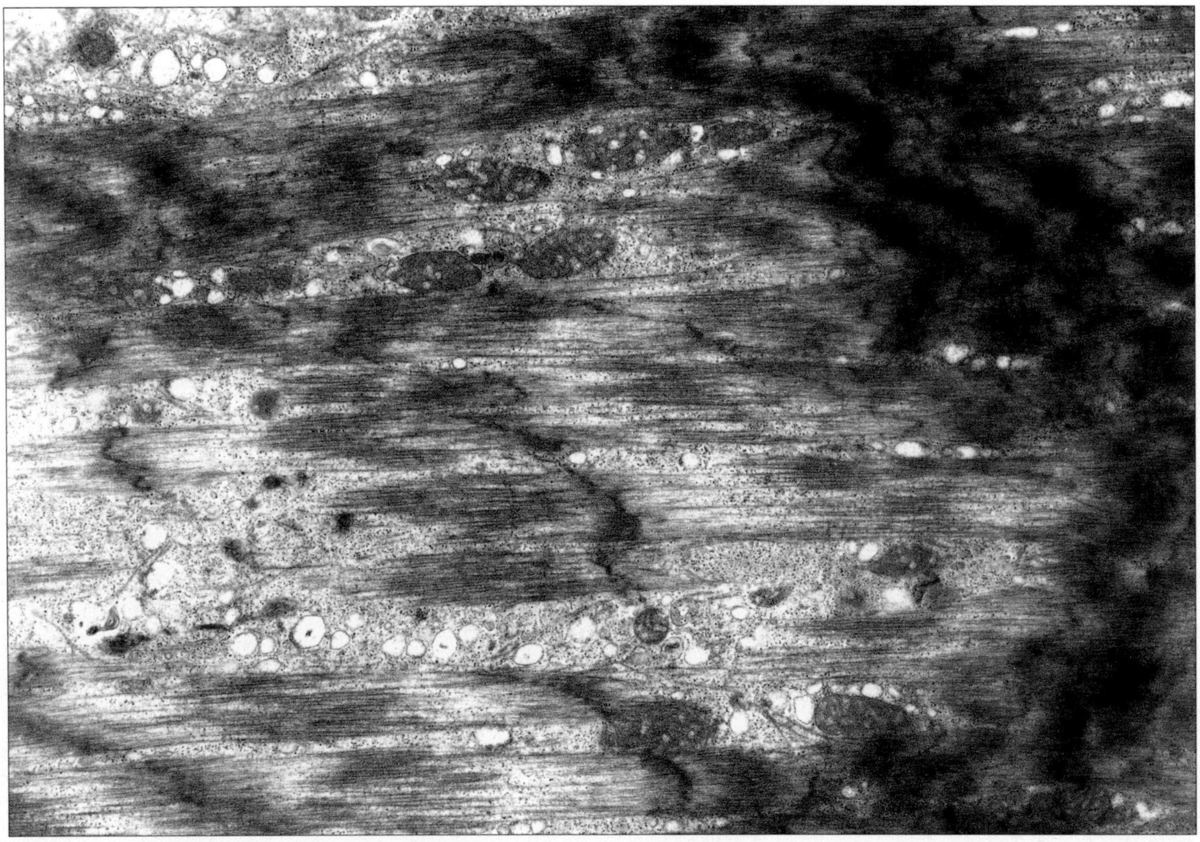

4.29b The separation of the A-bands and I-bands of the fibrils at the edge of such a lesion can be seen higher magnification. The filaments have been replaced by granular debris. (×15000)

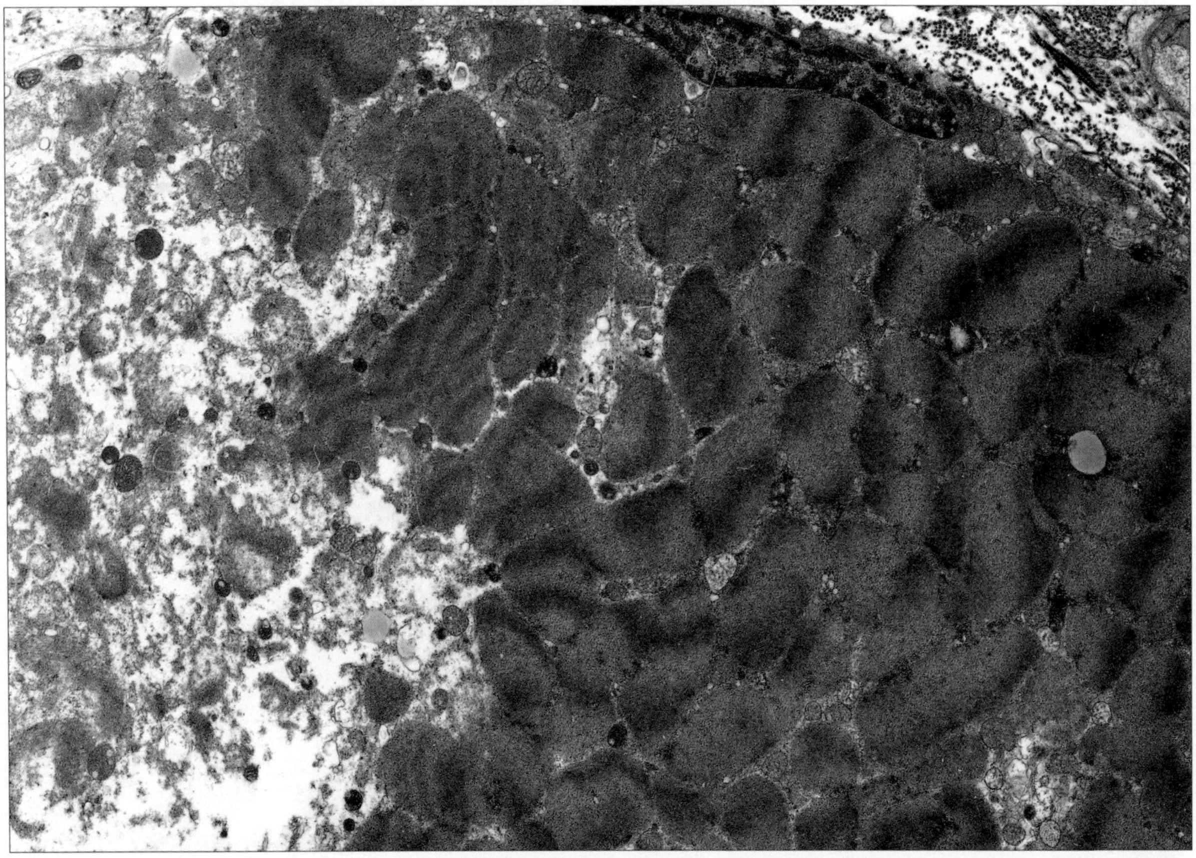

4.30 Fibres that appear to be hyaline by light microscopy contain clumped and disorientated myofibrils that are electron-dense. In and around these clumps, there are regions which are devoid of myofibrils. (×7950)

The hypercontraction bands are the basis for the 'hyaline' fibres that are seen with the light microscope in transverse sections from Duchenne Muscular Dystrophy (DMD) biopsies, and to a lesser extent in other myopathies (**4.28**, **4.31–4.32**,**4.29a**, **5.3**). As the hypercontraction bands form, a number of other changes occur in the contractual apparatus. In particular, there is the loss of I-bands (**4.33**) and the Z-band streaming, overstretching of the sarcomere, and the degeneration of the filaments to a granular debris.

The light micrograph correlate of **4.33** and **4.35–4.36** is **4.34**.

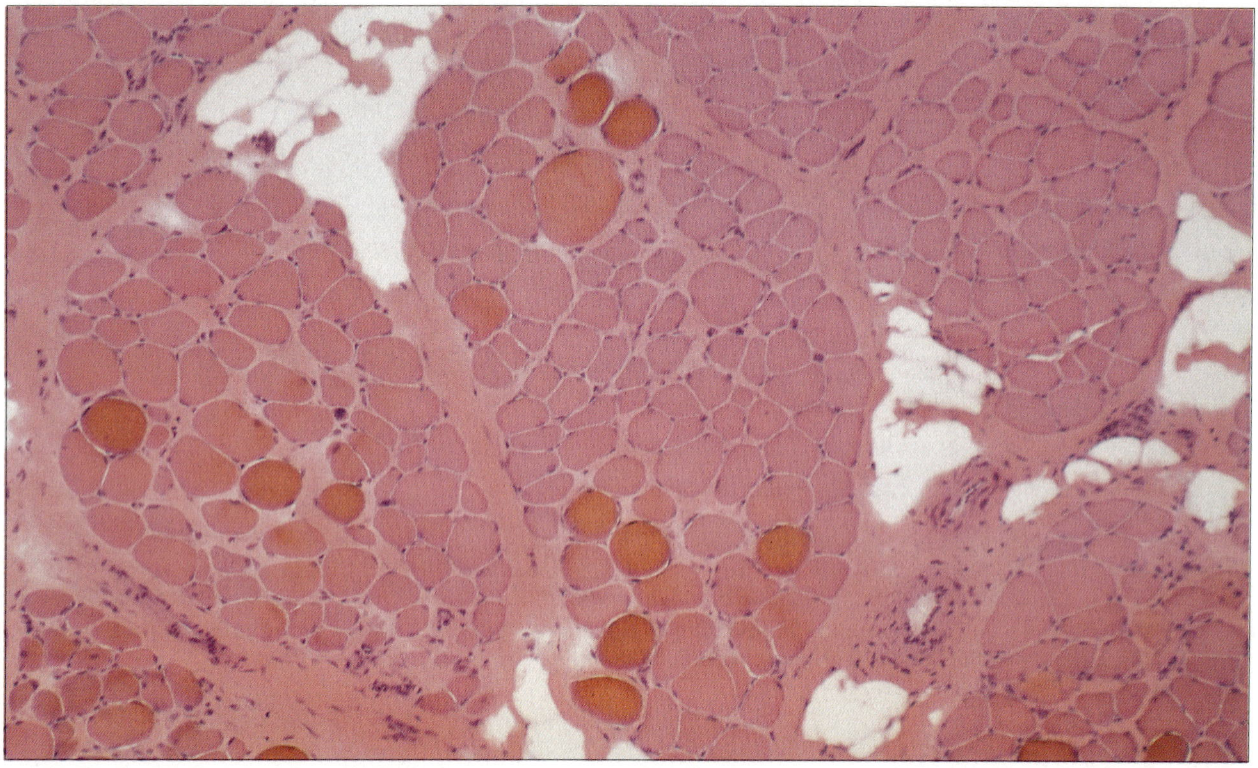

4.31 Numerous heavily stained, rounded, hyaline fibres are easily discernible in this section, which shows the typical appearances of Duchenne muscular dystrophy. (H&E, ×118)

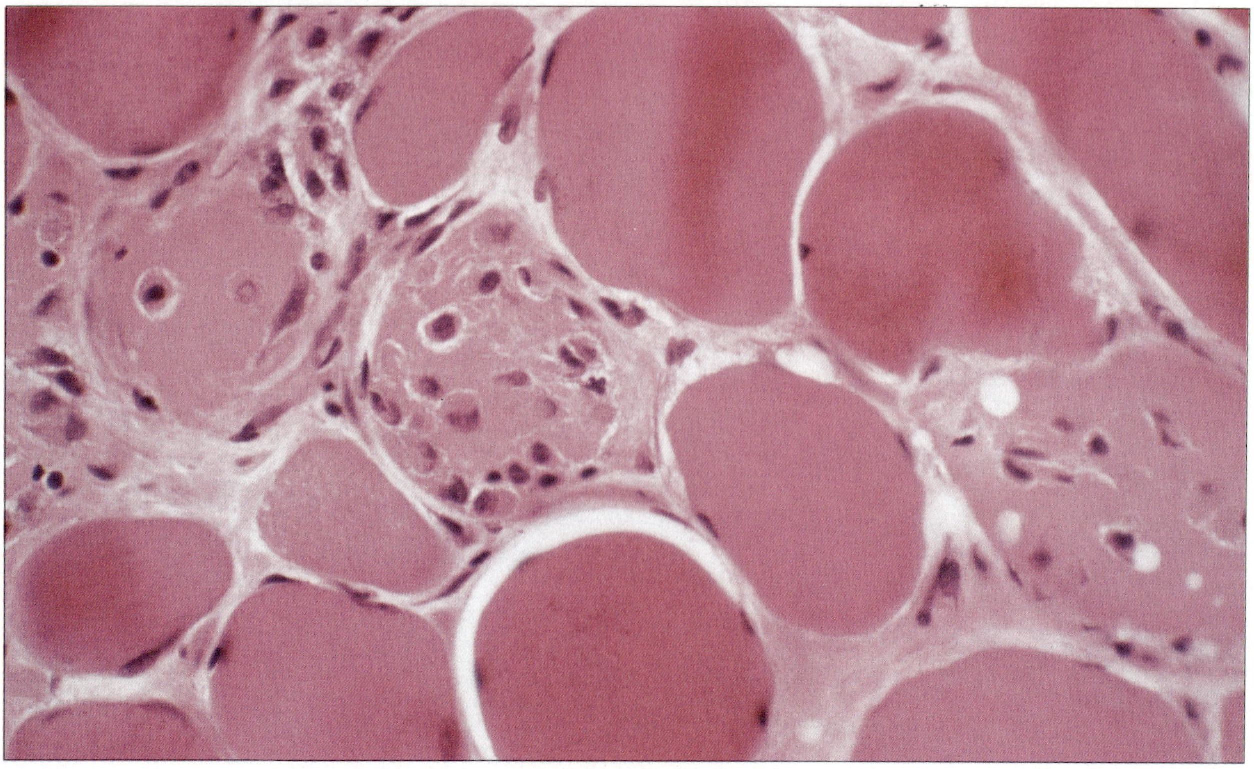

4.32 Hyaline fibres can be seen adjacent to fibres undergoing necrosis. These features may be present within the same muscle fibre, occurring at different regions along its length. (H&E, ×473)

Irrespective of the cause of the necrosis, this debris is subsequently removed by phagocytic cells (**4.22–4.24**, **4.32**, **5.8–5.9**, **7.22**, **7.25–7.27**).

Regeneration has been shown to be initiated by the differentiation of the satellite cells, which are normally found between the plasma and the basement membranes of the muscle fibres (**3.31**, **4.37–4.39**). These mononuclear cells are considered to have myogenic potential.

The regenerating muscle fibres are identified by their high RNA content (basophilia) and their large vesicular nuclei (**4.38–4.42**, **5.13–5.14**, **5.19**, **7.33**).

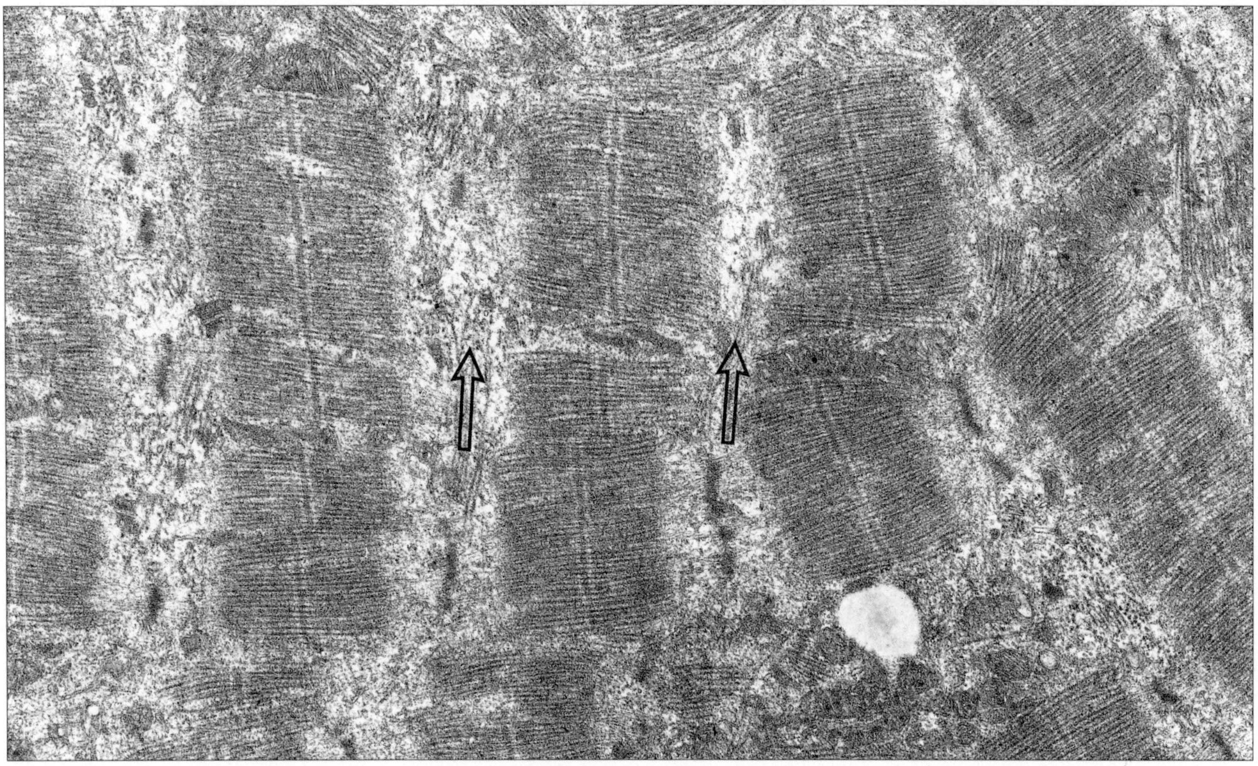

4.33 The loss of the Z-line (arrowed) and the breakdown of the I-band, with survival of the A-band, is a sign of muscle fibre degeneration that may be observed in muscular dystrophy and myositis. (×26250)

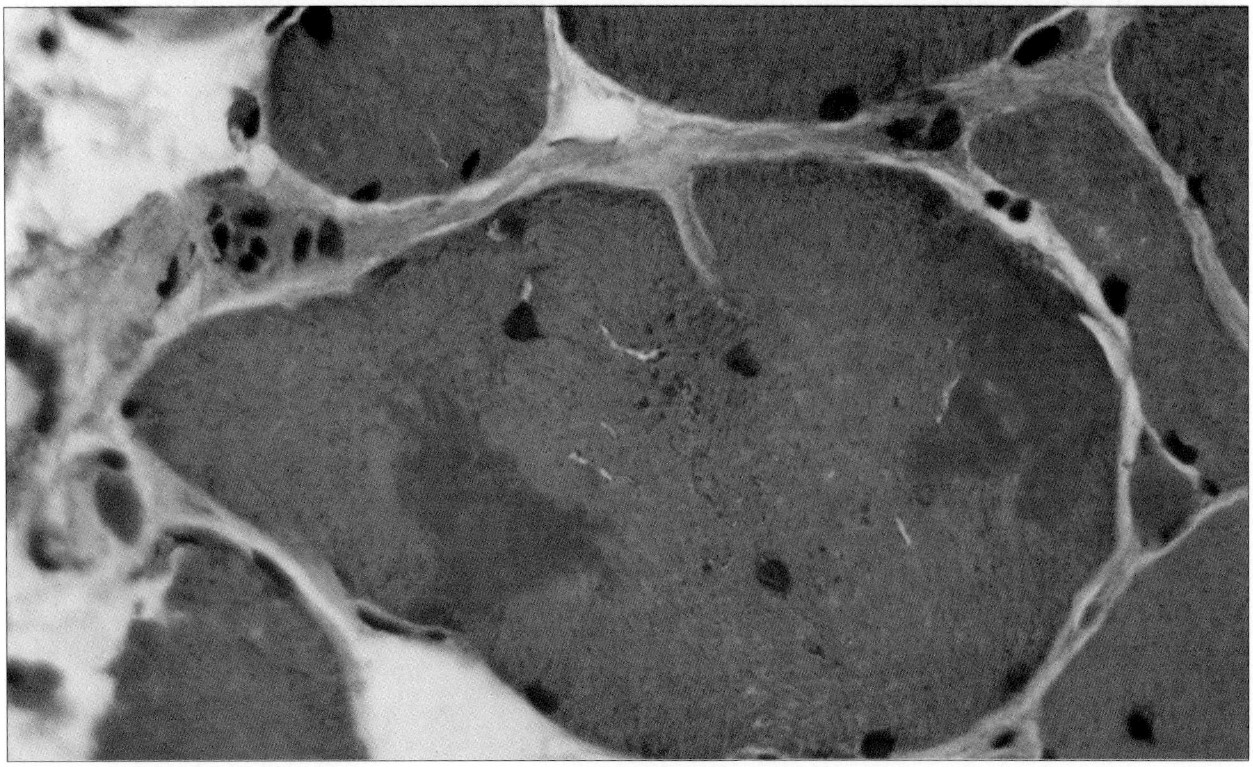

4.34 The large muscle fibre shown here has patches of 'fuzzy' densely-staining material. Such areas may represent zones of Z-band streaming that can be seen with the electron microscope (see **4.35b**). (H&E, ×638)

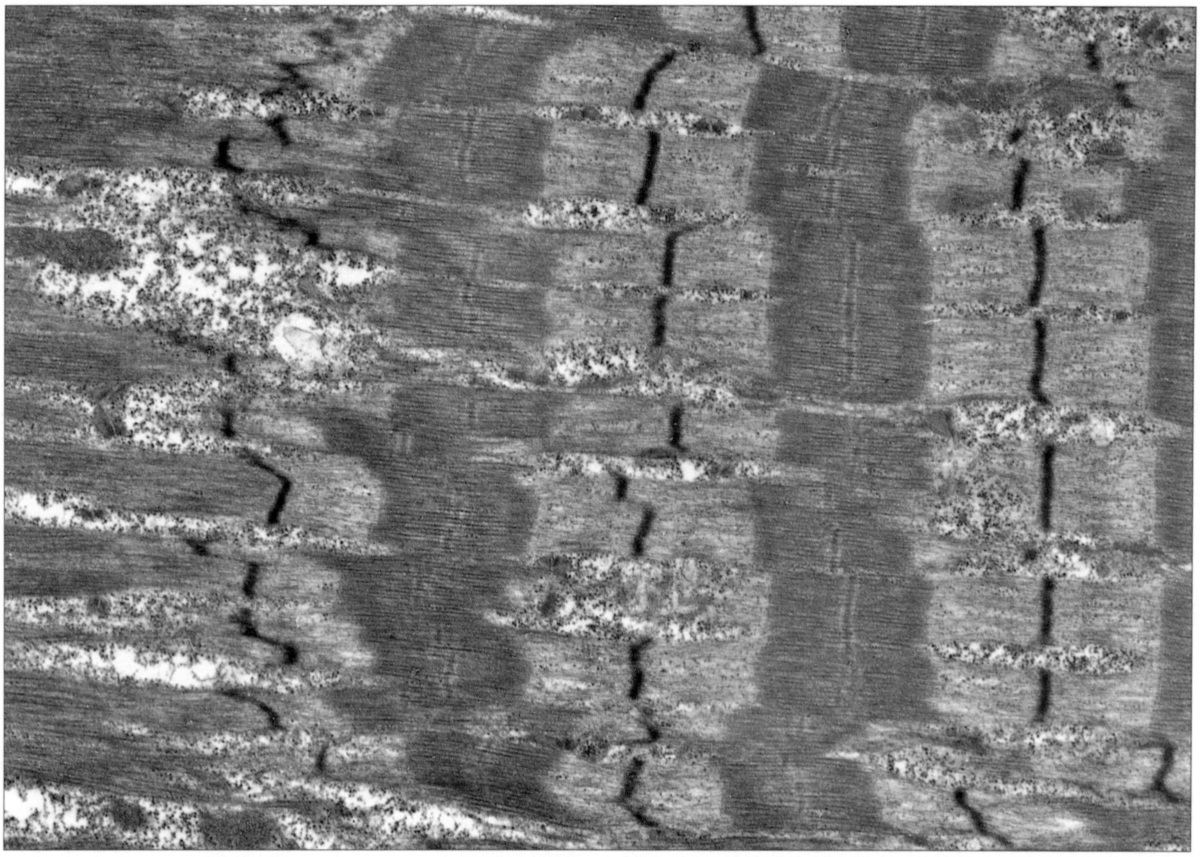

4.35a Streaming of the Z-lines is an early sign of fibre breakdown, and may be seen to affect single sarcomeres, or many. The Z-lines may be affected in sarcomeres adjacent to contraction zones. (×12000)

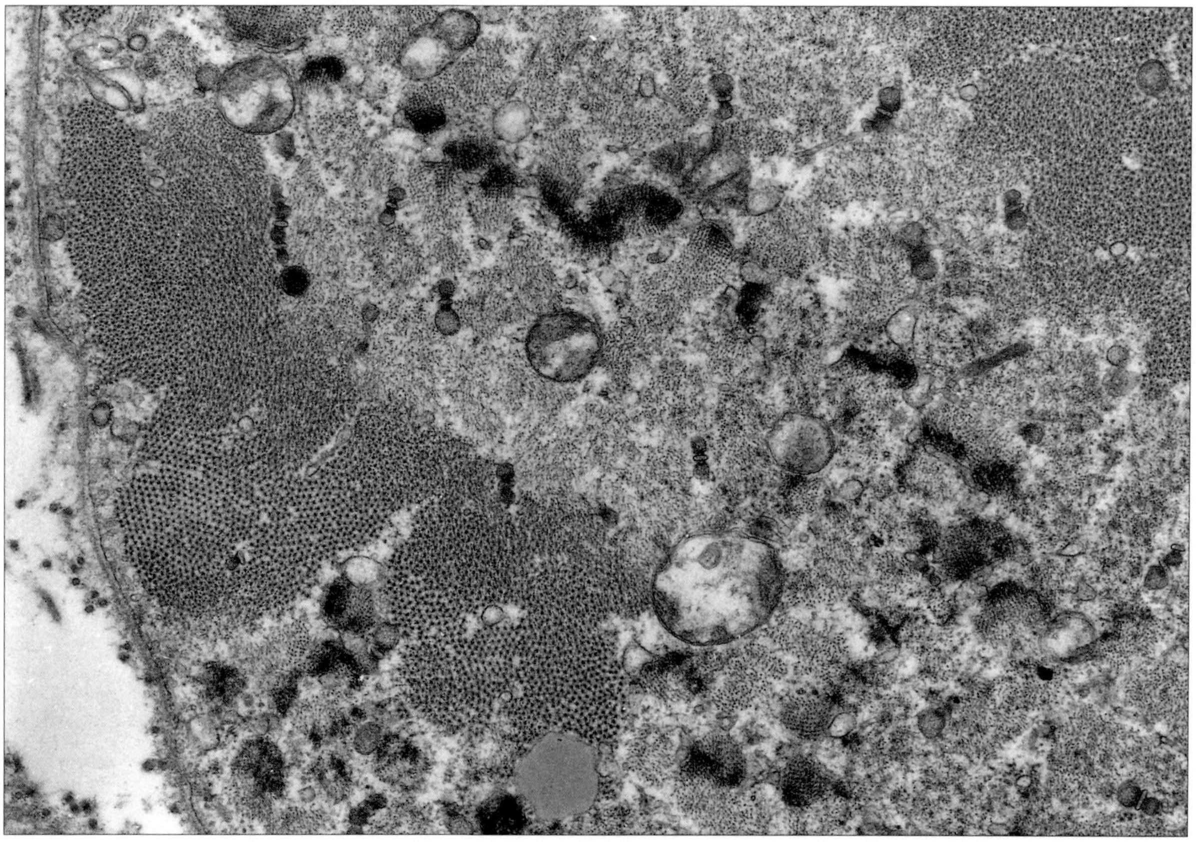

4.35b In transverse section, streaming causes the Z-disc to be seen as small heterogeneous structures lacking a regular position within the fibre. (×25000)

4.35c More extensive streaming may be seen to extend over several sarcomeres. (×22750)

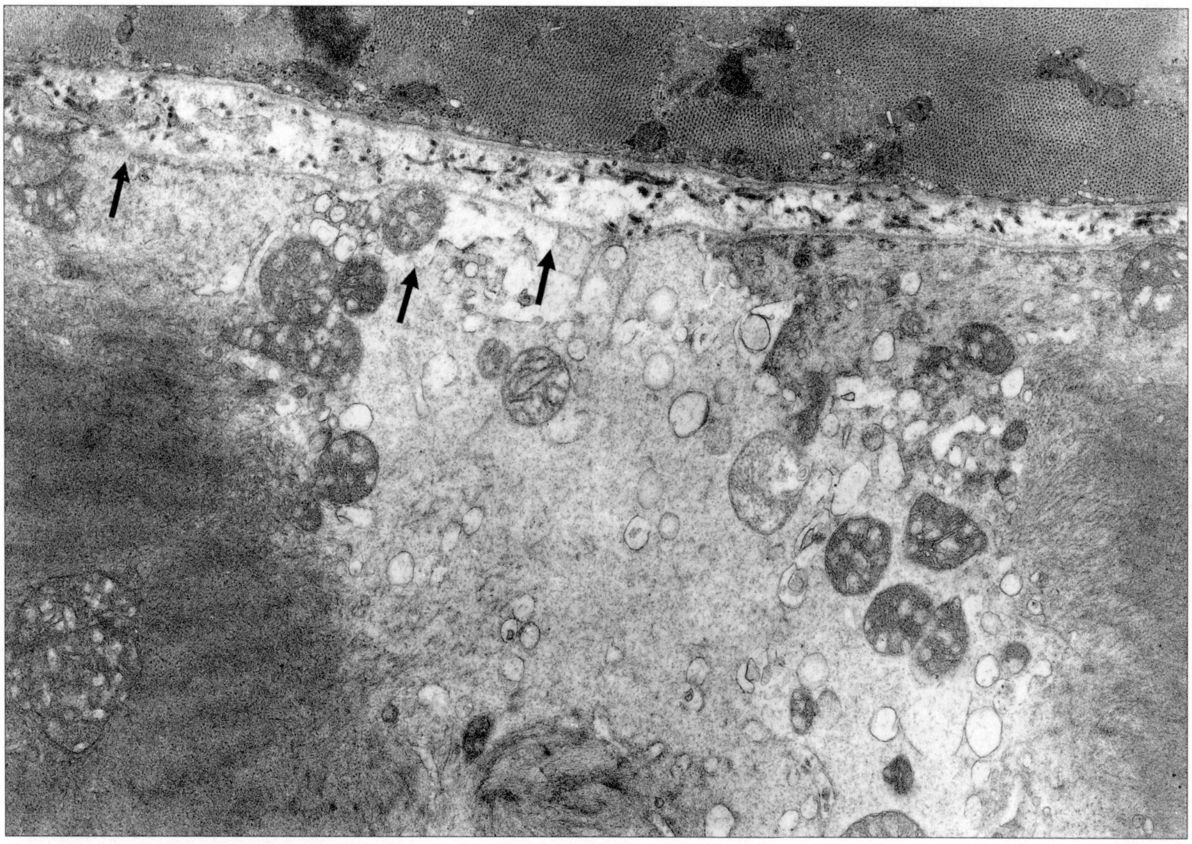

4.36 Necrosis of the muscle fibre, with focal loss of myofilaments and disruption of the plasma membrane (arrowed), is referred to as a 'delta' lesion. The basal lamina is seen to remain intact. (×18250)

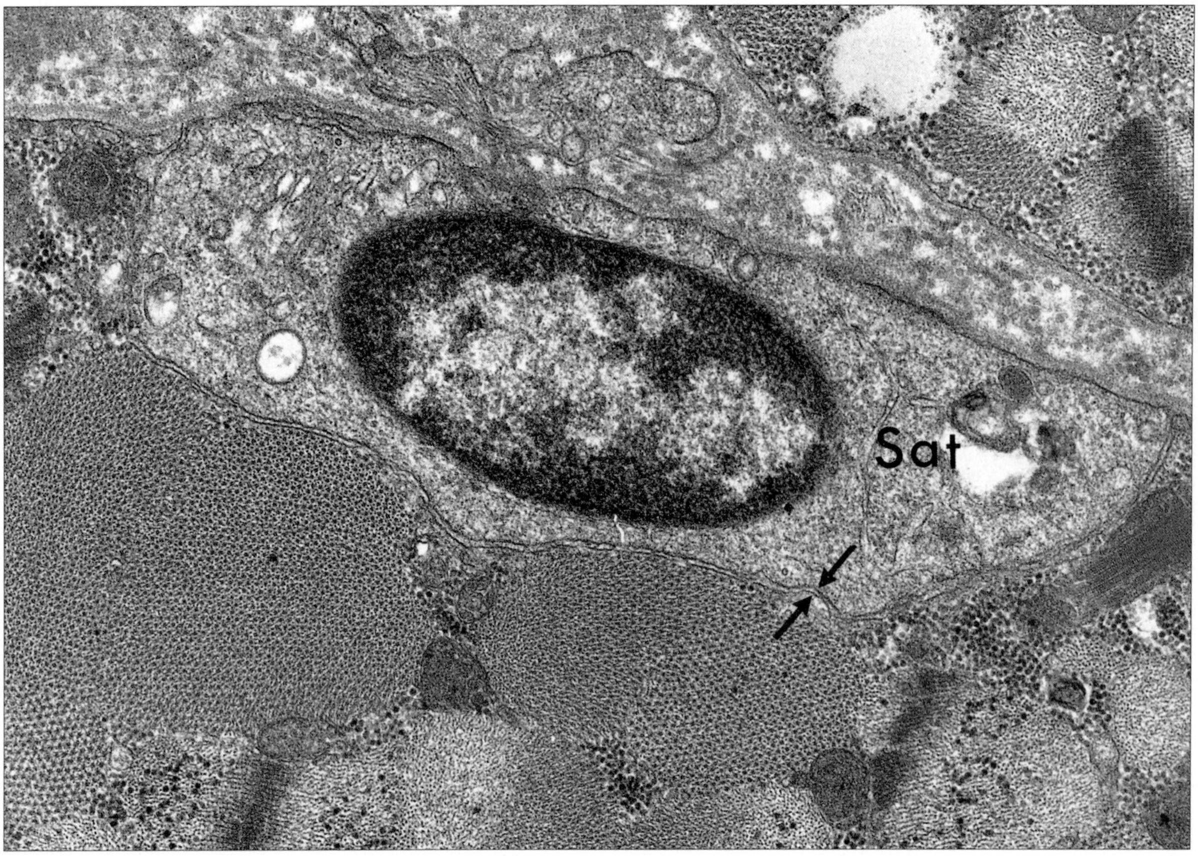

4.37 The satellite cell (Sat) lies beneath the basal lamina of the fibre but external to the plasma membrane of the muscle fibre. The cells are separated by their plasma membranes (arrowed). (×30000)

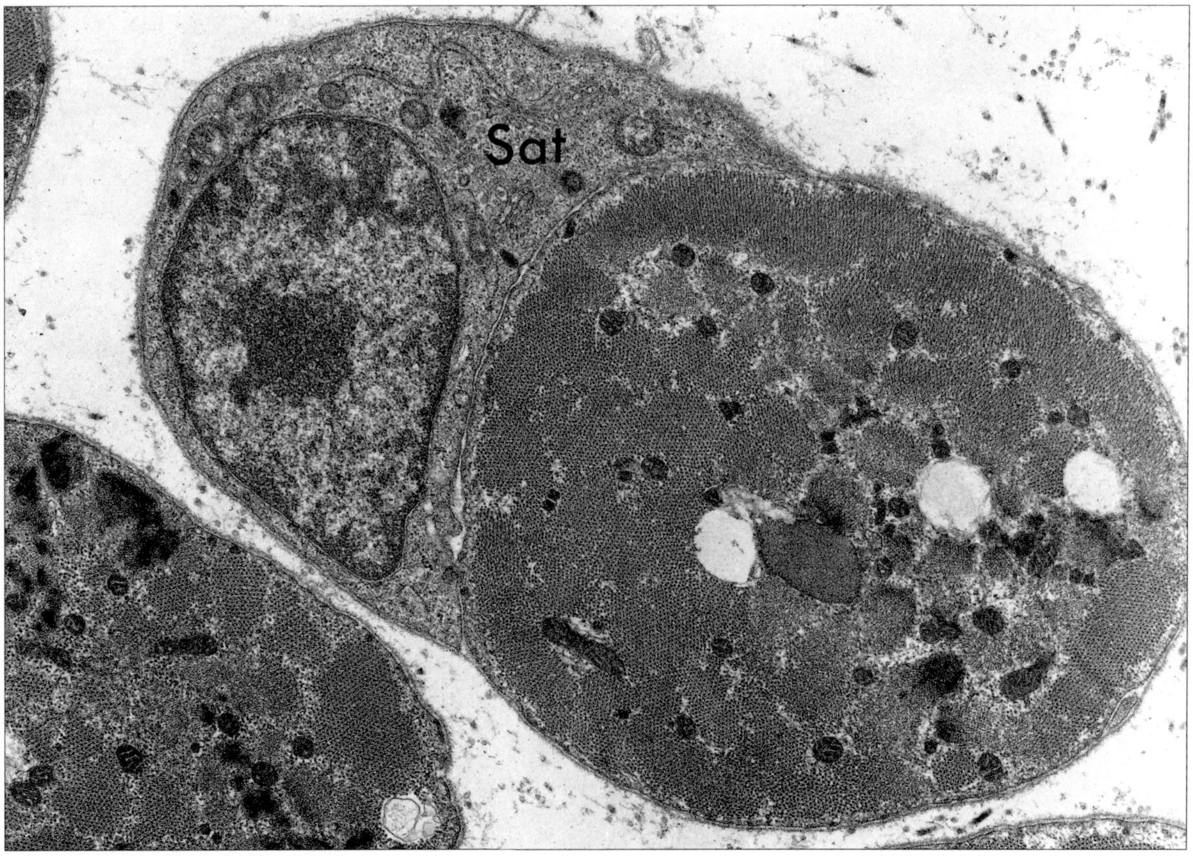

4.38 Satellite cells (Sat) are frequently seen in neonatal and atrophic muscle, where the numbers of fibres sampled is correspondingly greater than normal. They appear large with respect to the small muscle fibres. (×18000)

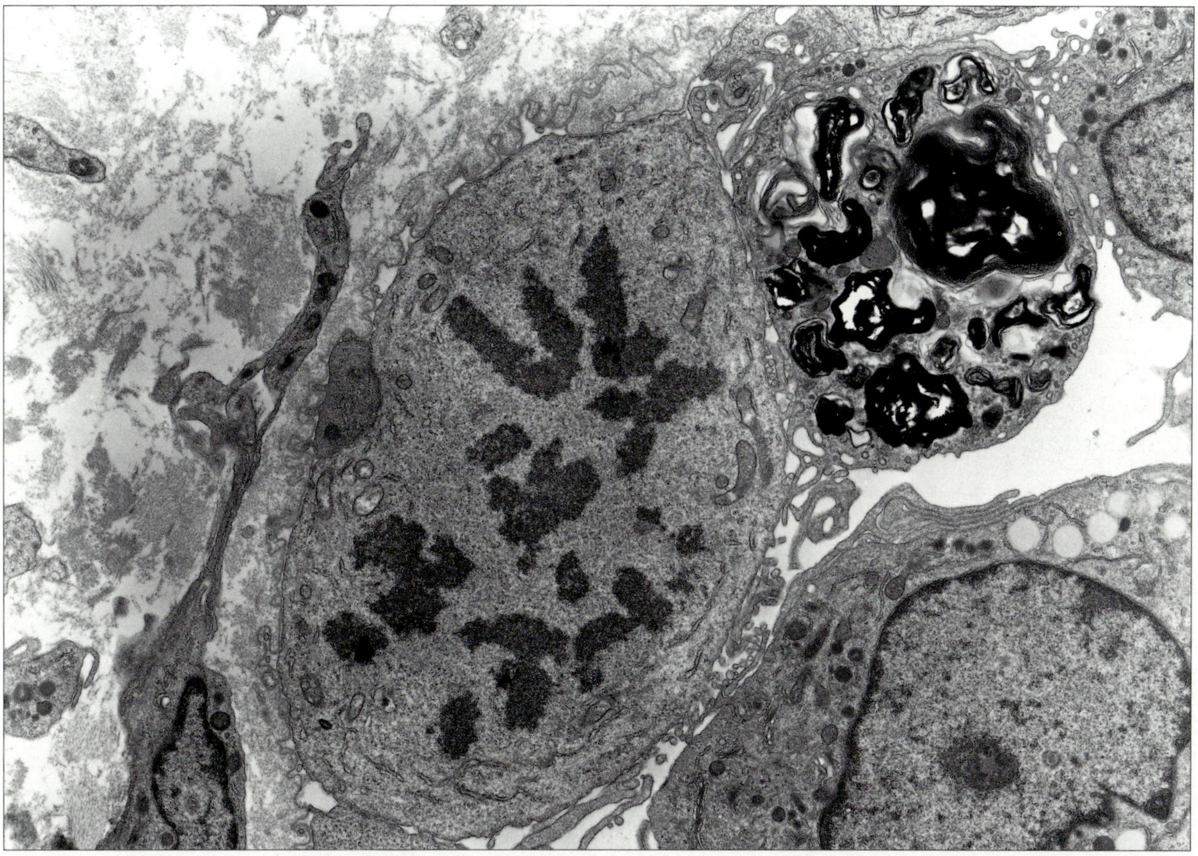

4.39 Mitotic figures are occasionally seen in satellite cells. The nuclear membrane is absent, and the chromosomes appear as darkly stained irregular structures within the cytoplasm. (×15000)

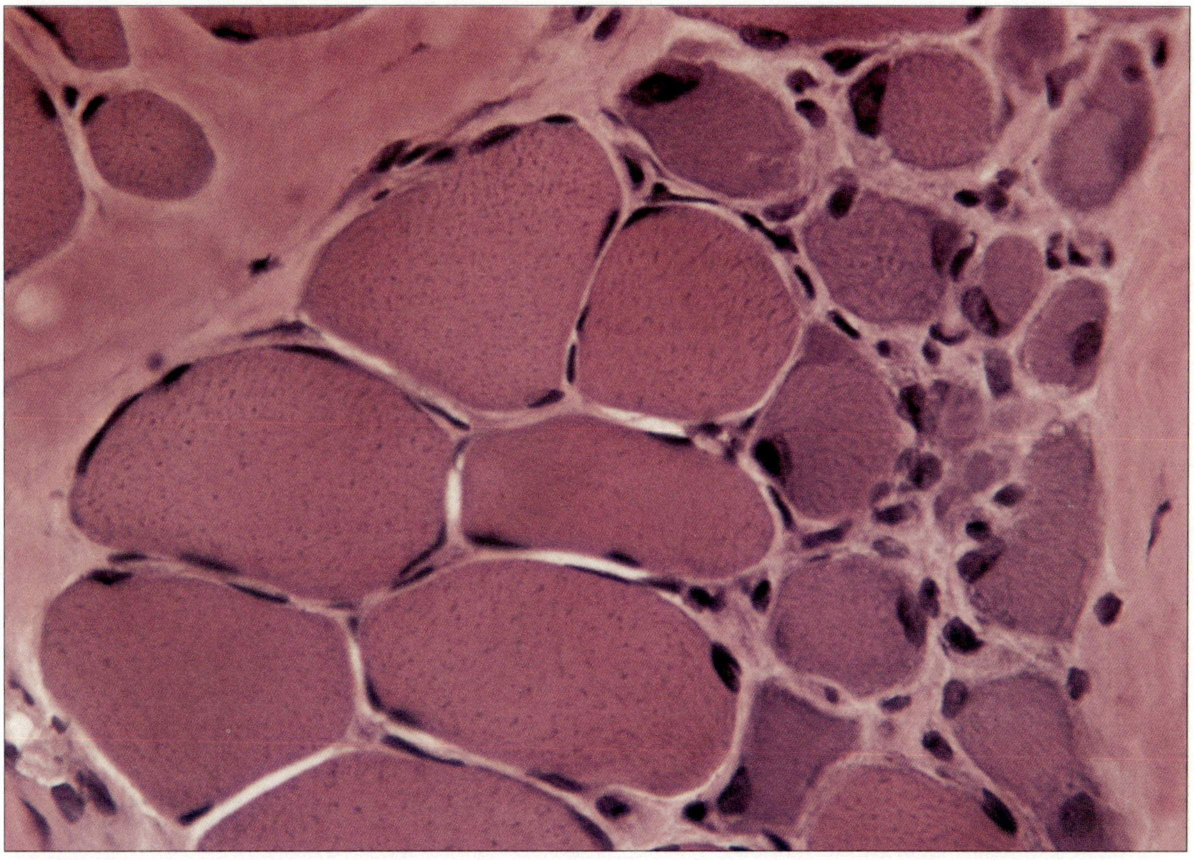

4.40 This group of small basophilic muscle fibres with large nuclei is indicative of regeneration. (H&E, ×638)

In DMD particularly, regenerative changes are profuse in the early stages of the disease, but the distribution tends to be patchy (**4.41**). The advanced pathology of DMD is illustrated in **4.16**. It will be clear that the abnormalities are essentially the same as in any other chronic myopathy. In sharp contrast, **4.43** illustrates the relative integrity of many of the muscle fibres in another necrobiotic myopathy, in this case polymyositis. This emphasises the patchy nature of the pathological process in the inflammatory myopathies, in comparison with the more generalised involvement of muscle tissue in DMD.

The relative amounts of degeneration and regeneration seen will vary depending on the nature of the insult. In chronic conditions it is common to see all stages of necrosis and regeneration within the same section (**4.41, 4.43, 7.17, 7.32**); phagocytosis may even be seen to occur concomitantly within the same basement membrane (**4.24, 4.39**).

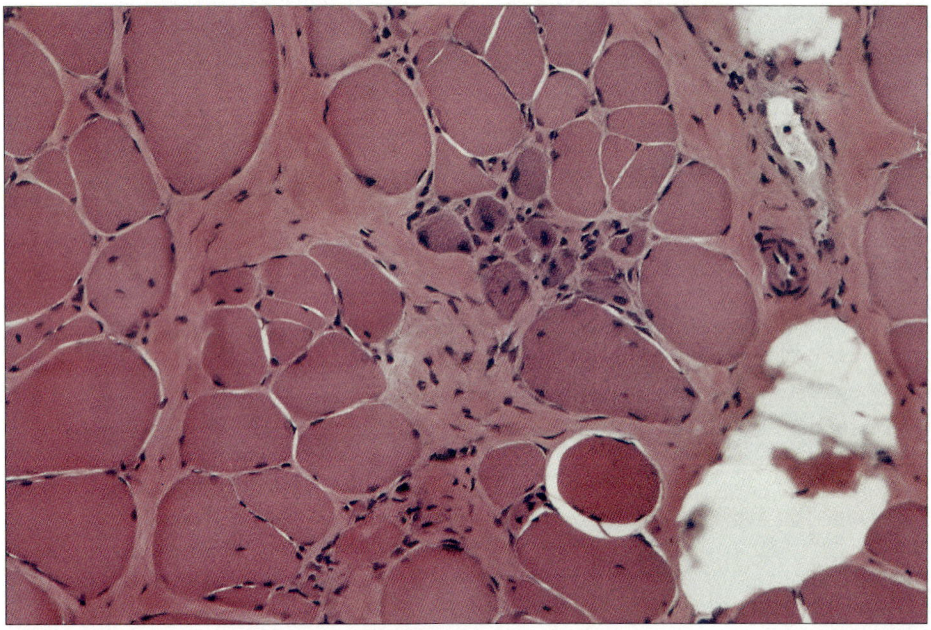

4.41 A combination of myopathic features is evident in this biopsy. Note the excess connective tissue, the presence of some fatty replacement, and the great variation in muscle fibre size and shape.
In addition to a prominent hyaline fibre, many muscle fibres show evidence of splitting or centralised nuclei. A group of regenerating fibres is also present near the centre of the field.
(H&E, ×217)

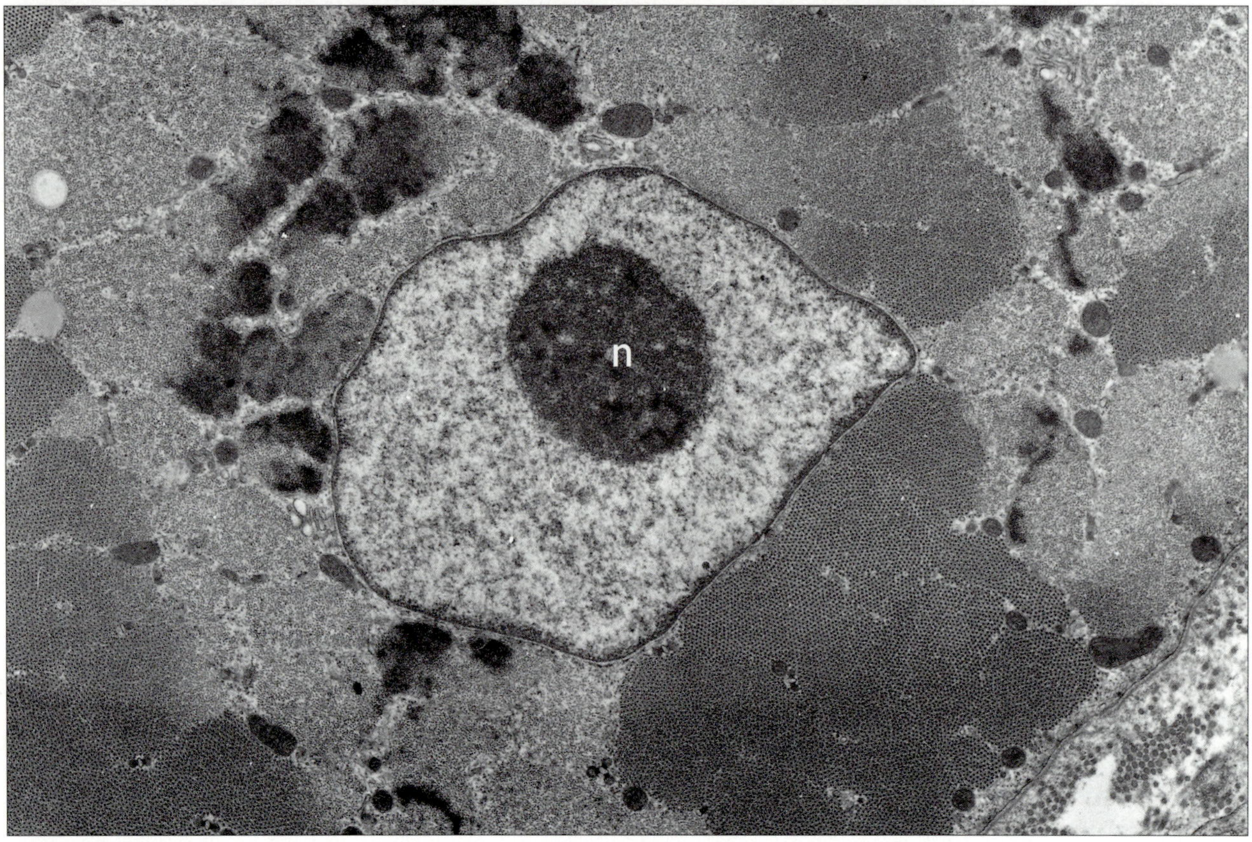

4.42 In fibres that have undergone regeneration, the nucleus is centrally located, smooth in outline, and contains a prominent nucleolus (n). (×15750)

Vacuoles and storage products

Many of the metabolic myopathies produce histological changes within the muscle fibre. In some instances these take the form of vacuoles that may appear to contain nothing, or in which there are accumulations of either lipid or polysaccharide. In others, there are accumulations of normal or abnormal mitochondria, or subsarcolemmal accumulations of carbohydrate.

The pathological reactions of muscle seen in systemic metabolic and endocrine disorders are referred to in Chapter 10.

Glycogen

In some types of abnormal glycogen metabolism, the storage material may be membrane bound, as well as free within the sarcoplasm (**4.44**, **9.1–9.9**, **9.11–9.12**, **9.14–9.18**).

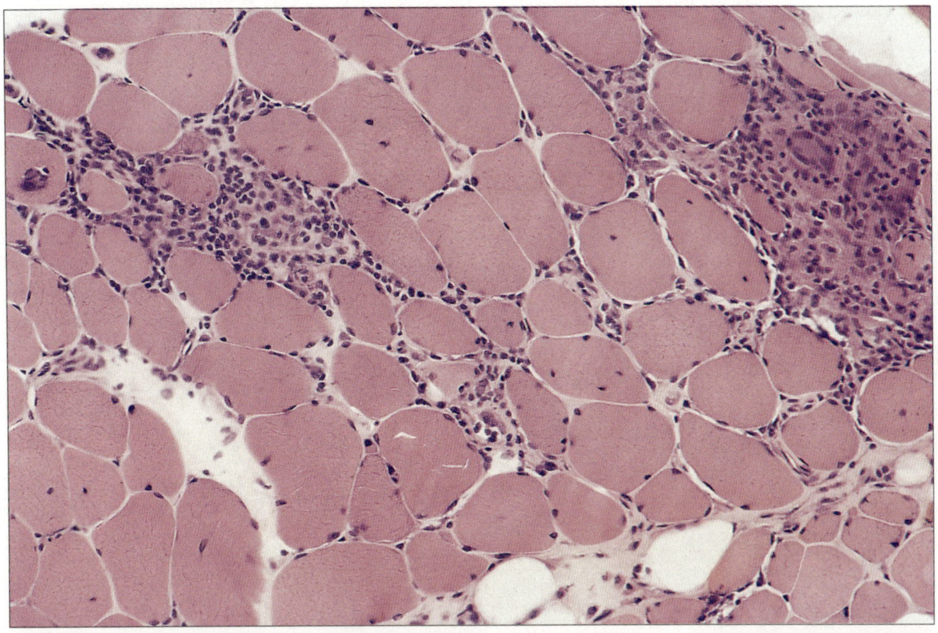

4.43 An inflammatory condition, showing focal lymphocytic infiltrates and extensive myophagocytosis. Unlike other cases shown previously, many of the muscle fibres remain relatively normal in appearance and arrangement. (H&E, ×217)

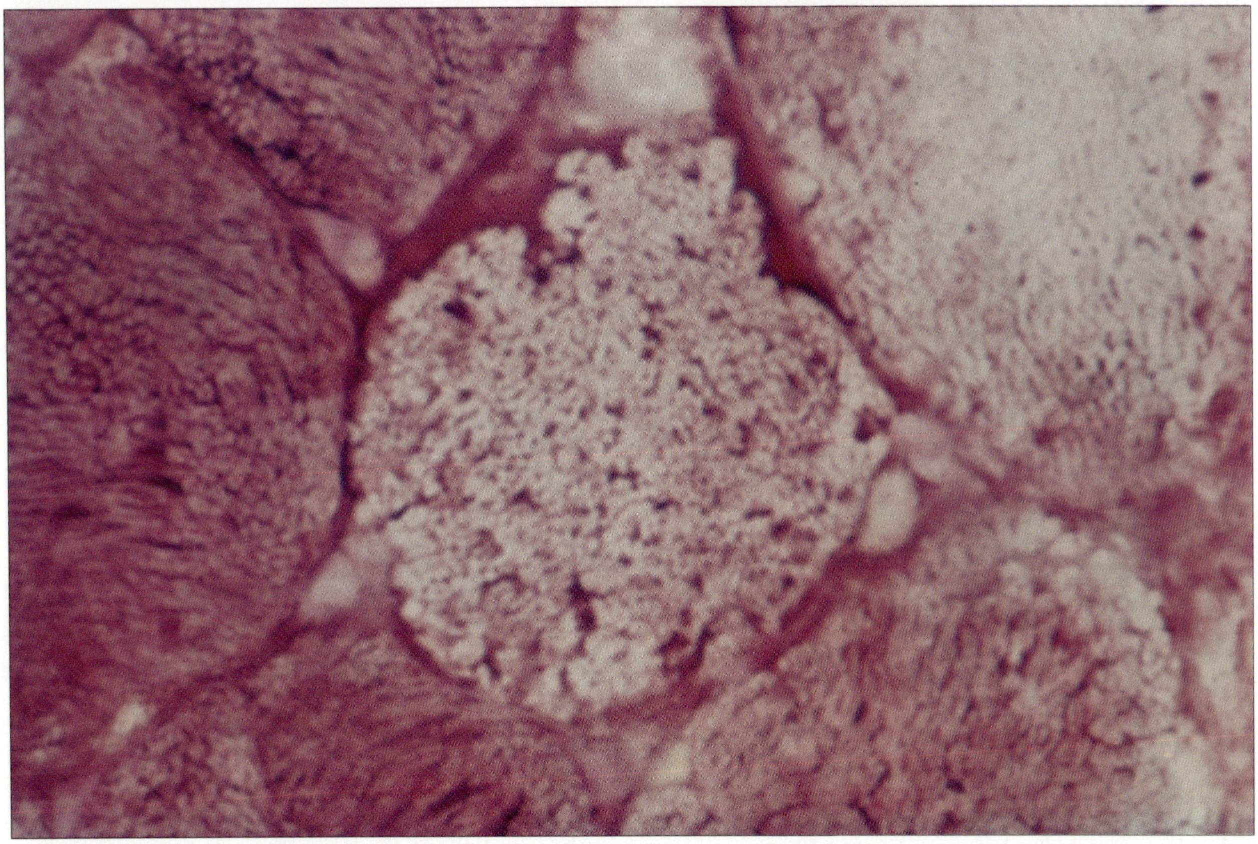

4.44 Storage of excess glycogen, especially in subsarcolemmal regions, is typical of many of the glycogenoses affecting muscle. It appears as empty vacuoles with other stains. (PAS, ×732)

Lipid

Recent evidence suggests that disorders of lipid metabolism are secondary to defects in mitochondrial function, and that they may or may not be associated with evidence of structural alterations of mitochondria. However, the cardinal histopathological feature of the 'lipid myopathies' is the presence of excess lipid droplets between the myofibrils.

The abnormalities seen in biopsy material from patients with lipid storage disease are illustrated in **4.45–4.47**, **4.54** (see also **9.19–9.23**). These abnormalities are most obvious in Type I and Type IIa fibres (**4.46**, **9.20**), and correspond to the neutral lipid deposits seen in **4.47**.

Electron microscopy shows rows of lipid droplets lying between the myofibrils. These droplets are confined by a single limiting membrane and are often in close proximity to the mitochondria (**4.48**).

In H&E stained sections, the sarcoplasmic vacuoles seen in lipid storage myopathy should not be confused with those occurring in other storage disorders.

In glycogen storage, these vacuoles tend to be less discrete and more variable in size. In the specific instance of myophosphorylase deficiency, the vacuoles are confined to the subsarcolemmal region (**4.44**).

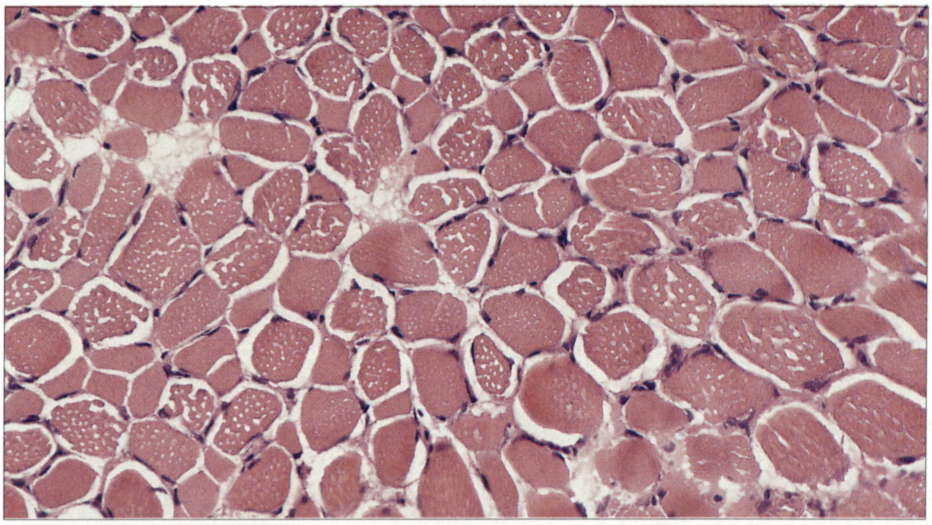

4.45 Vacuoles are present within the fibres from this biopsy.
Subsequent histochemical staining of serial sections is needed to reveal their content (see **4.46–4.47**).
(H&E, ×210)

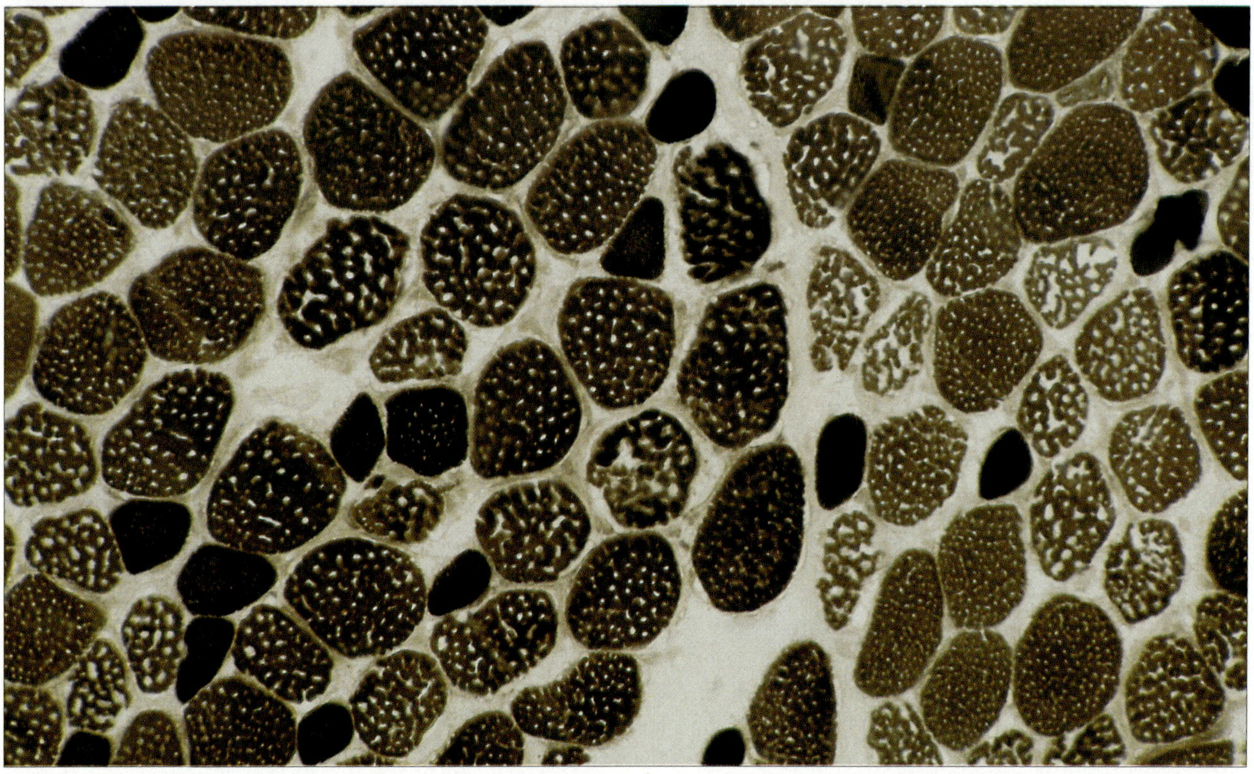

4.46 Lack of myofibrillar ATPase staining corresponds to the vacuoles shown in **4.45**. (Myofibrillar ATPase, pH 10.2, ×284)

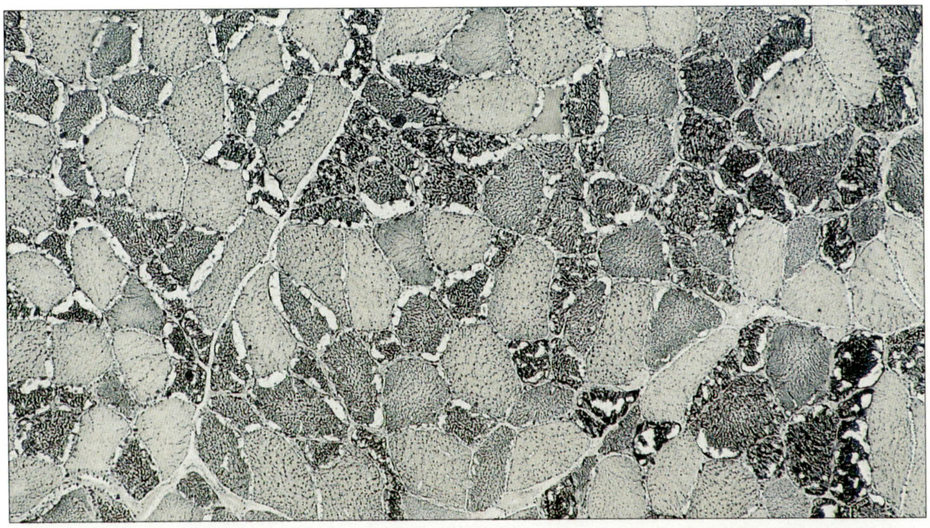

4.47 Staining reveals that the vacuoles, which are predominantly in Type I fibres (see **4.45–4.46**), contain stored lipid.
a) Sudan Black B, ×88;

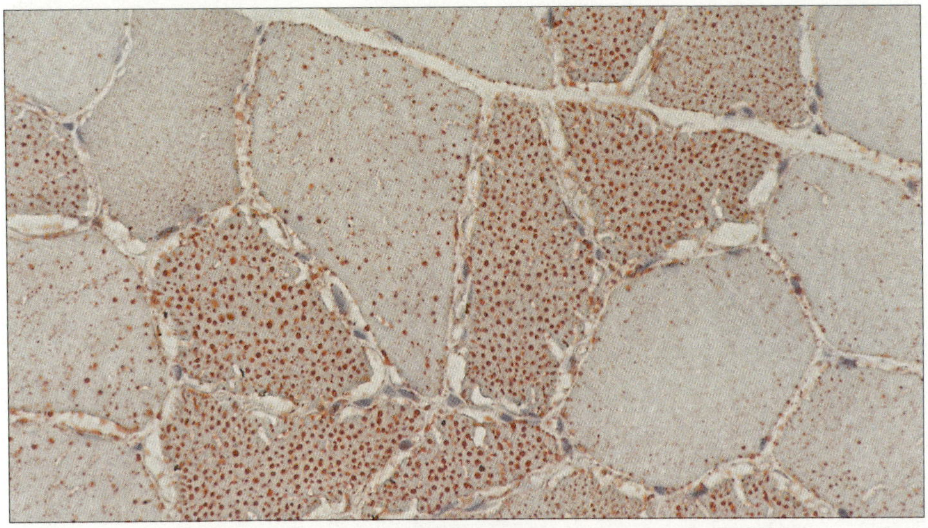

b) Oil red O, ×350.

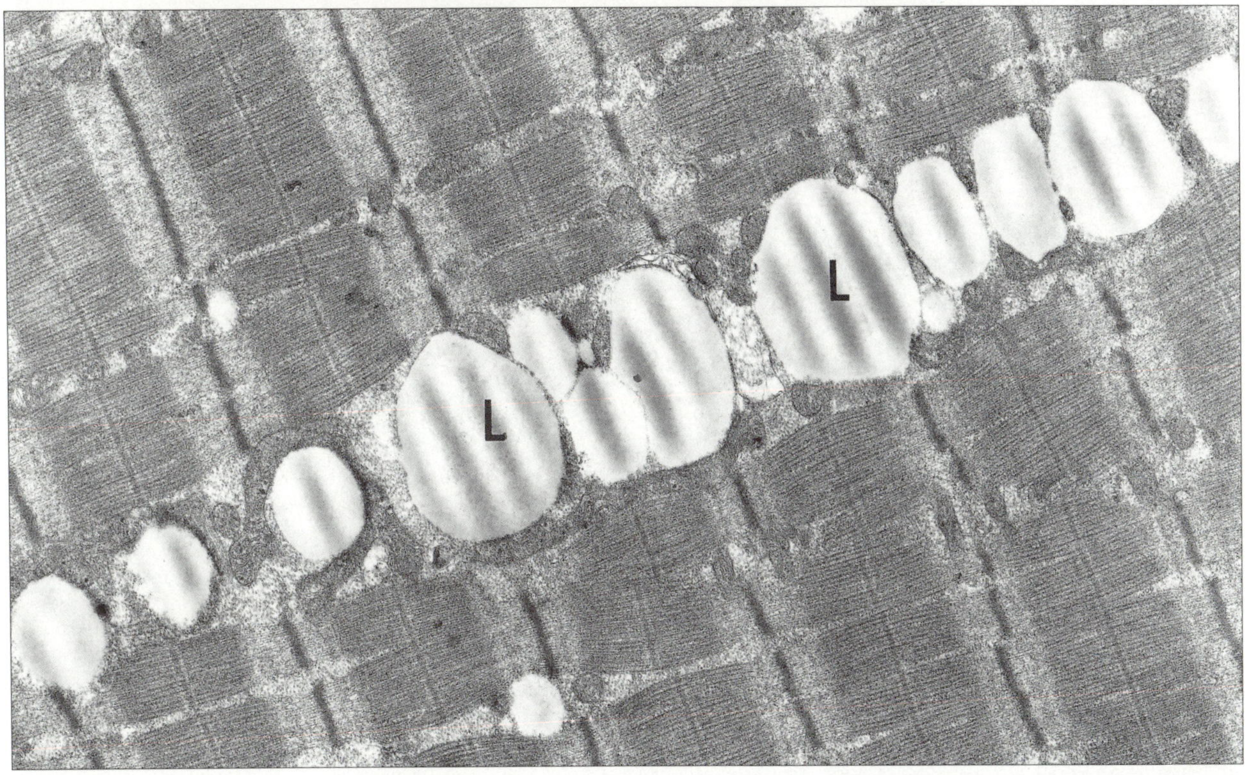

4.48 Single lipid droplets (L) are occasionally seen between the myofibrils at the level of the I-band. In certain diseases, they may be seen to accumulate between the myofibrils, varying in size and exhibiting a characteristic banded appearance which is an artefact of either fixation or microtomy. (×12600)

Mitochondrial changes

In mitochondrial myopathies, the histopathological features are diverse, and range from apparent normality to gross morphological changes. Because of this, histochemical investigation is essential, requiring confirmation with electron microscopy, and, where available, with biochemistry.

The H&E may show completely normal appearances, granular, diffuse or subsarcolemmal basophilia or the presence of structurally abnormal fibres (**4.49–4.50**, **9.28**). The oxidative enzyme reactions (NADH-Tr, SDH and cytochrome oxidase) frequently demonstrate fibres in which there are subsarcolemmal accumulations of reaction product, or the presence of fibres which are structurally abnormal and show increased staining intensity (**4.51–4.52**, **4.61**, **9.31–9.32**). In addition, the cytochrome oxidase reaction may be negative in some fibres (**9.32**).

The modified Gomori trichrome stain may show fibres with a pattern of granular reddish cytoplasmic staining, or one in which there are subsarcolemmal aggregates. There may also be fibres that are structurally abnormal, with intense heterogeneous staining for which the term 'ragged-red' was coined (**4.53–4.54**, **9.30**).

At the ultrastructural level, the appearances range from large subsarcolemmal accumulations of mitochondria, which may appear normal or contain bizarre inclusions, to fibres containing degenerate mitochondria, vacuoles and disorders of myofibrils (**4.54–4.56**).

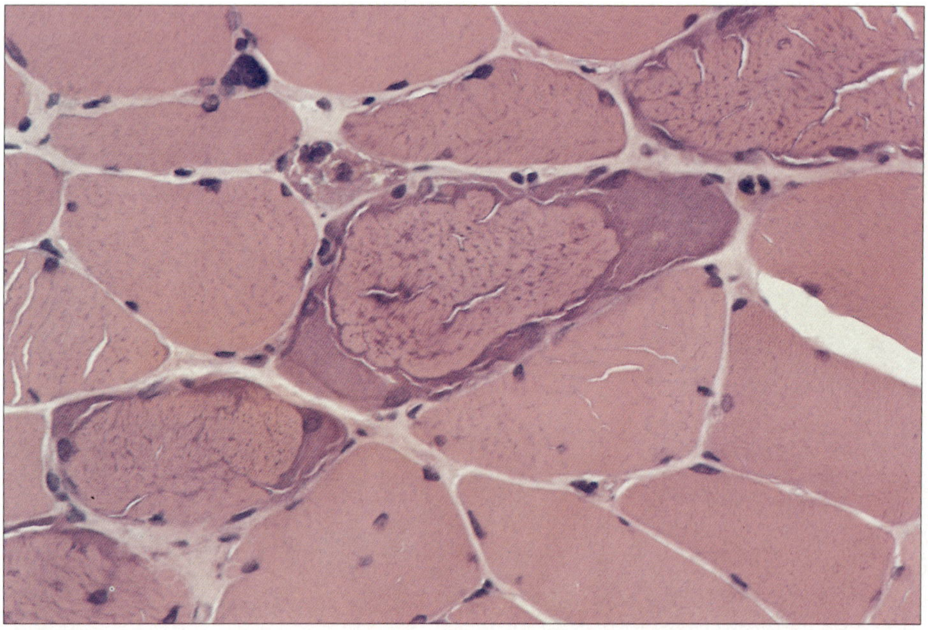

4.49 Prominent subsarcolemmal basophilic staining, is shown, in addition to increased intermyofibrillar basophilia. Such accumulations may be abnormal mitochondria ('ragged-red' fibres), and must be confirmed by subsequent histochemical staining of serial sections. (H&E, ×543)

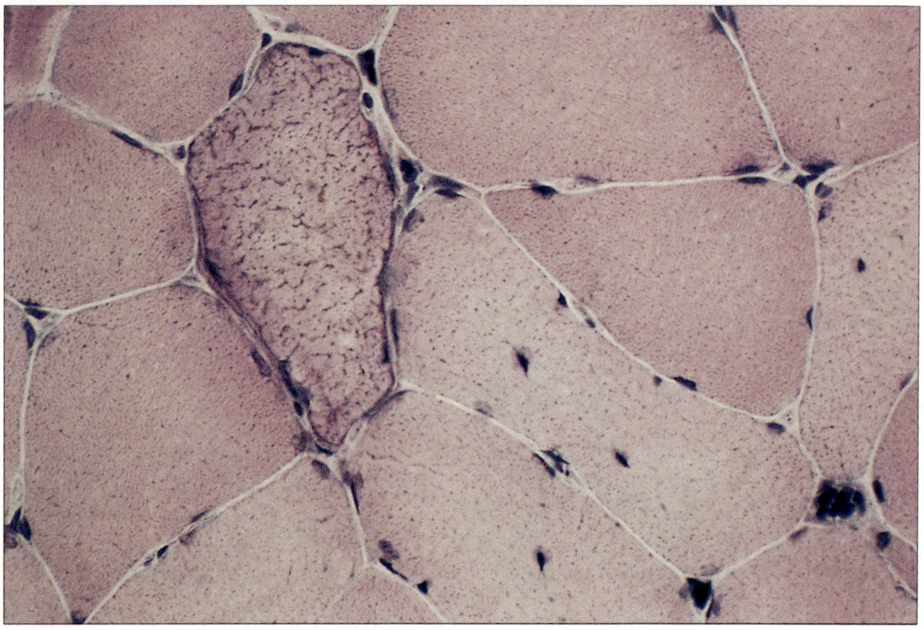

4.50 Isolated fibres that show a granular and dense peripheral basophilic staining with haematoxylin often correspond to the typical ragged-red fibres, and subsequently reveal mitochondrial abnormalities upon histochemistry or electron microscopy. (H&E, ×350)

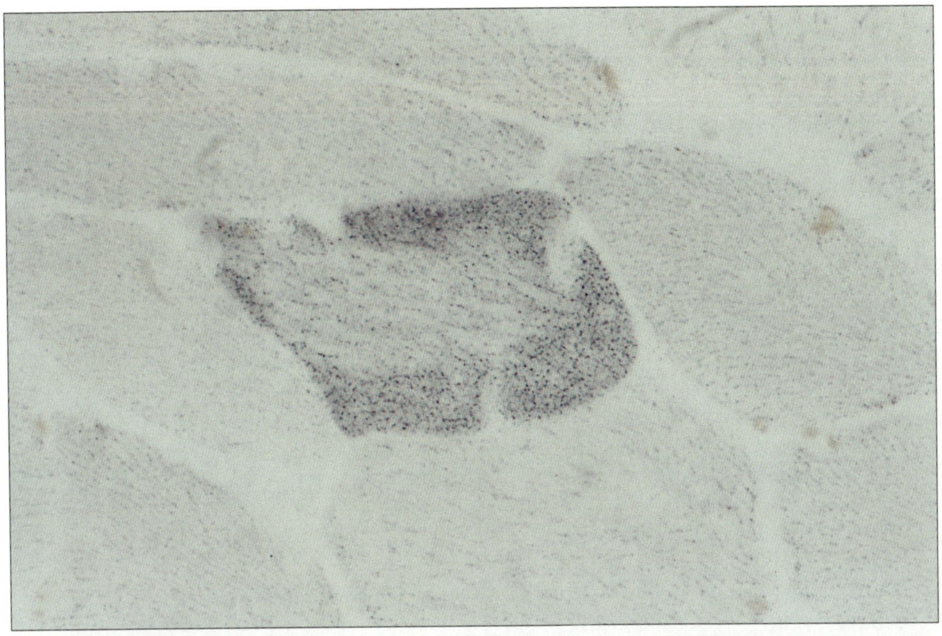

4.51 Abnormally dense subsarcolemmal accumulations of mitochondria from the same biopsy as shown in **4.49**, **4.53**. (SDH, ×560)

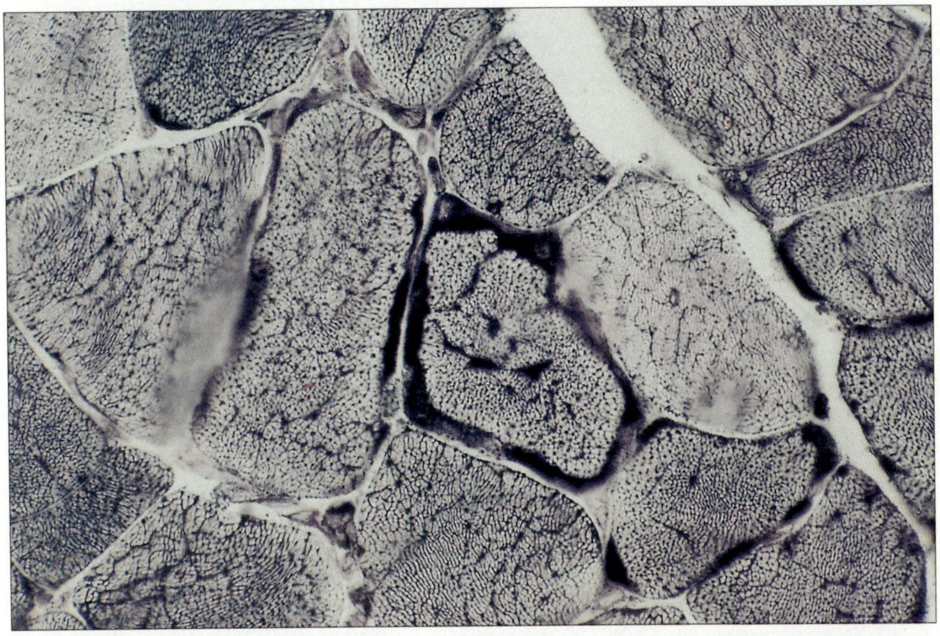

4.52 Subsarcolemmal mitochondrial accumulations in the same biopsy as in **4.49**, **4.51**–**4.52** are stained densely with the NADH-Tr reaction. (NADH-Tr, ×438)

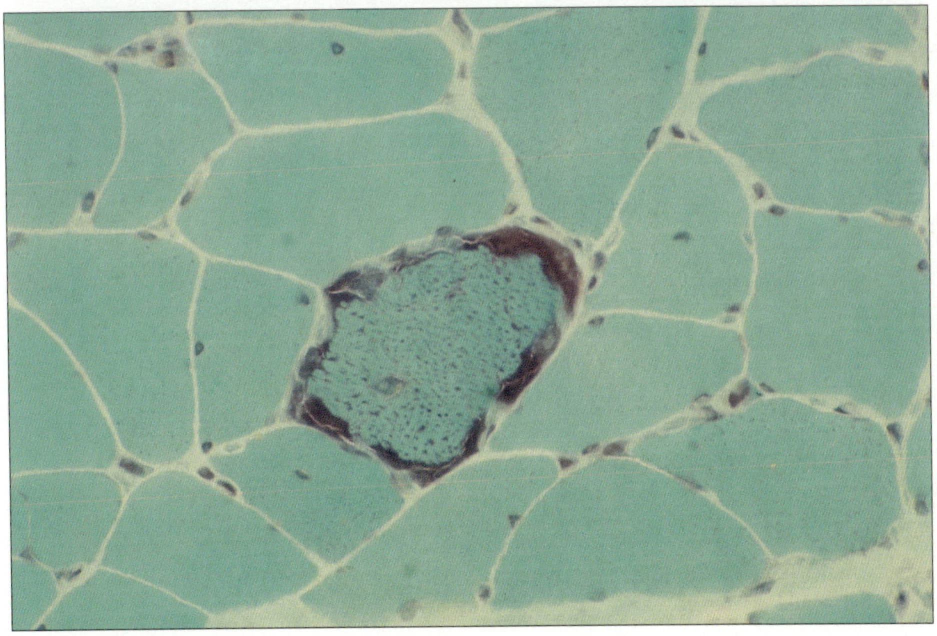

4.53 A ragged-red fibre with characteristic red peripheral staining can be seen (see **4.47**). Such fibres are often 'cracked' in appearance.
(Modified Gomori trichrome, ×543)

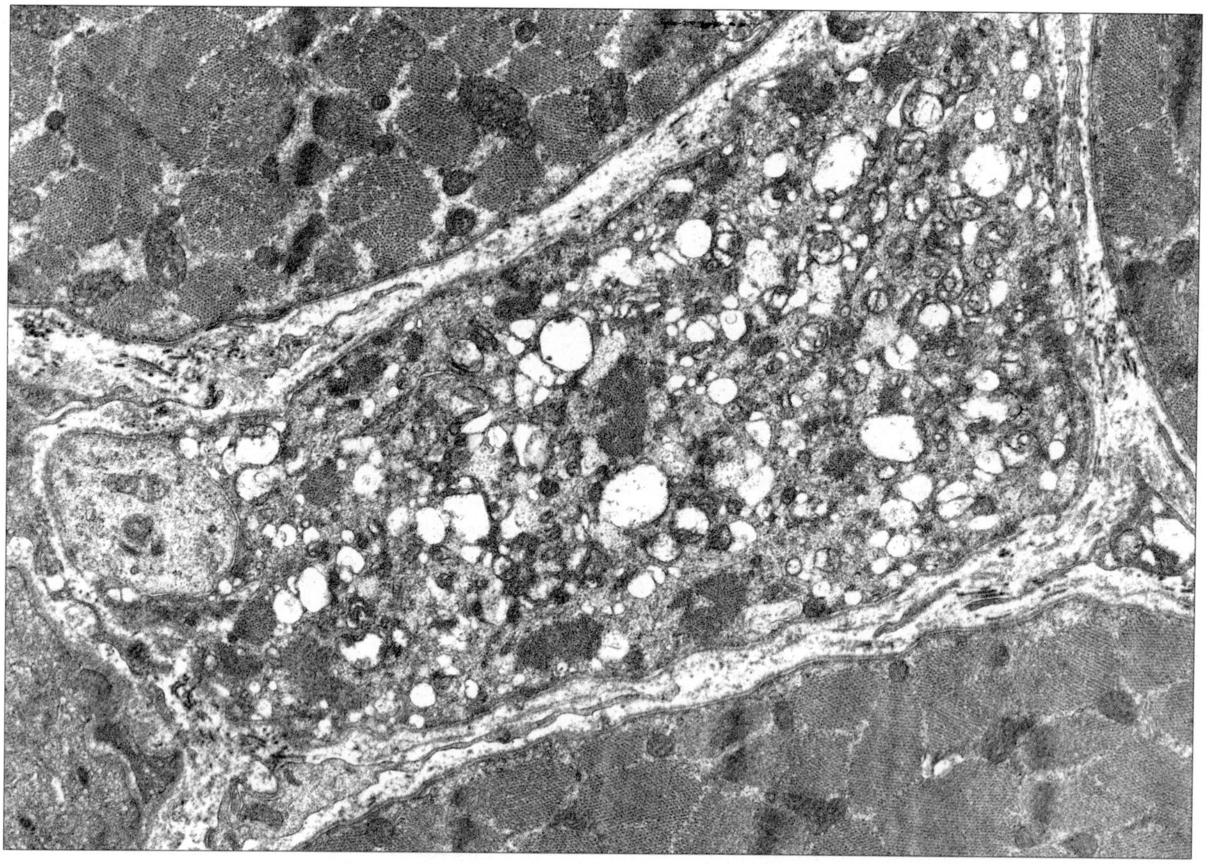

4.54 The myofibrillar architecture of a ragged-red fibre is severely disrupted. The myofilaments are sparse, and the sarcoplasm contains abnormal mitochondria and many vesicles devoid of recognisable contents. (×12500)

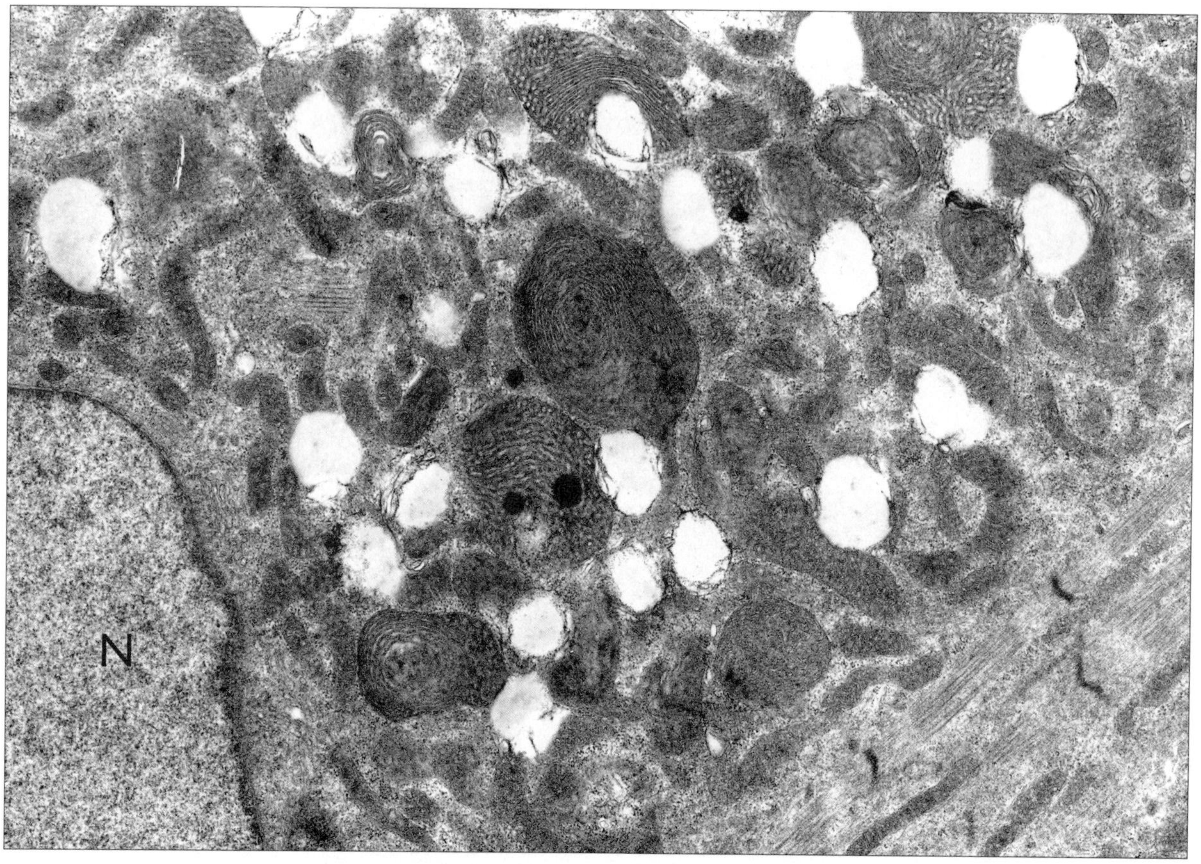

4.55 Abnormal mitochondria, sometimes found together with lipid droplets, are often located in numbers close to nuclei (N) and on the edge of the muscle fibre. (×15000)

4.56a Loss of myosin filaments from the centre of the fibre produces a 'core'. The mitochondria and the Z-lines are still present in what is known as a 'structured core'. (×8000)

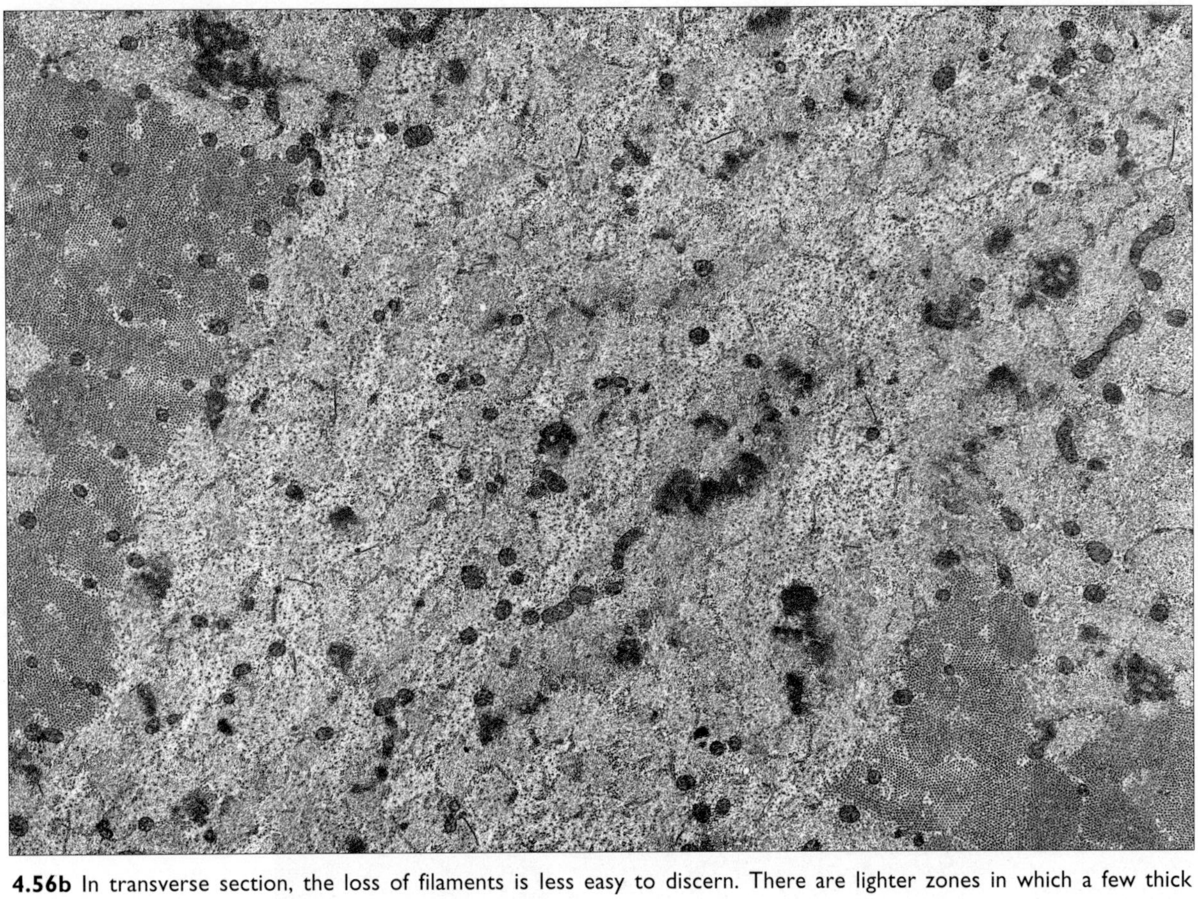

4.56b In transverse section, the loss of filaments is less easy to discern. There are lighter zones in which a few thick filaments can be seen. (×12000)

Miscellaneous structural abnormalities

There are a variety of miscellaneous structural abnormalities which may be observed within the biopsy. Some are encountered frequently, whereas others are derived from single case reports or individual families. We will consider only those which are seen most commonly in practice.

Core myopathies

Central core disease was the first 'specific' pathological entity described in the benign congenital myopathy (infantile hypotonia) syndrome. It should be noted that examination of the H&E preparation may not reveal these abnormalities.

The cores consist of zones devoid of reaction product in the oxidative enzyme preparations (**4.57**). These cores may be structured or unstructured, depending upon the integrity, or otherwise, of the contractile apparatus in that region (**4.58**).

Since the original description of core myopathy, a number of variations on this theme have been described; a particular instance of this is mini/multicore disease (**4.59–4.60**, **8.13**).

A 'lobulated' appearance of the mitochondrial enzyme distribution may occur in mitochondrial disorders, or as a non-specific change in other neuromuscular diseases, e.g. FSH syndrome (**4.61**).

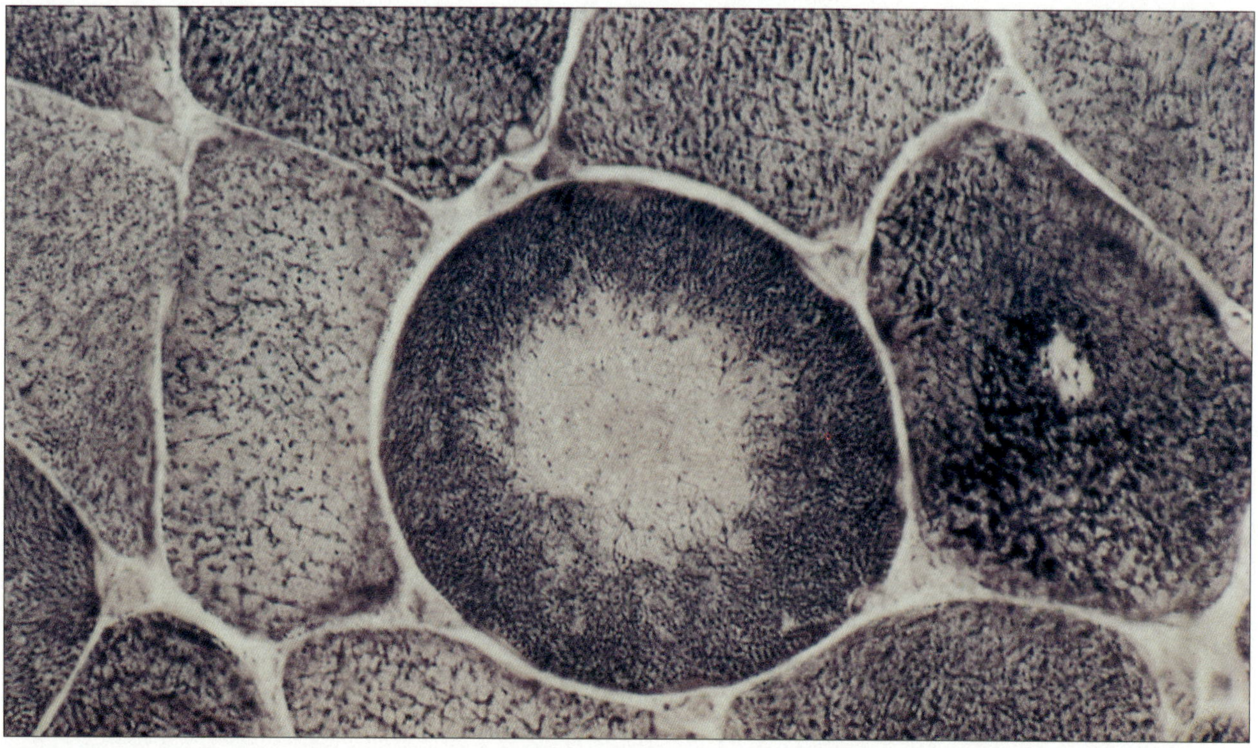

4.57 A central core lacking mitochondrial enzyme activity, surrounded by a dense staining rim. Central cores are usually restricted to Type I fibres. (NADH-Tr, ×732)

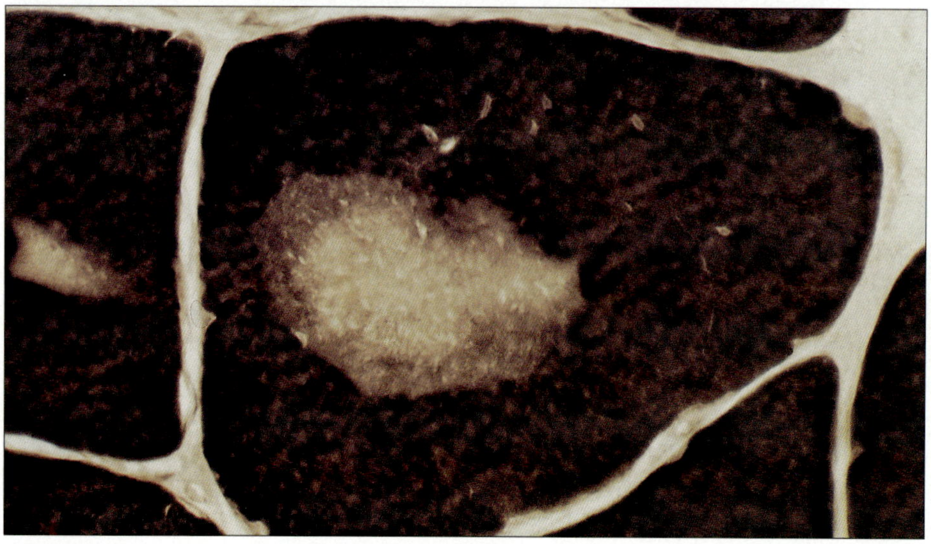

4.58 An unstructured central core lacking myofibrillar organisation is present within this muscle fibre. (Myofibrillar ATPase, pH 4.35, ×543)

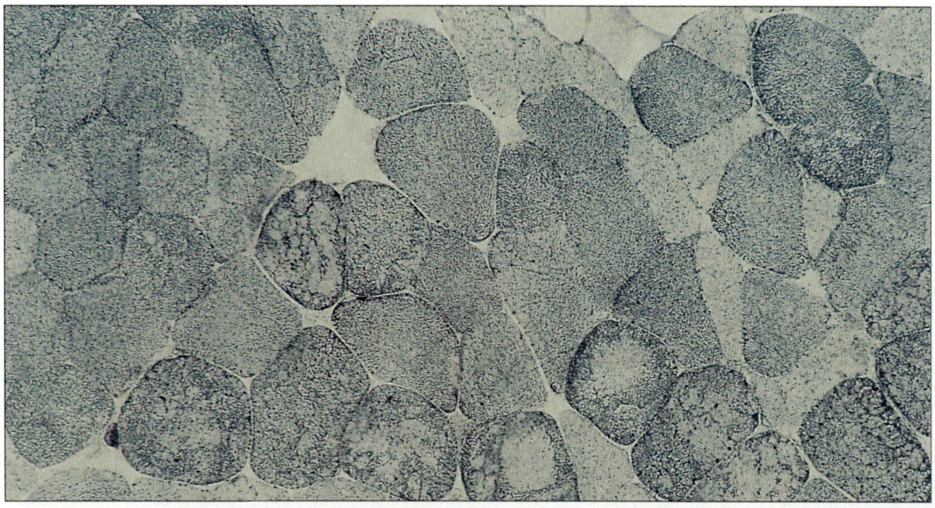

4.59 This section shows disturbance of the oxidative enzyme staining pattern. This appearance is described as multicore or minicore. (NADH-Tr, ×350)

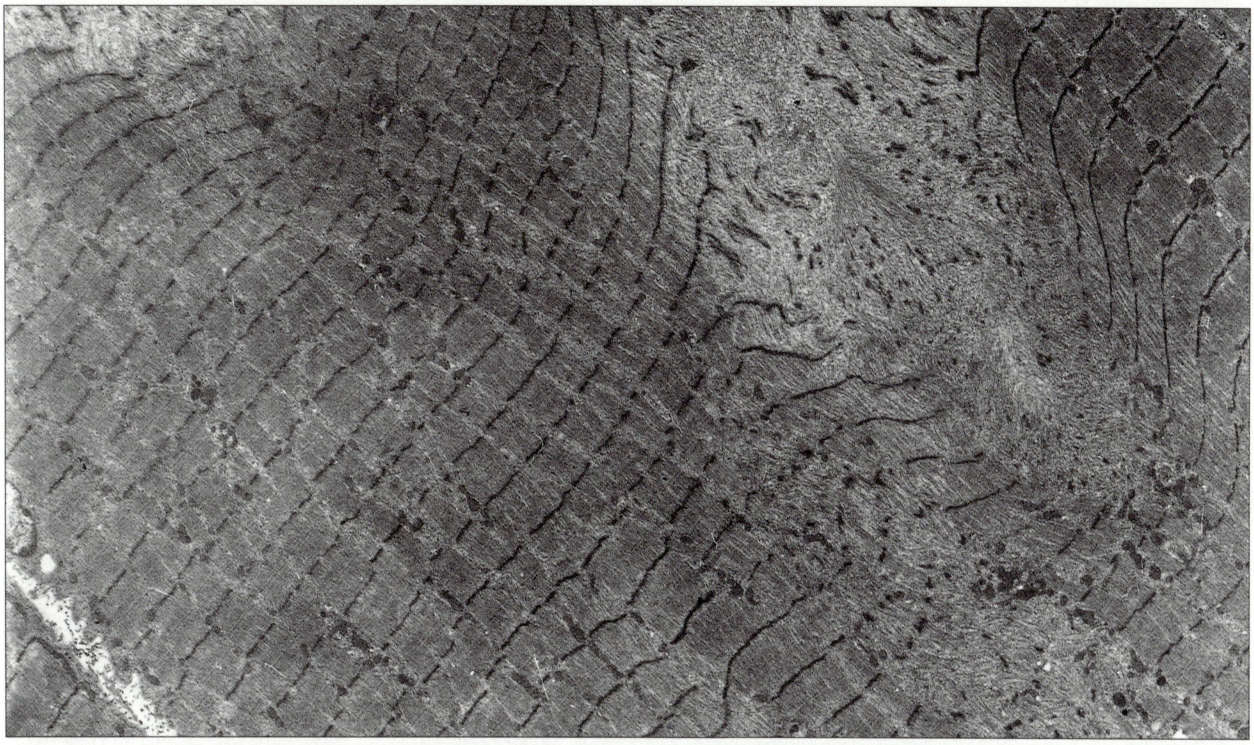

4.60 Disruption of the myofibrils may be focal, and may occur in more than one region of the fibre, resulting in the formation of what are termed 'mini cores'. These are 'unstructured', as the cross striations have been destroyed. (×5775)

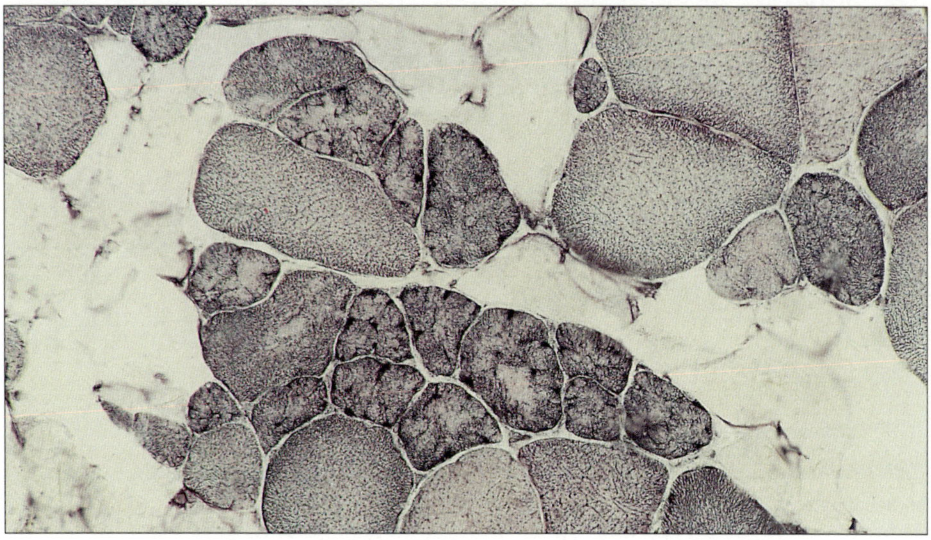

4.61 Fibres showing a 'lobulated' appearance. (NADH-Tr, ×350)

Nuclear myopathies

Nuclear changes occur as a consequence of diverse pathological reactions, such as fibre atrophy, necrosis (which produces pyknosis and disintegration), and in regeneration where nuclei are enlarged and show prominent round nucleoli and dense chromatin (**4.62**-**4.69**, **4.42**). Only rarely are mitotic figures seen. In ischaemia, the nuclear staining becomes pale and acidophilic (**7.19**).

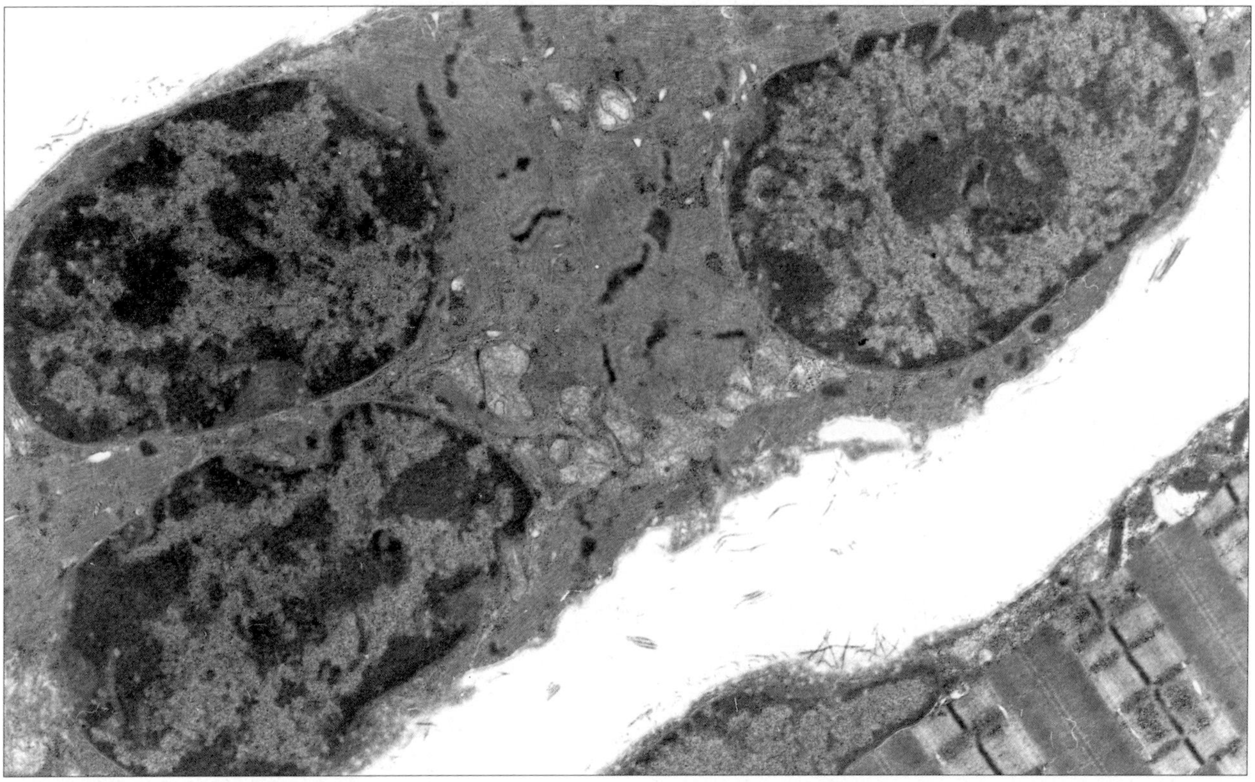

4.62 The nuclei of atrophied fibres are rounded in outline, and contain many dense chromatin accumulations. They may be seen as aggregates in severely atrophic fibres, sometimes refered to as 'nuclearknochen'. (×10500)

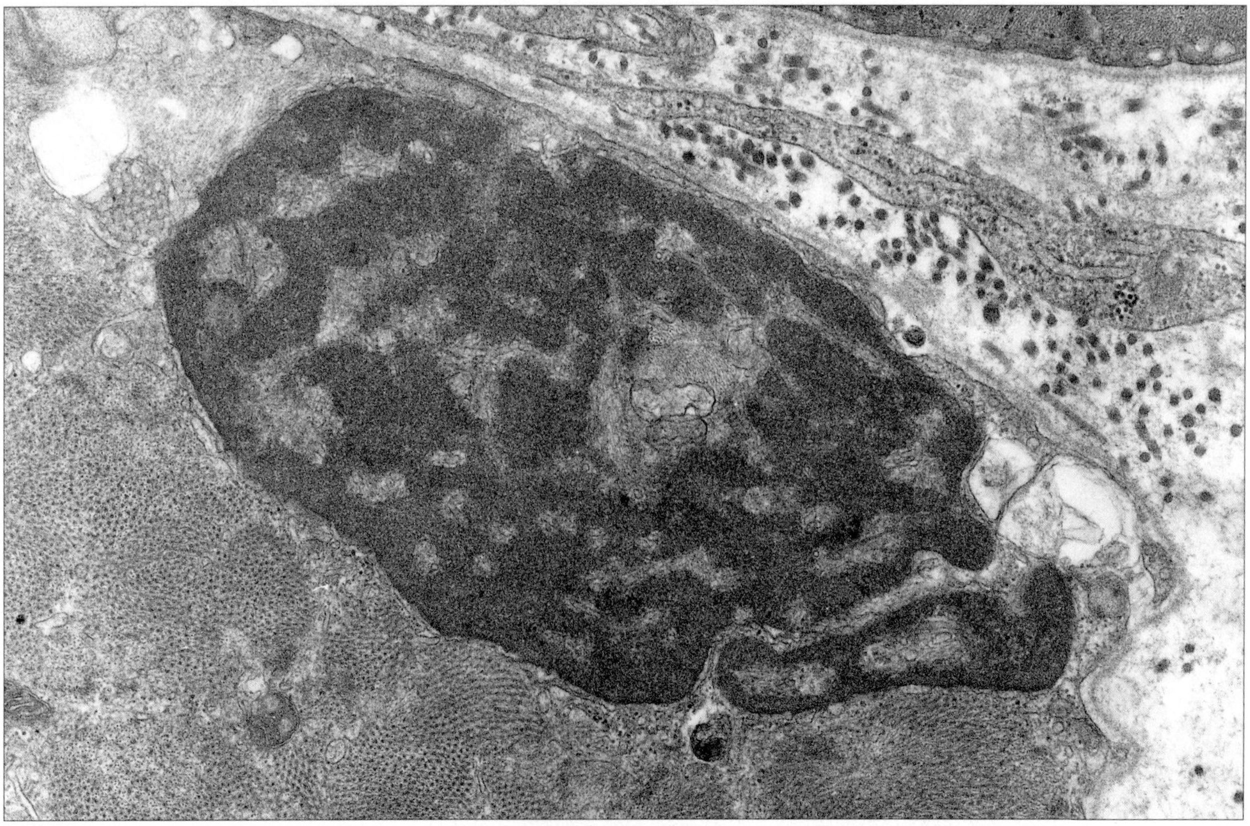

4.63 Lobulated nuclei that show abnormal patterns of chromatin distribution may be found. (×39375)

Electron microscopy is essential to distinguish myonuclei from both satellite cells (**4.37**) and from the nuclei of invading lymphocytes and phagocytic cells.

Centronuclear myopathies are discussed in Chapter 8. Nuclei may appear abnormal or altered in many disorders; this may be in response to denervation, necrosis, inflammation or regeneration of the muscle fibre.

The nuclei seen within fibres may also be those of satellite or invading inflammatory cells.

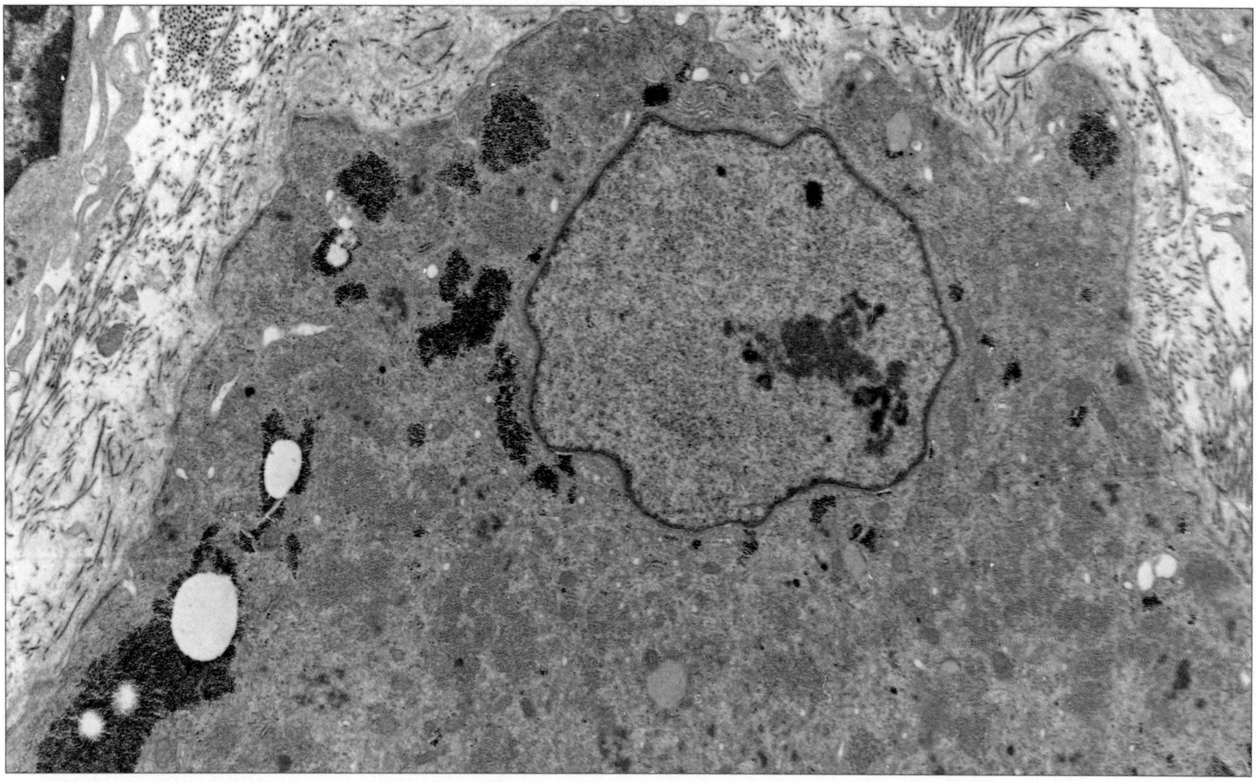

4.64 Enlarged nuclei, angular in outline and containing very little chromatin may be found in structurally abnormal fibres. (×5775)

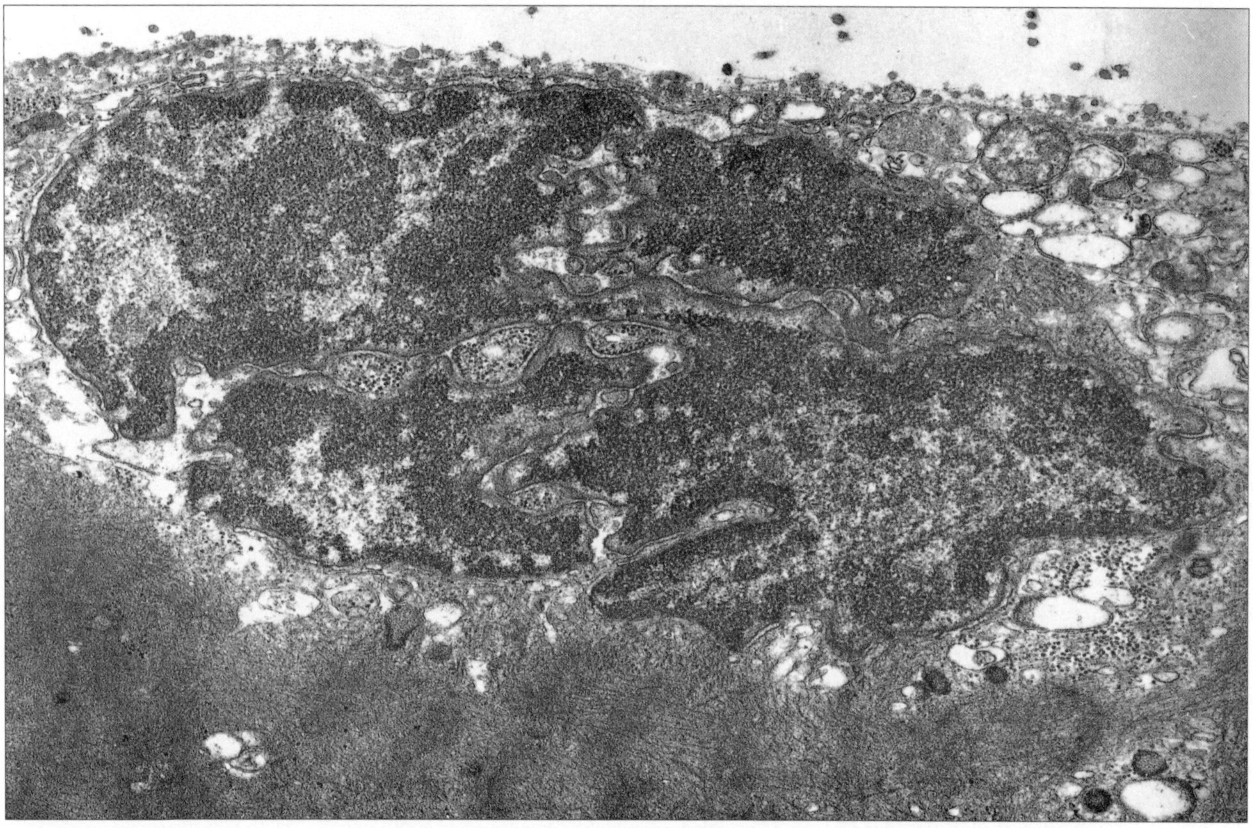

4.65 Nuclei in hyaline fibres also undergo degeneration, with the formation of lobules and aggregation of the chromatin. (×23888)

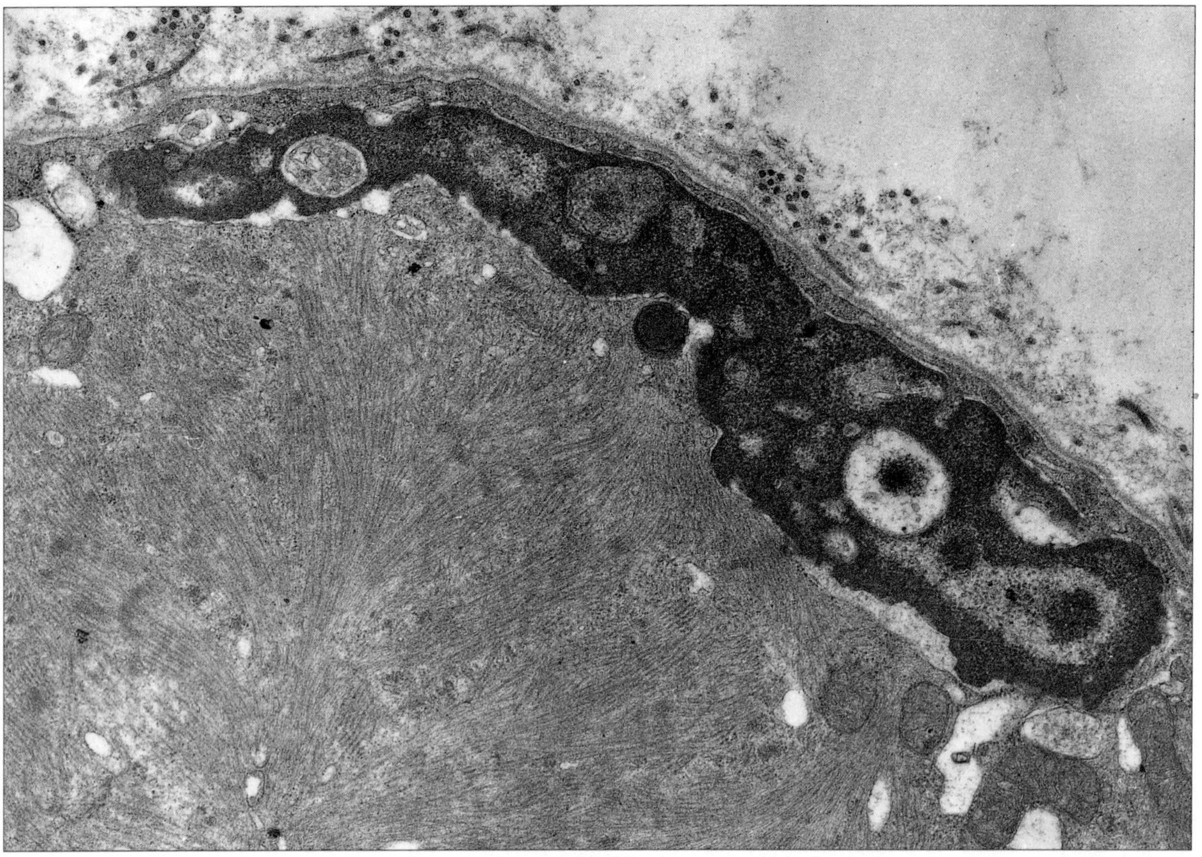

4.66 Intranuclear inclusions occur in IBM, in the form of vesicles containing a variety of structures. (×26250)

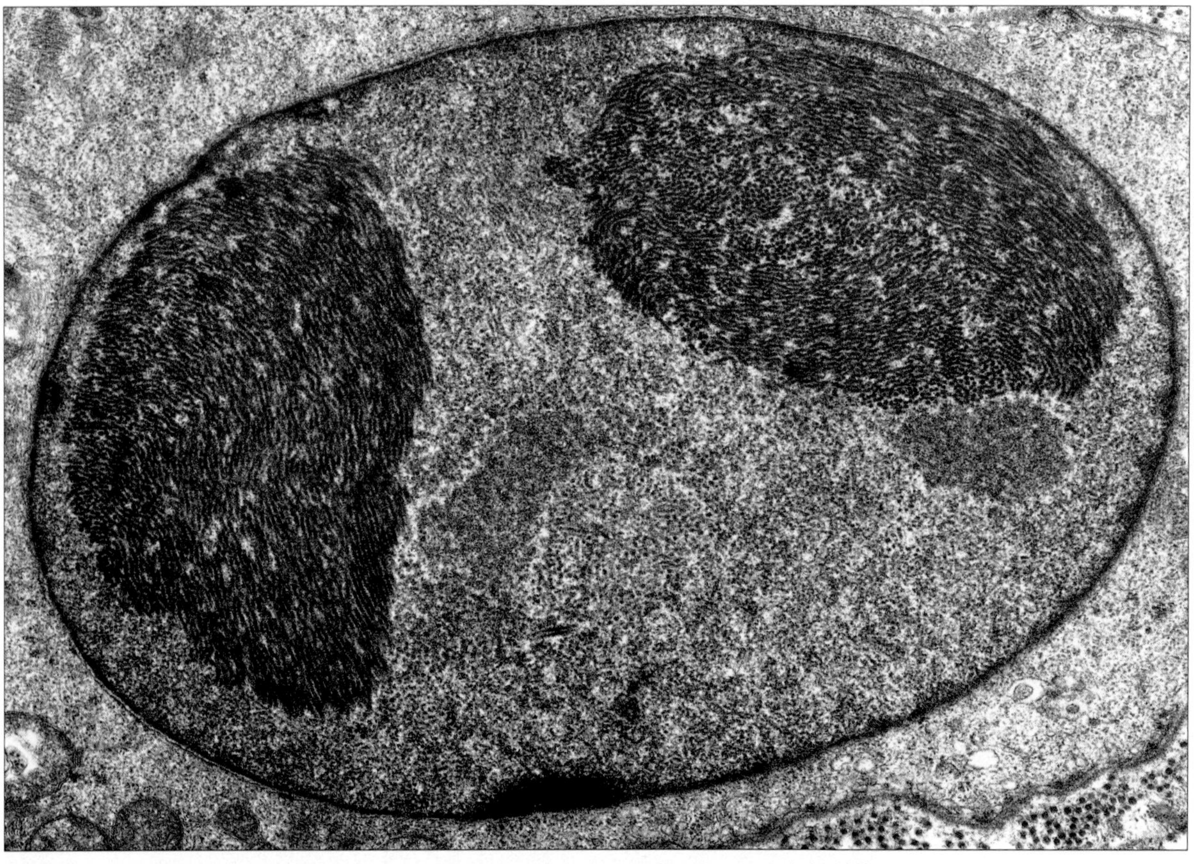

4.67 Filamentous nuclear inclusions with no limiting membranes may be observed. (×21000)

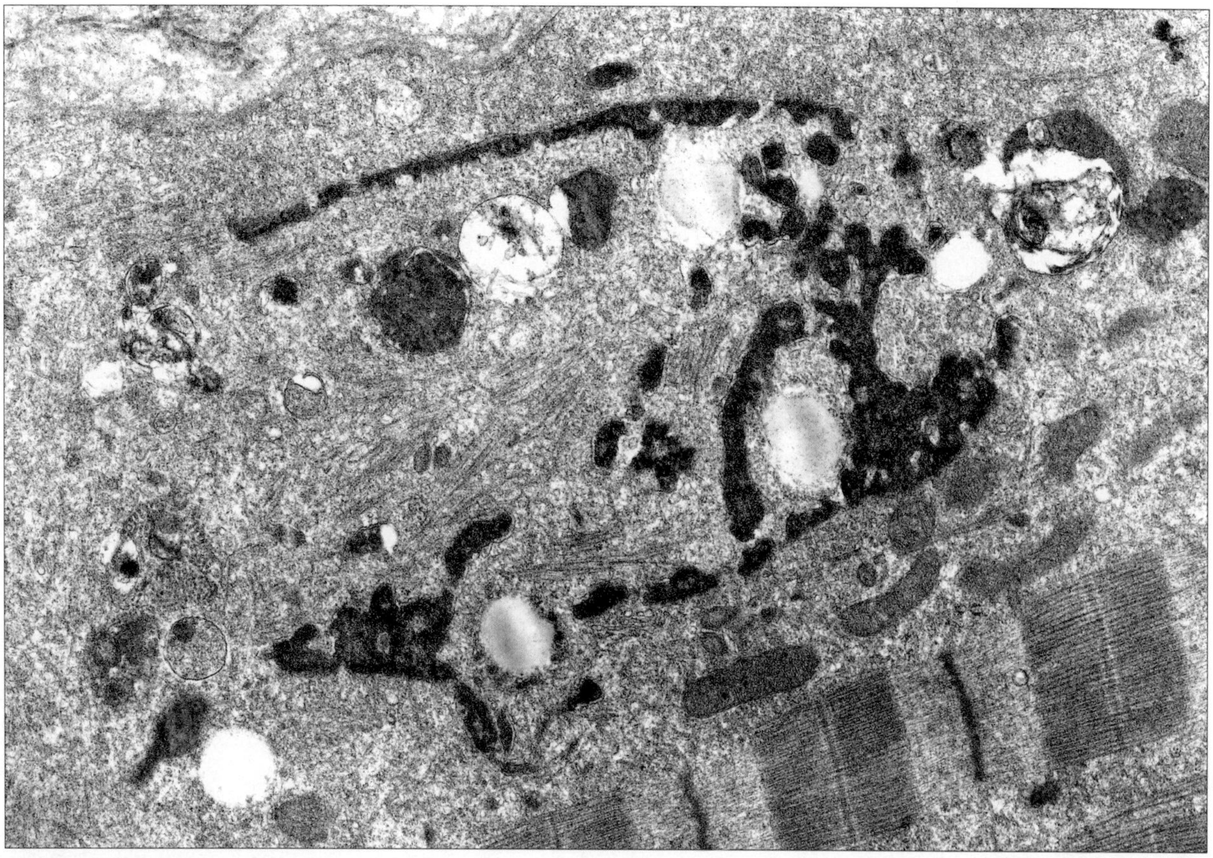

4.68 Remnants of necrosed nuclei appear as electron-dense fragments of chromatin, retaining the outline of the original structure. (×14750)

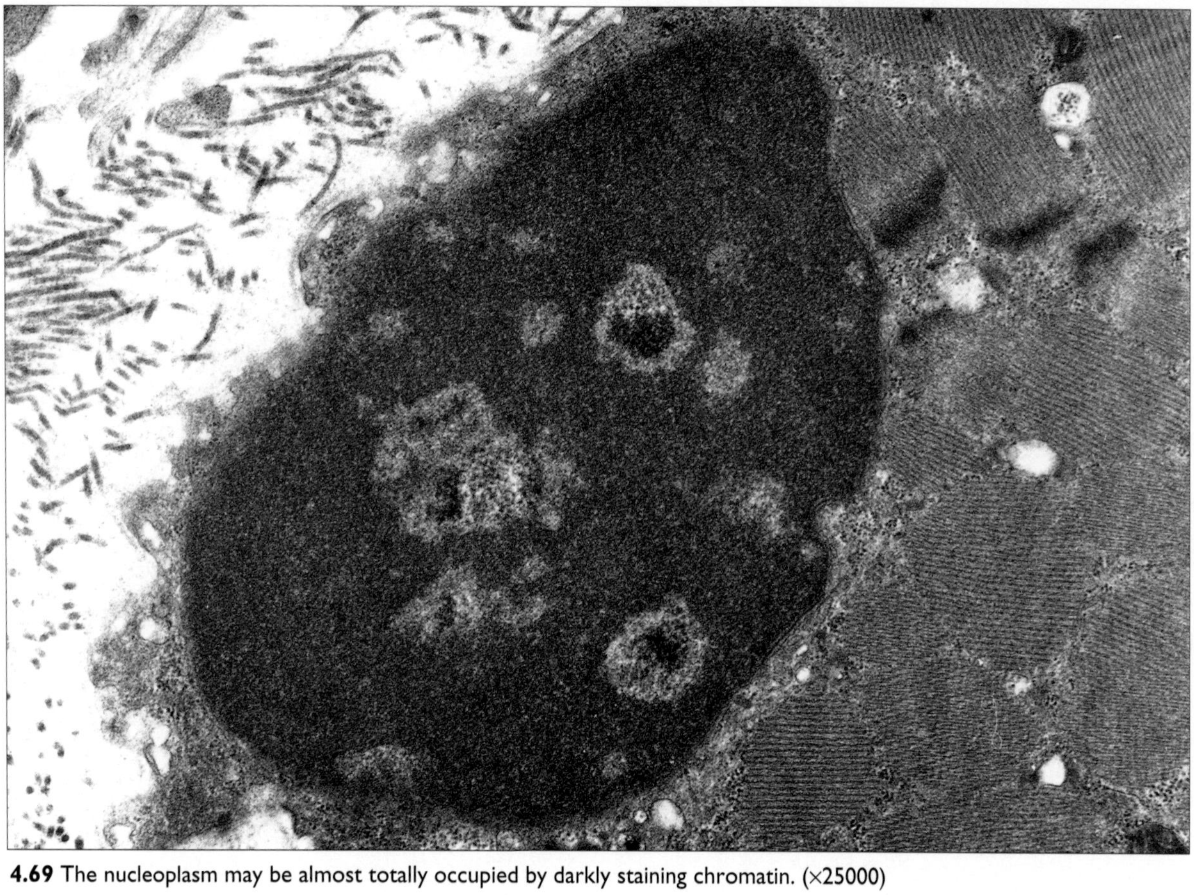

4.69 The nucleoplasm may be almost totally occupied by darkly staining chromatin. (×25000)

Nemaline rods

Nemaline rods are seen within the cytoplasm, and are best demonstrated with the modified Gomori trichrome (**8.8**). They are the predominant feature of nemaline myopathy where they are principally found in Type I fibres (Chapter 8), but have also been reported as a non-specific reaction of fibres of all types in other conditions. They are derived from the Z-band material (**4.70–4.72, 8.9–8.11**), and have been labelled immunocytochemically with anti-α-actinin.

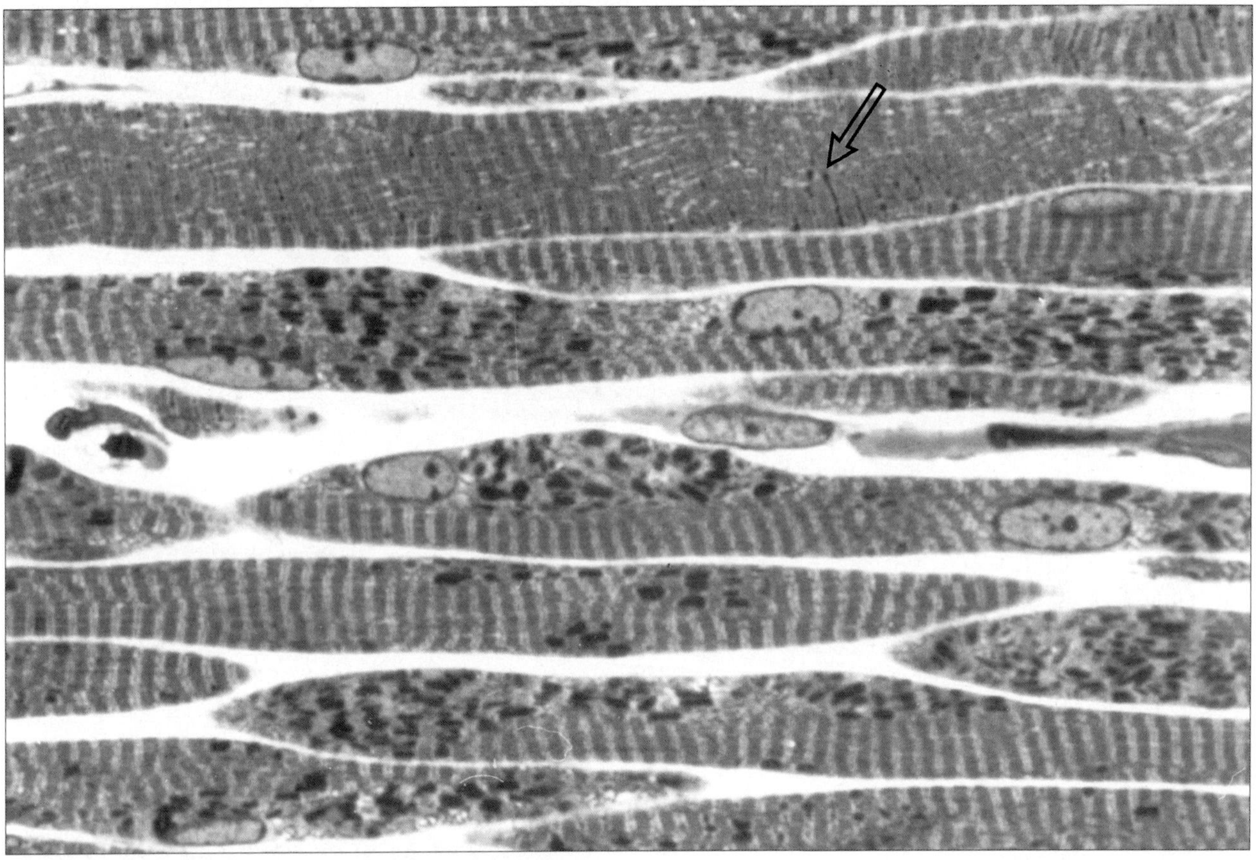

4.70 Nemaline rods arise from thickening of the Z-line (arrowed), and are identifiable in 1 μm thick, toluidine blue stained sections. (×1575)

4.71 The thickening of the Z-lines, which gives rise to the formation of nemaline rods is seen more clearly with the electron microscope. (×37500) (see also **8.9–8.10**.)

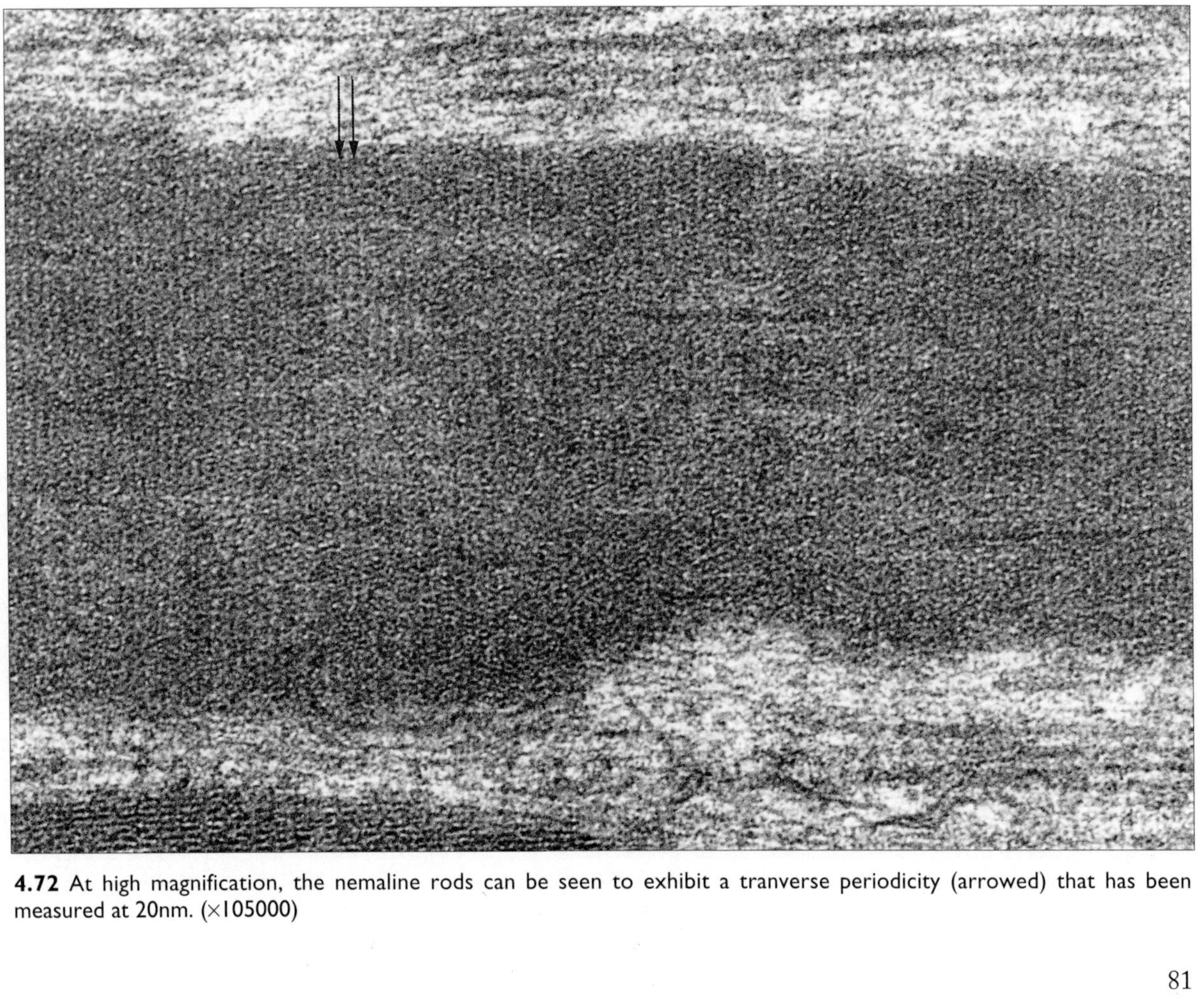

4.72 At high magnification, the nemaline rods can be seen to exhibit a tranverse periodicity (arrowed) that has been measured at 20nm. (×105000)

Tubular aggregates

Tubular aggregates are difficult to discern in the H&E stained section. However, they appear as distinct peripheral accumulations of reaction product with NADH-Tr (**10.6**), and are SDH and myofibrillar ATPase negative (**10.5**). Their presence may be confirmed by electron microscopy (**4.73–4.74**, **10.7**), whereby their distribution throughout the cytoplasm can be seen to be wider than by light microscopy. Tubular aggregates occur most commonly in the periodic paralyses and myotonic disorders (Chapter 10), but have been reported in other conditions (e.g. in association with myalgia) and may occur as a non-specific finding.

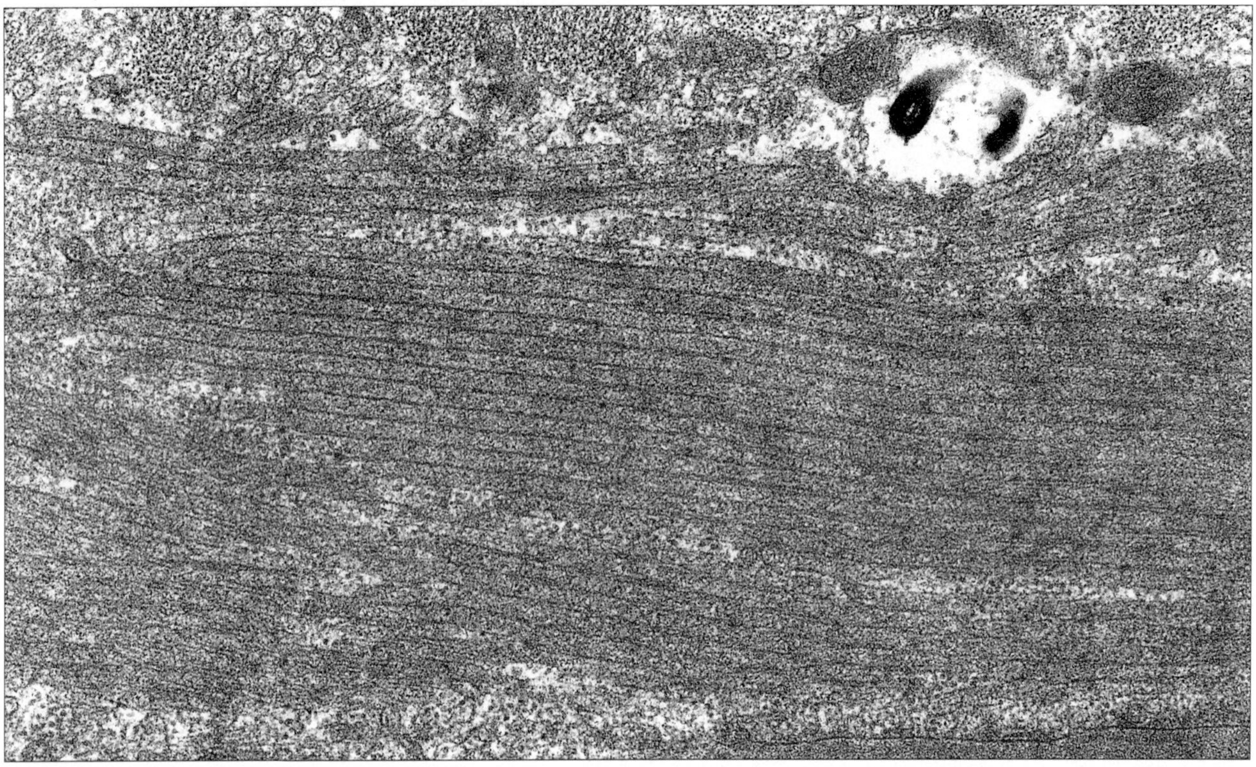

4.73 Tubular aggregates are composed of tightly packed, long, thin-walled tubes approximately 30nm in diameter. They have a granular core, giving the appearance of a translucent wall. (×50400)

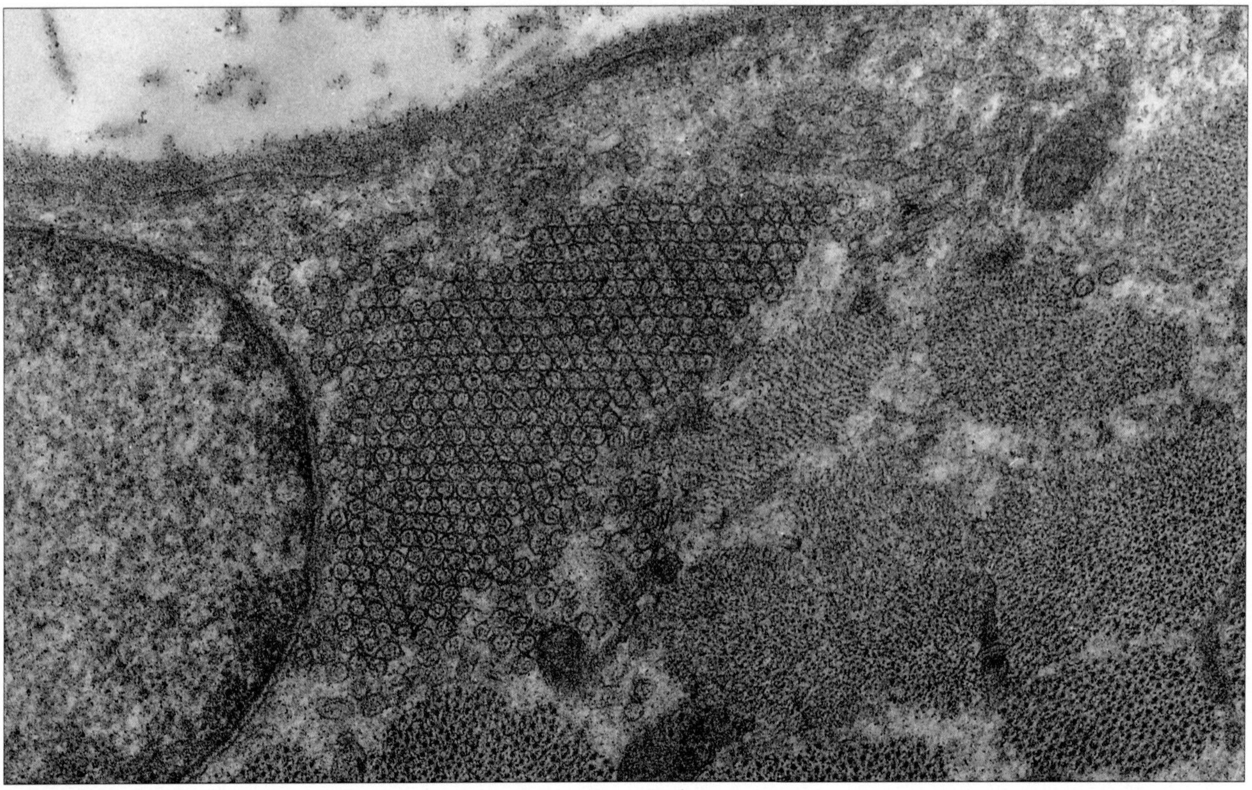

4.74 Tubular aggregates may vary in size and distribution within the fibre. In transverse section, the closely packed tubules can be seen to have a thin, dense outer wall, and a dense central core separated by a clear zone. (×49350)

Miscellaneous ultrastructural disorders

In addition to the various abnormalities illustrated and discussed previously, there are a number of morphological changes seen only at the electron microscopic level that do not appear to be specific to any one pathological process. These include duplication and folding of the basement membrane (**4.75–4.77**), dilatation of the terminal cisterns of the SR and the T-tubules (**4.78–4.79**), and the dissolution of the A-bands, which occurs in several necrobiotic myopathies, particularly DMD and PM/DM (**4.33**).

Other non-specific changes are seen in such conditions, including the presence of honeycomb inclusions (**4.80–4.81**), intramyofibrillar vacuoles (**4.82**), cytoplasmic bodies (**4.83**), concentric laminated bodies (**4.84**) and cylindrical spirals (**4.85**). The pathogenetic significance of these features is unclear, but they are seen sufficiently frequently to merit inclusion in any comprehensive discussion of the ultrastructural pathology of skeletal muscle.

In addition to these features, myelin whorls and lipofuscin deposits (**4.85–4.86**) are seen as late-stage changes in virtually any form of myopathic degeneration. These features are generally held to be the end product of the breakdown of intracellular phospholipid membranes, with lipofuscin being regarded as a lysosomal remnant.

There have been a number of case reports of unusual inclusions, the pathophysiological significance of which remains obscure.

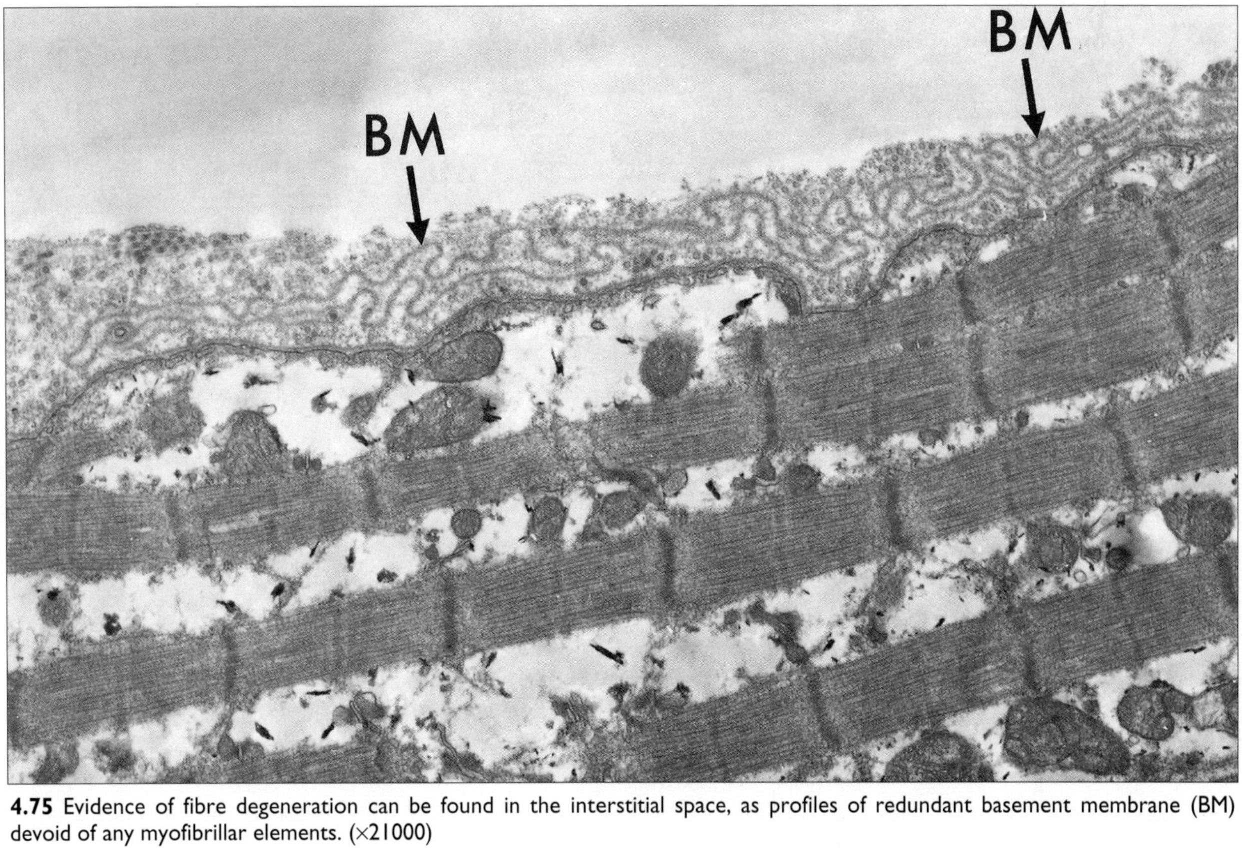

4.75 Evidence of fibre degeneration can be found in the interstitial space, as profiles of redundant basement membrane (BM) devoid of any myofibrillar elements. (×21000)

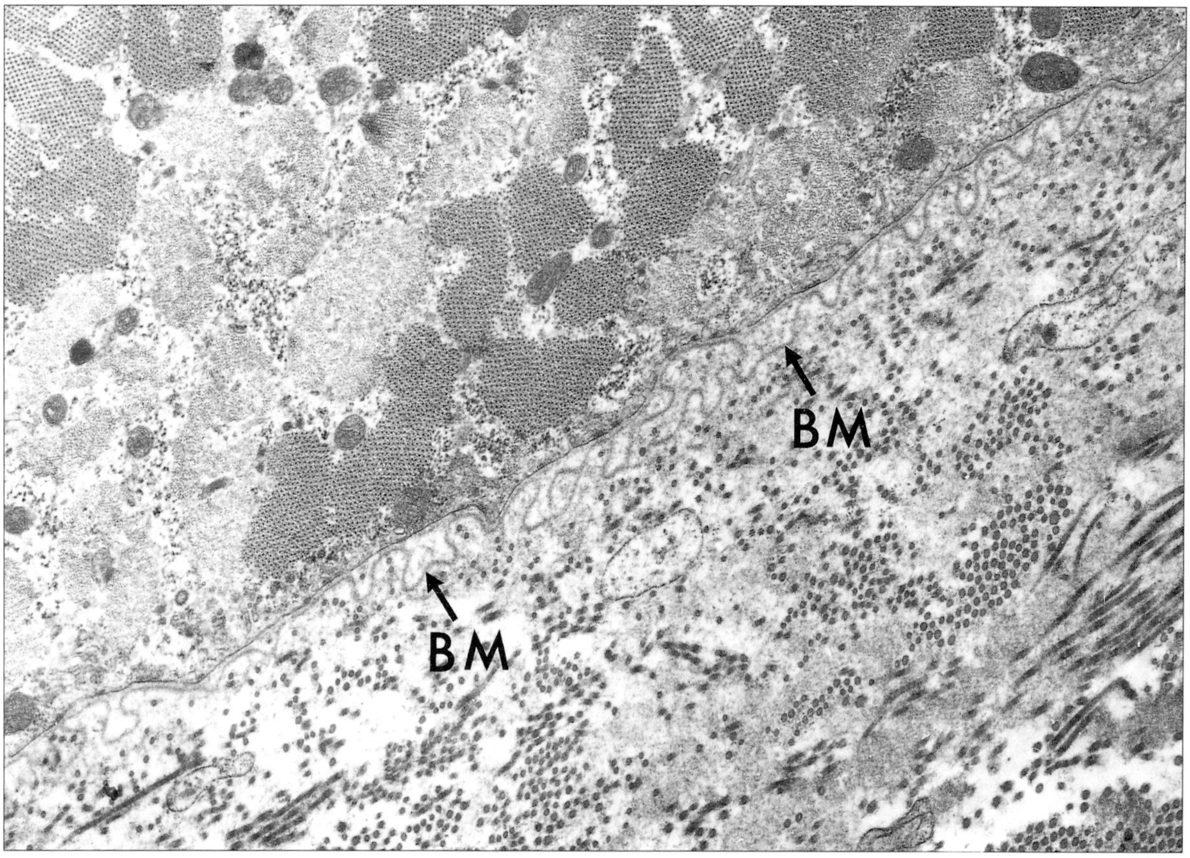

4.76 Thebasement membrane (BM) may show detachment from the plasma membrane and duplication. This may be as a result of regeneration of the fibre within its original basement membrane. (×20000)

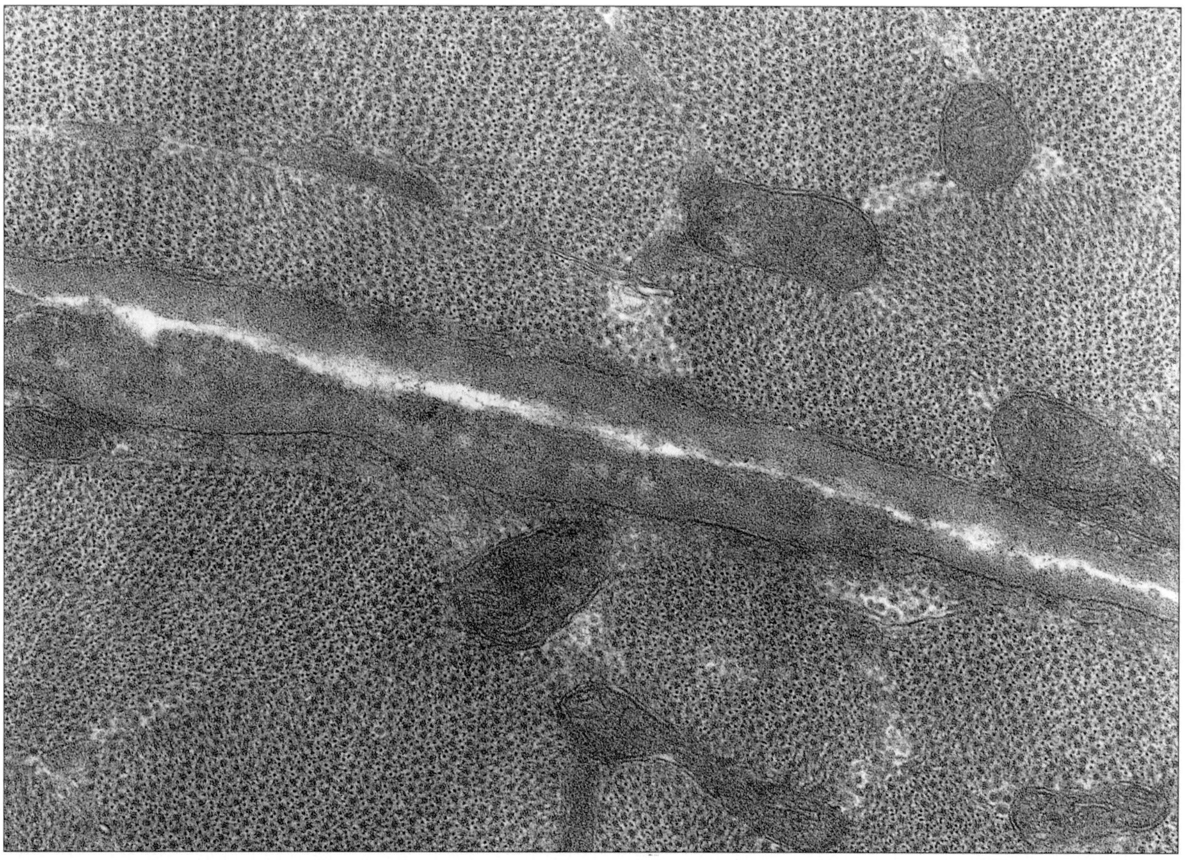

4.77 Thickening of the basement membrane, as illustrated, has been reported in several unrelated conditions. The cause is not known. (×60000)

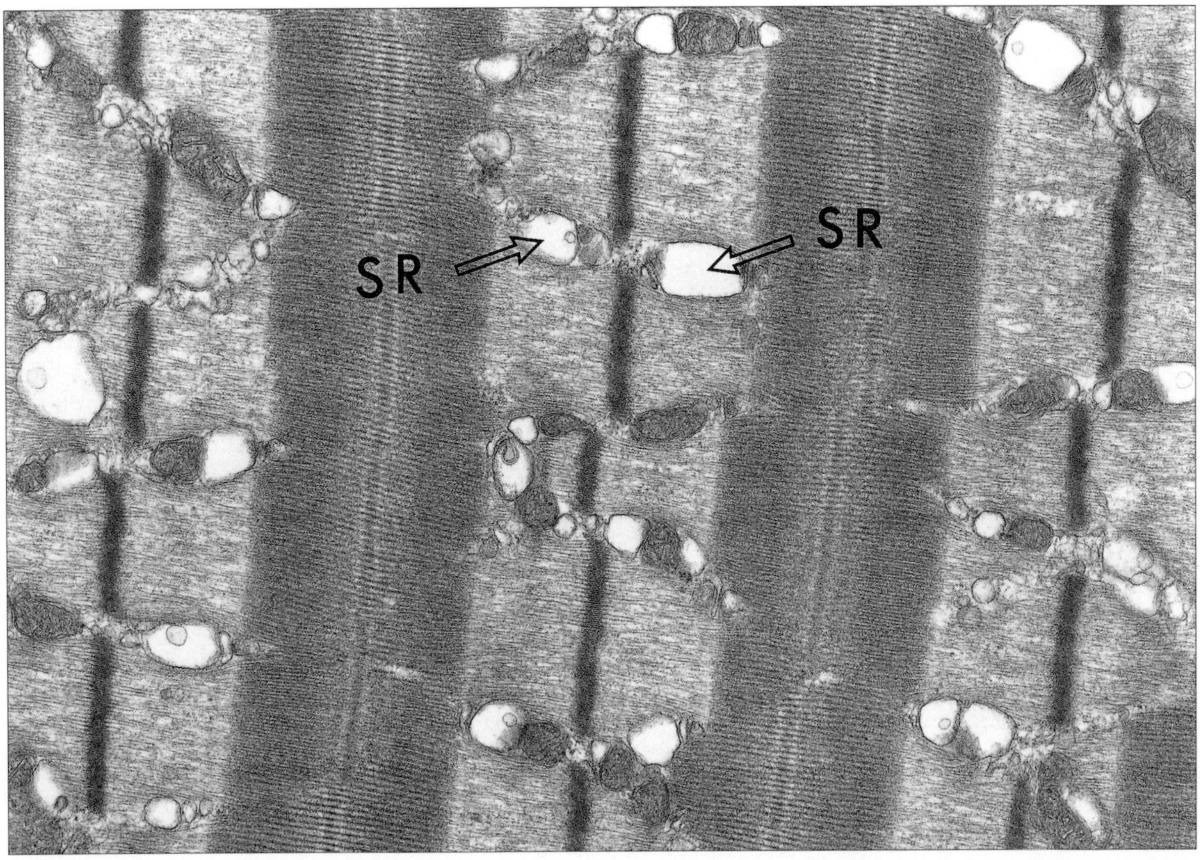

4.78 Dilatation of the sarcoplasmic reticulum (SR) is a nonspecific phenomenon seen in several disorders. It may also occur as a consequence of inadequate fixation. (×30000)

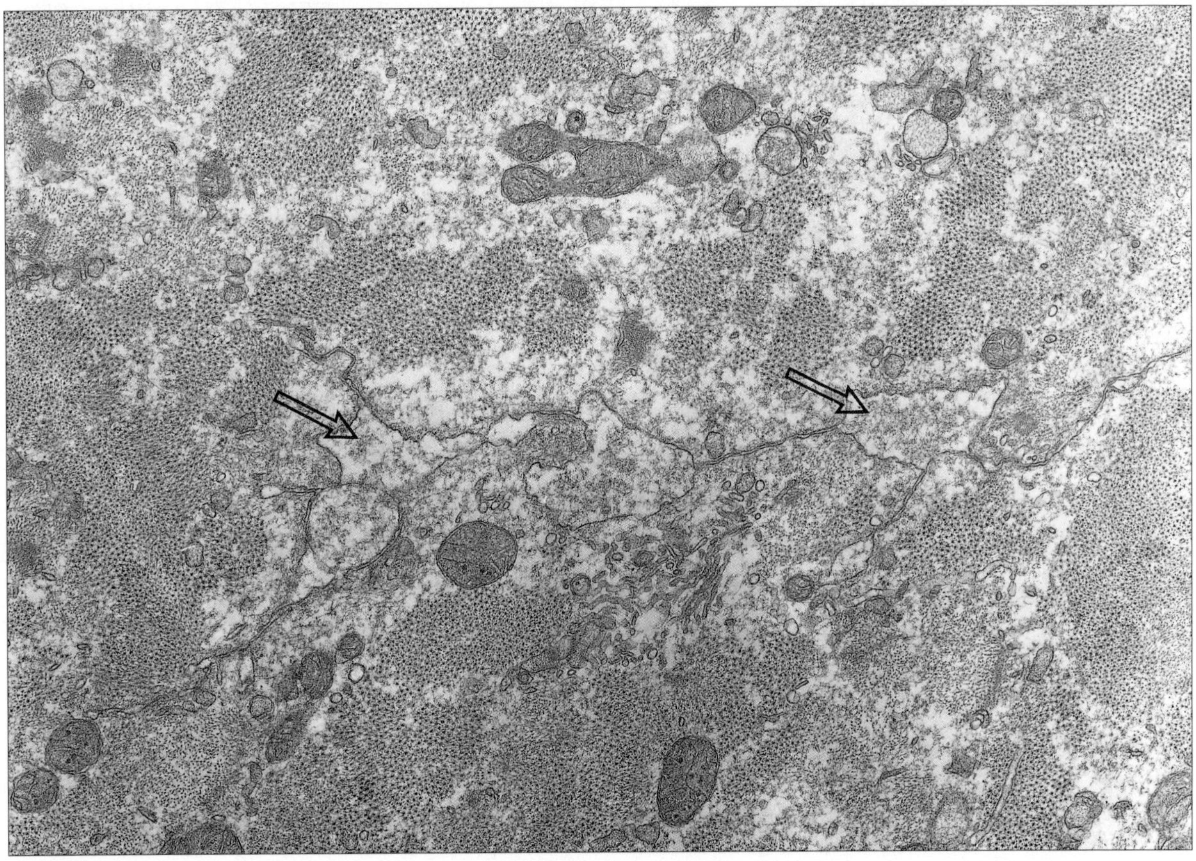

4.79 Dilatation of the T-tubules can lead to the formation of vacuoles (arrowed) that may contain unidentified granular material. Similar vacuoles have been reported in the periodic paralyses. (×26250)

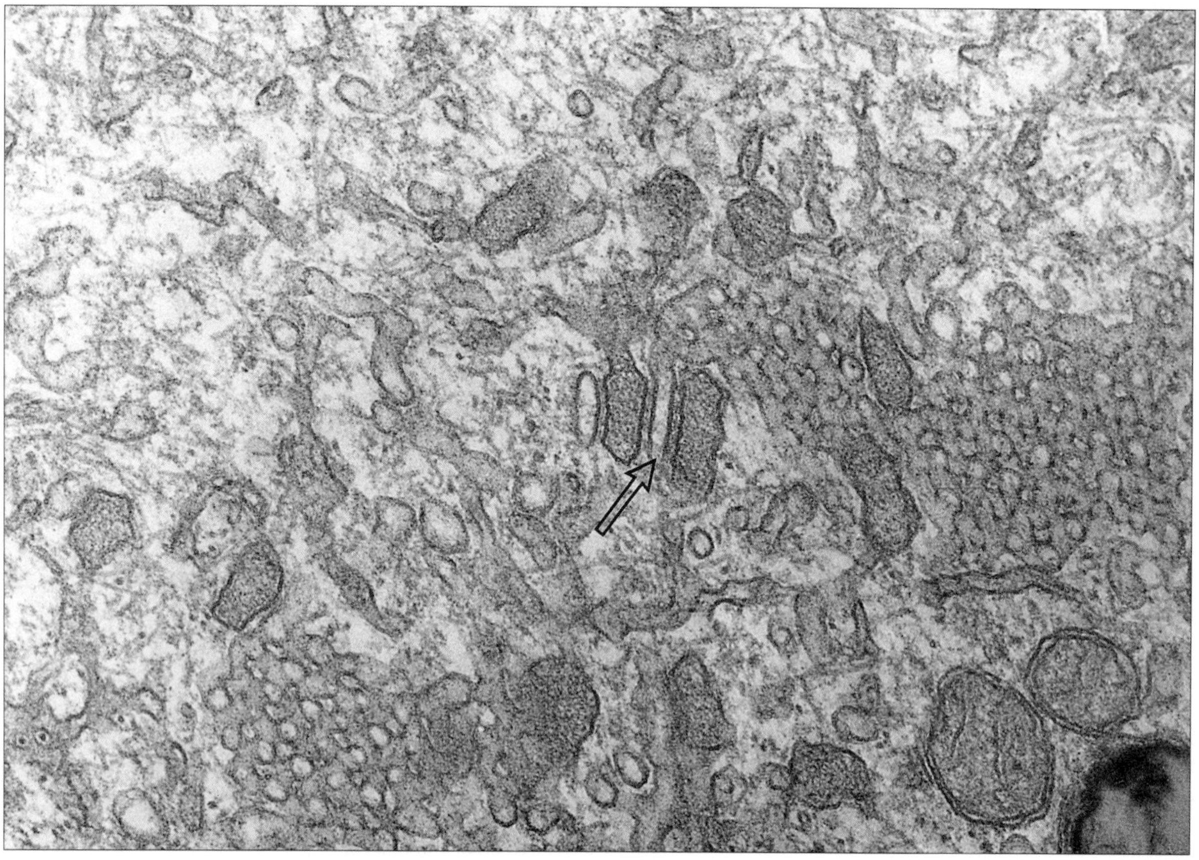

4.80 Proliferation of the T-tubules produces honeycomb-like inclusions. Such structures (arrowed) may be associated with the tubular portion of the triads. (×90000)

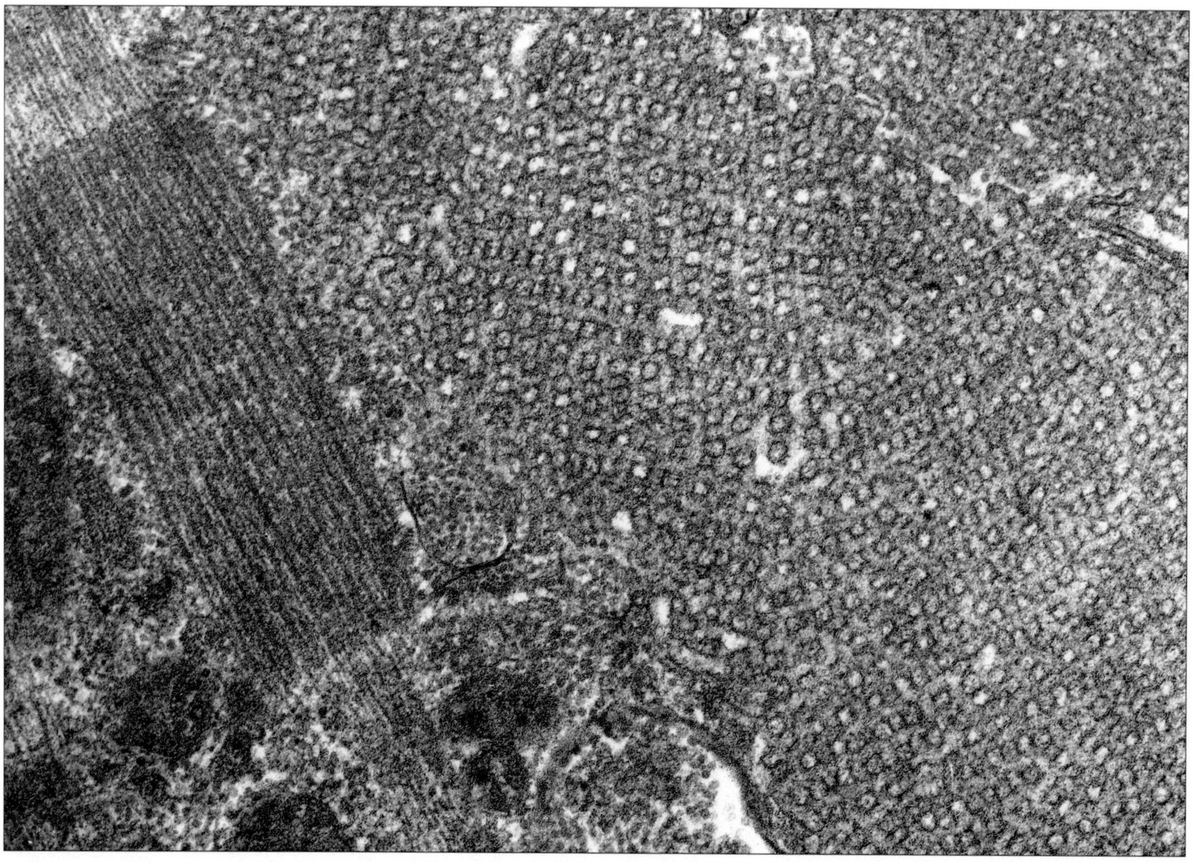

4.81 Honeycomb inclusions can vary considerably in size and distribution, from a small cluster of spheres to extensive interfibrillar masses. (×60000)

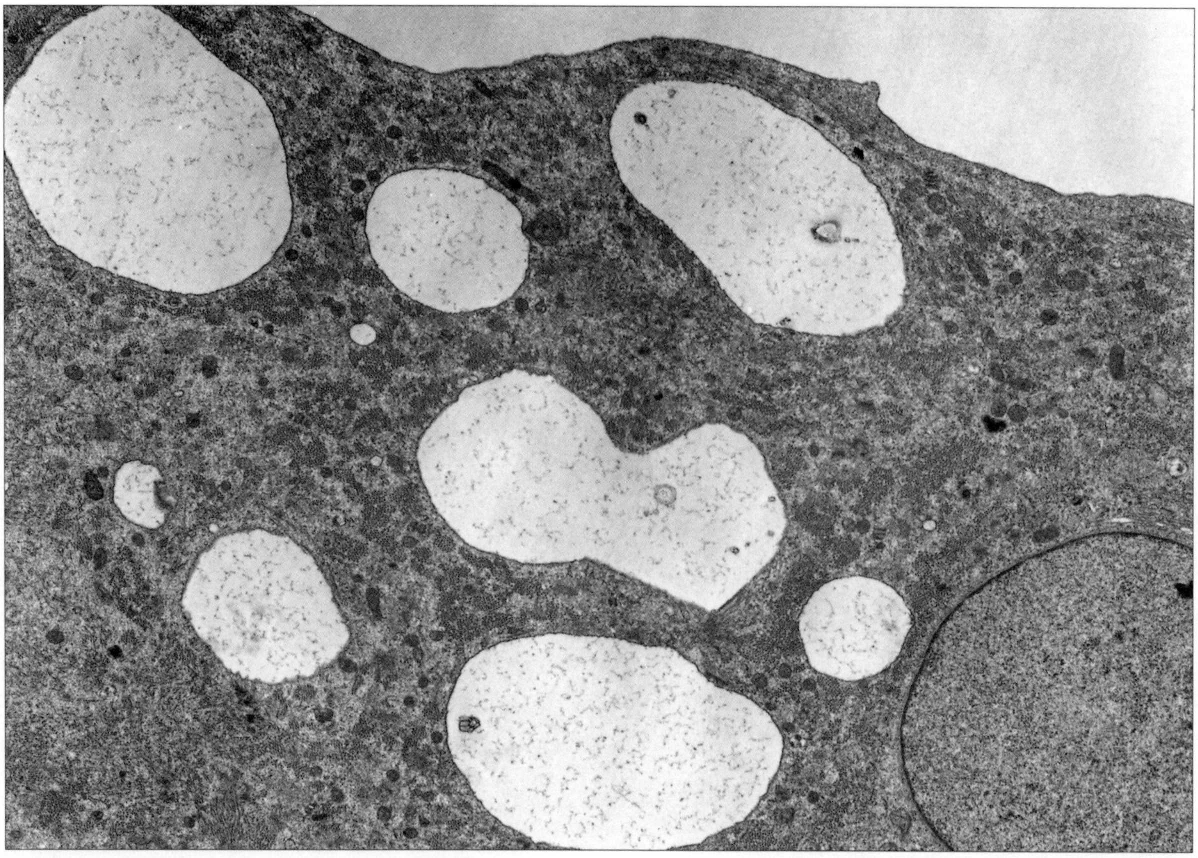

4.82 Intrafibrillar vacuoles lacking identifiable contents may be observed in many unrelated disorders. Unlike those seen in storage diseases, they may be empty or contain only a few membranous remnants. (×33000)

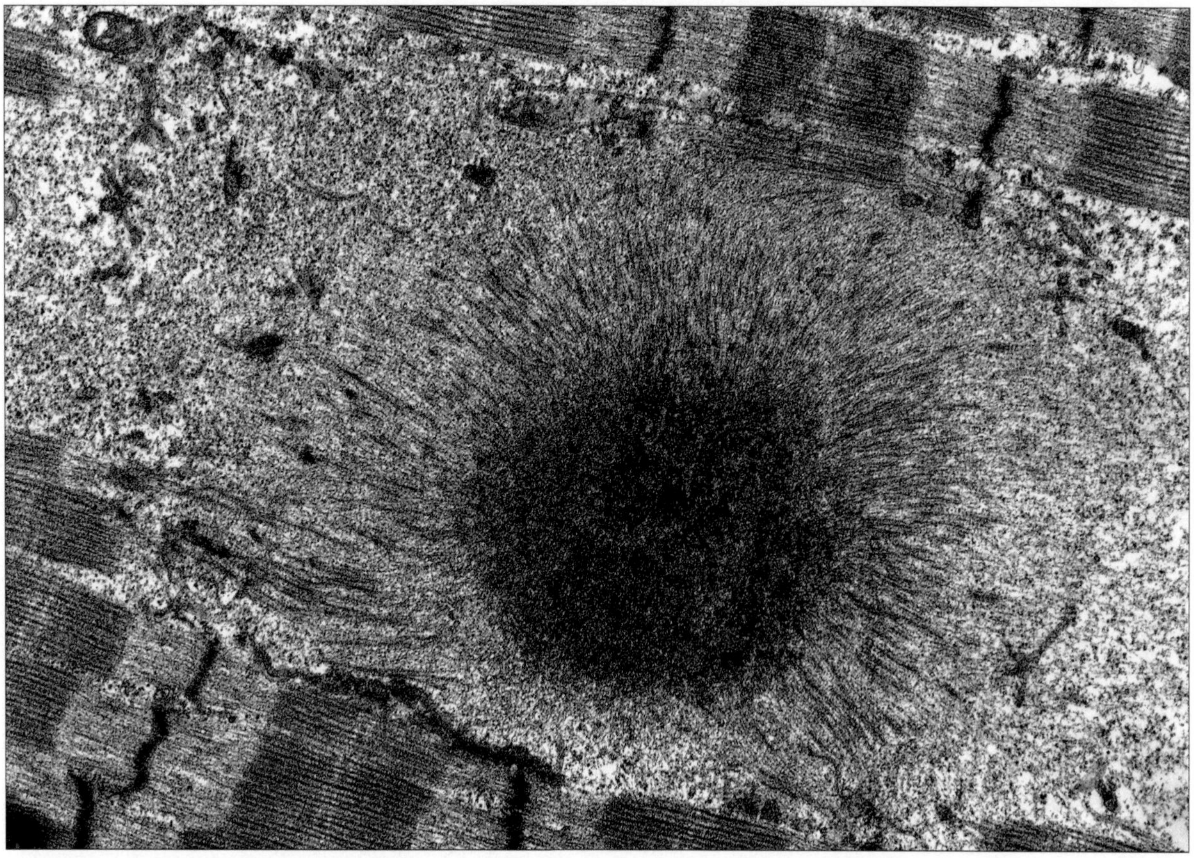

4.83 Cytoplasmic bodies are round or ovoid inclusions, with thick and thin filaments radiating from a dense central structure. They may be related to nemaline rods or they may be thickenings of the Z-disc from which actin filaments are seen to extend (see **8.10**). Both have been described in several unrelated disorders. (×22000)

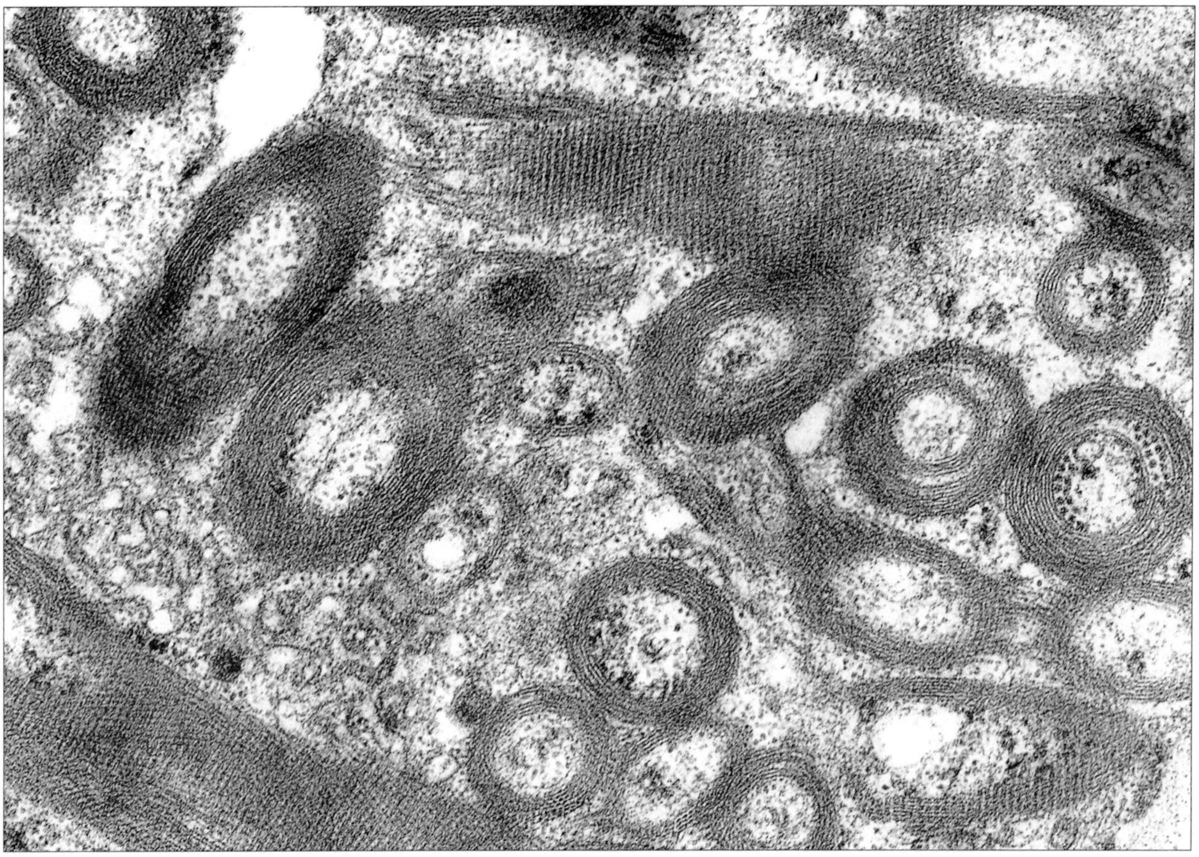

4.84a Concentric laminated bodies are tubular inclusions possessing a regular longitudinal periodicity. They have been observed in many muscle diseases, especially those associated with polysaccharide or mucopolysaccharide storage. (×45000)

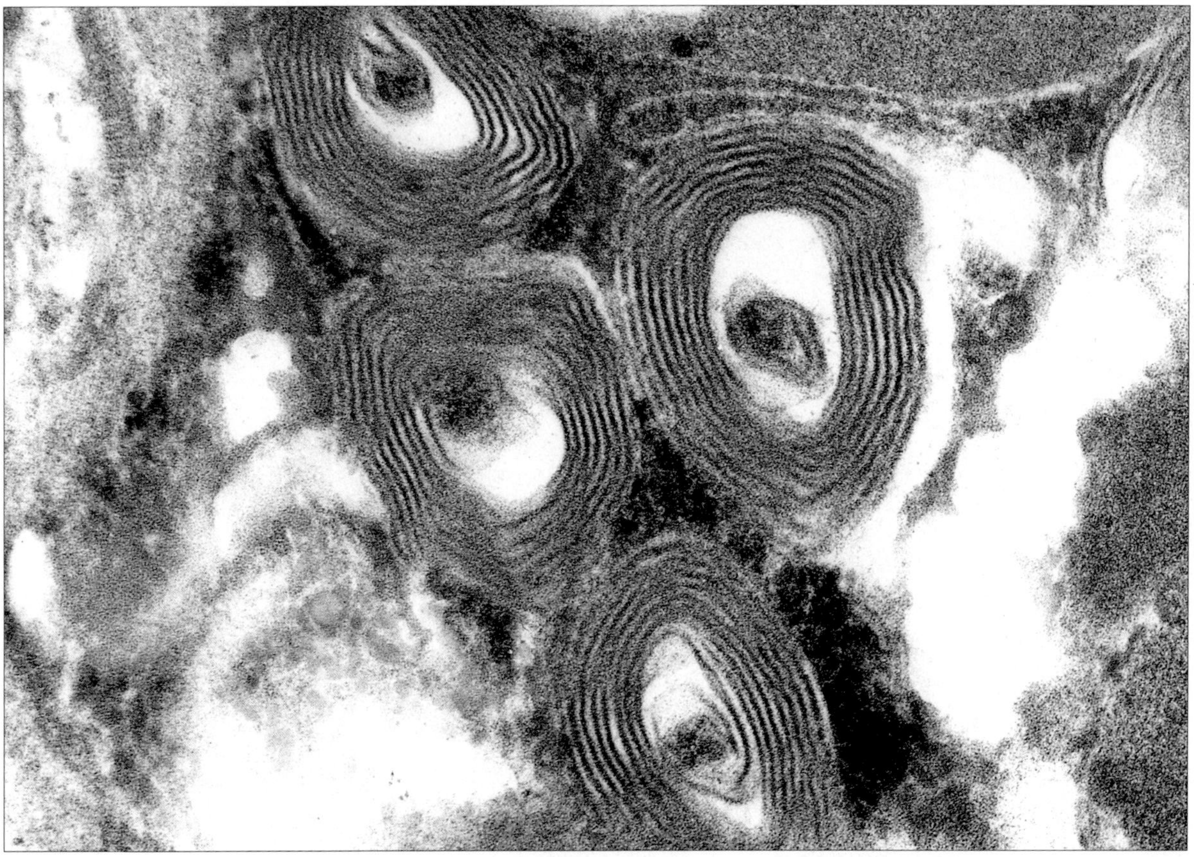

4.84b Cylindrical spirals are composed of concentric lamellae surrounding a granular core. (×100000)

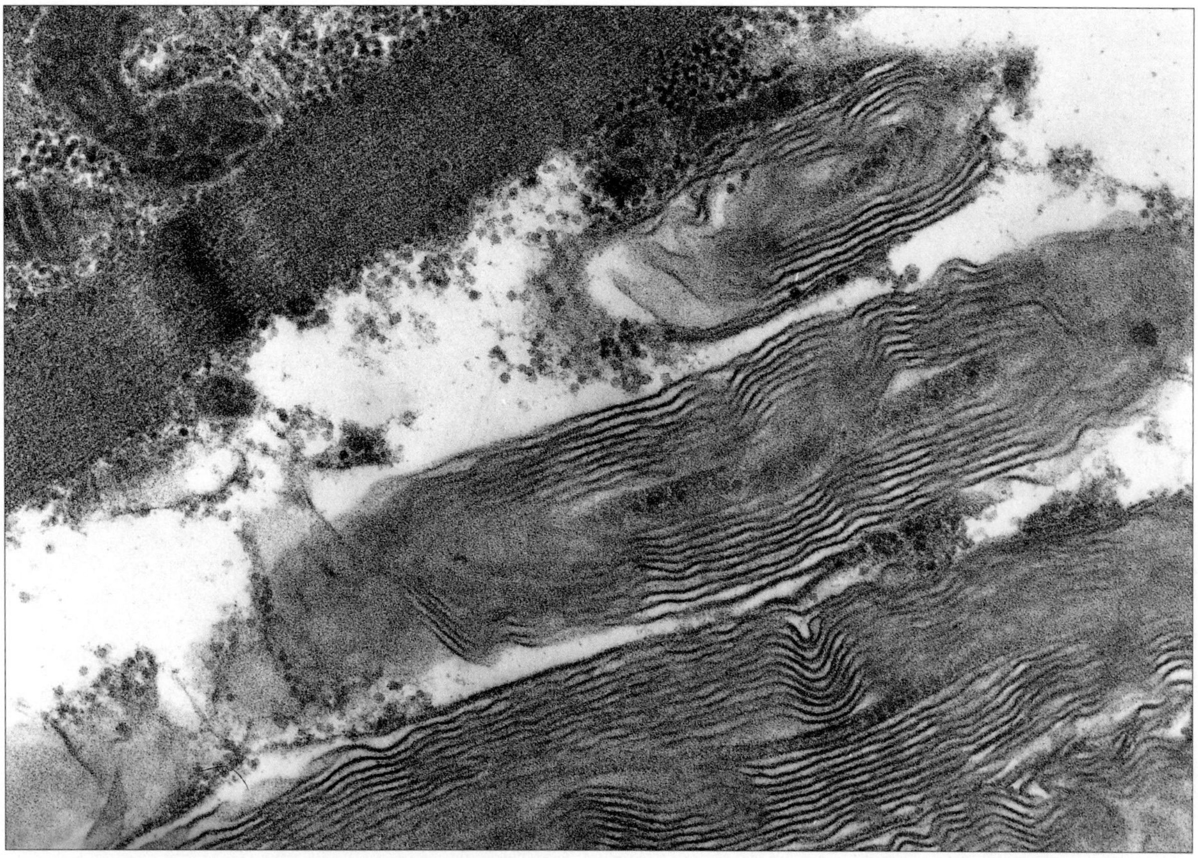

4.84c In longitudinal section, these structures can be seen to contain a core of glycogen-like granules. They lack the cross-striations seen in concentric laminated bodies. (×62500)

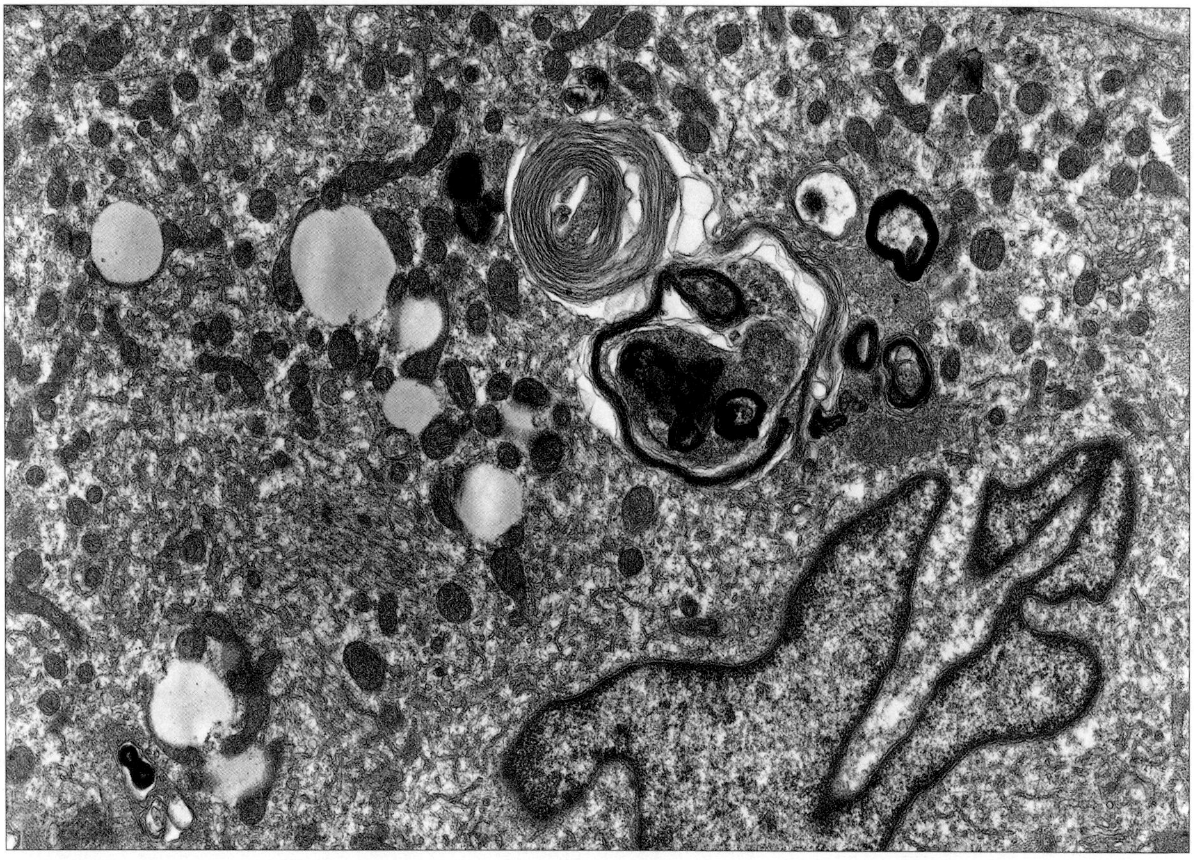

4.85 Myelin figures are electron-dense, laminated structures that vary considerably in size, shape and numbers within a fibre. Most often found in fibres undergoing degeneration, they are degradation products of intracellular phospholipid membranes. (×22000)

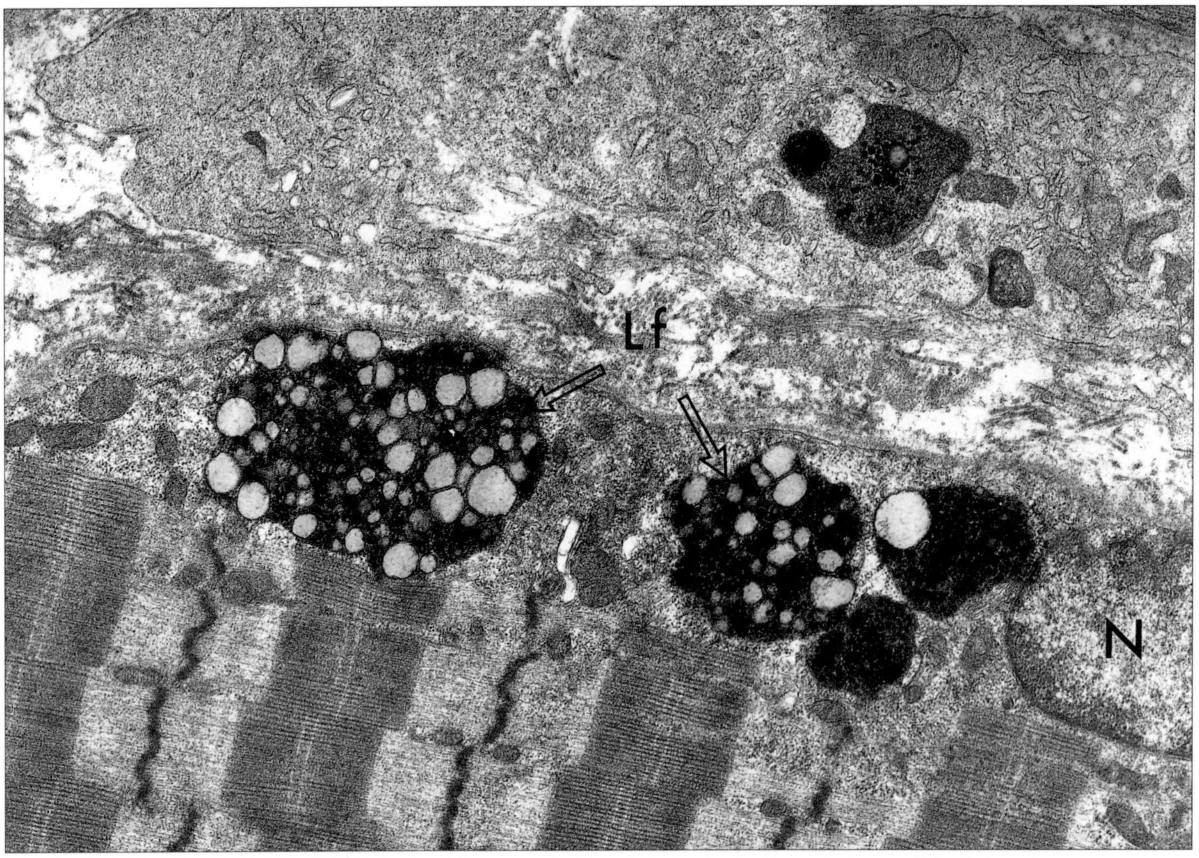

4.86 Lipofuscin granules (Lf), seen as electron-dense vesicular bodies, are common in the muscles of elderly patients. They are usually adjacent to the nucleus (N), and are the only manifestation of lysosomes seen in normal muscle. (×20000)

5

X-linked recessive myopathies (Xp21.2 myopathies)

The impact of molecular biology on the clinical nosology of neuromuscular disease has been profound, most particularly in what are now called the Xp21.2 myopathies. Although it has long been recognised that both DMD (affecting 1 in 3500 live born males) and BMD (affecting 1 in 30000 live born males) are X-linked, recessive disorders, they were for some time considered to be distinct diseases.

In DMD, the clinical presentation is in the first five years of life, with an inexorable decline into wheelchair dependence by the early teens, and to death in the late teens or early twenties. In contrast, in BMD, the age of onset is later and less predictable, varying from the teens to midlife. The rate of progression of BMD was considered to be much slower than that of DMD, with many patients becoming wheelchair dependent in their twenties (or even later), and some patients having a normal life span.

However, the identification of 'outliers' in DMD (i.e. patients whose course is initially rapid and who then appear to progress more slowly with long survival), and early-onset disease in otherwise 'typical' BMD, started to blur the distinction between the two disorders. An arbitrary cut off was developed: if the patient was wheelchair confined by the age of 12, the disorder was considered to be DMD, if after the age of 12, it was considered to be BMD.

Genetic studies led to the localisation of the DMD gene to the short arm of the X-chromosome at site Xp21.2. Subsequent linkage analysis showed that the BMD gene mapped to the same locus, thus indicating that the diseases were allelic (for detailed review see Emery, 1993). The protein product of the Xp21.2 gene was identified (Hoffman *et al.,* 1987) and was named dystrophin.

Dystrophin is a cytoskeletal protein (**5.1**) that is absent or barely detectable in DMD, and of reduced abundance in BMD. It shows variable expression in carriers. Thus, although histological analysis of biopsies from DMD and BMD remains of major importance, the final confirmation of diagnosis rests in dystrophin studies (*see below*).

The clinical and pathological characteristics of these conditions are detailed below.

DMD

The clinical features of DMD were described in detail by Meryon (1852). Nonetheless, the condition was misdiagnosed more often than not until the clinical nosology of the myopathies was placed on a firm footing for the first time by Walton and Nattrass (1954).

Occasionally however, diagnostic difficulty will still arise, most often in the sporadic cases which account for approximately one third of all new presentations. In addition, other neuromuscular problems developing in children during the first decade of life continue to be labelled 'muscular dystrophy' erroneously, on both clinical and pathological grounds. Accordingly, the following cardinal points concerning the clinical features of DMD should be borne in mind:

- The median age of presentation of DMD is 3.5 years, although parents often notice that their son's motor function is not entirely normal before that time. In particular, boys with DMD tend to walk later than their normal peers, and they fall frequently when attempting to run.
- As the classical waddling gait of DMD becomes obvious, the boys start to employ Gowers' manoeuvre when rising from the floor.
- Examination at this stage will demonstrate the characteristic pattern of weakness and wasting described in detail by Walton and Nattrass (1954), and re-analysed by many authors since then. By this time, the boys will have obvious winging of their scapulae and calf hypertrophy. The parents often notice minor, usually asymmetrical signs of both these features at an earlier age.
- From the age of 7–8 years onwards, DMD progresses inexorably, and the boy becomes wheelchair dependent by the age of 10–12 years in the majority of cases. For some time thereafter, he may be able to lead a relatively independent life, although his incapacity will increase with deteriorating muscle strength in his upper limbs. This phase of the boy's life will also be punctuated by respiratory difficulties that are usually triggered by upper respiratory tract infections and made worse by increasing kyphoscoliosis. During this time the cardiomyopathy, which is part of the disease process, may become manifest clinically and will certainly be evident on the ECG and echocardiogram.
- The quality of life for boys with DMD in their latter years is undoubtedly much better now than previously. Nevertheless, their life expectancy has only been increased marginally and death may be anticipated between the ages of 16–21 years in the majority of cases.
- Many boys with DMD manifest mild to moderate degrees of intellectual impairment. However, they are all educable, and it is of the utmost importance that this aspect of their management is not overlooked, particularly with reference to the acquisition of computer literacy which will be invaluable to them with increasing muscle weakness.

5.1 Dystrophin and related proteins at the muscle fibre sarcolemma

Dystrophin is a large, rod-shaped cytoskeletal protein (molecular weight 427 000 Daltons) transcribed from a 14 000 base mRNA, that is encoded by 79 exons within the enormous 2.4 million base DMD/BMD gene located at Xp21.2. Although dystrophin accounts for only 0.002% of total muscle protein (2% of sarcolemmal protein or 5% of sarcolemmal cytoskeletal protein) its absence, reduction or abnormality, and changes in its associated membrane proteins (see below) appear to be the primary lesion in a number of related myopathic or dystrophic disorders, resulting in variable phenotype of muscular wasting (see main text).

Dystrophin is localised to the cytoplasmic face of the sarcolemma in normal skeletal cardiac and smooth muscle fibres (see **5.1a** and **5.1b**), where it forms an interlacing network approximately 15 nm from the plasmalemma. It is most abundant at costameres (cytoskeletal associations with the Z-line) and at neuromuscular and myotendinous junctions. Associations with the T-tubular system have also been reported.

Dystrophin first appears in trace amounts after about 9 weeks of foetal life, and is fully expressed in differentiating myotubes and immature muscle fibres by 17 weeks of gestation. It also appears in small amounts and in a slightly different form (Dp71, Dp116) and arrangement in other tissues, most notably that of the brain. It is not usually present in muscle in DMD and is reduced in amount and size in BMD.Complete sequencing of the human DMD cDNA has predicted dystrophin to be a 3685 amino acid protein that is approximately 125–150 nm in length. It is composed of four structural domains (**5.1c**):

- The N-terminal (amino or 5') globular domain of 240 amino acids with high homology to actin binding proteins such as α-actinin and β-spectrin. This region associates with F-actin.
- The rod domain, consisting 25 repeats of 109 amino acids in a triple helix resembling spectrin (approximately 2725 amino acids).
- The cysteine rich domain that binds to sarcolemmal proteins (approximately 300 amino acids).
- The C-(carboxy) terminal (unique to dystrophin), which is leucine rich and contains 420 amino acids.

Molecular genetics of dystrophin

The massive DMD/BMD gene is particularly prone to deletions, which are particularaly prone to occur around exons 44 and 45 (as is observed in 40% of affected individuals). In 94% of patients, in-frame gene deletions lead to a truncated or reduced abundance protein and a mild phenotype (BMD), whilst out-of-frame deletions give rise to a marked abnormality or almost complete loss of dystrophin and severe phenotypes (DMD). Exceptions occur mainly with respect to deletions in exons 3–7, where a near normal sized dystrophin and mild phenotype results.

Generally, deletions at the 5' end (which code for the N terminus) lead to a variable but severe BMD like phenotype. In contrast, the proximal rod region appears to be relatively dispensible, resulting in a very mild phenotype. Abnormalities of the distal rod give rise to more typical BMD, and variations in the C–terminus (3' end of the gene) produces a severe or DMD phenotype.

Dystrophin-glycoprotein complex (DGC).

This complex of four dystrophin-associated glycoproteins (156 kD, 43 kD, 35 kD, 50 kD DAGs) and two dystrophin-associated proteins (25 kD, 59 kD DAPs) appears to link dystrophin to the muscle fibre plasma membrane and the laminin component of the extracellular matrix (**5.1b**). Dystrophin and DGC also co-localise to the sarcolemma of intrafusal fibres, and the neuromuscular and myotendinous junctions.

The 156 kD and 43 kD dystroglycan are encoded by a single cDNA and are also expressed in nerve and other non-muscle tissues. Other proteins (58 kD, 87 kD, 94 kD) may also be associated with the complex, particularly at post-synaptic membranes.

There is a drastic reduction in the amount of the dystrophin-associated proteins in DMD and in the mdx mouse homologue, but levels of the latter may be restored in the laboratory in DYS treated transgenics. In severe childhood autosomal recessive muscular dystrophy (SCARMD, gene locus 13q12), there appears to be a particular deficiency of the 50 kD DAG. It is also possible that Fukuyama muscular dystrophy (9q31–q33) may be associated with deficiency of another protein from the complex (probably the 43 kD DAG).

Utrophin

Utrophin (also known as DMD-Like protein, or dystrophin related protein, DRP) is an autosomal (6q24) homologue of dystrophin. Its molecular weight, 395 000 daltons, is similar to dystrophin, and it is produced from a 13 000 base mRNA transcript, bearing about 83% homology to that of dystrophin.

Utrophin is expressed in all tissues, but is found in greater quantities in smooth muscle and nervous tissues. In adult skeletal muscle, it is localised exclusively to the neuromuscular junction where it is associated with DAPs. It is also expressed more widely during development and regeneration, and in DMD and in mdx mice, utrophin spreads beyond the neuromuscular junctions. As yet, no specific utrophin-deficient related neuromuscular myopathies have been identified.

Physiological role of dystrophin and the DGC

From the morphological and sequencing data, a number of proposed functions have been suggested for these proteins within muscle tissue. In particular, the complex is thought likely to provide structural support (stabilising the sarcolemma), to assist force or signal transduction between the sarcolemma and contractile apparatus, to organise membrane specialisations, and to regulate the intracellular calcium concentration. Thus, dystrophin, (a lack of which appears to lead to disruption of the DGC and loss of DAPs) may play a key role in the cascade of events leading to muscle cell necrosis in DMD.

The reader is referred to original and review articles by Ahn and Kunkel (1993), Koenig, Monaco and Kunkel (1988), Lindenbaum and Carbonetto (1993), Love *et al.* (1993), Matsumura and Campbell (1993, 1994) for a fuller account of the molecular biology of these proteins and their role in the 'dystrophinopathies'.

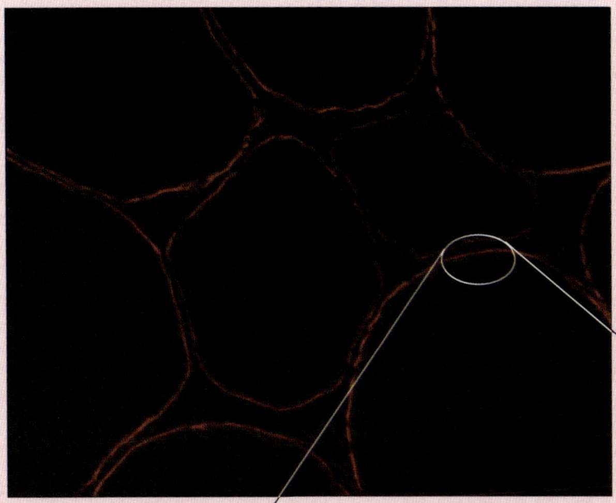

5.1a Muscle fibres showing peripheral localisation of dystrophin with fluorescent Texas-red labelled antibody.

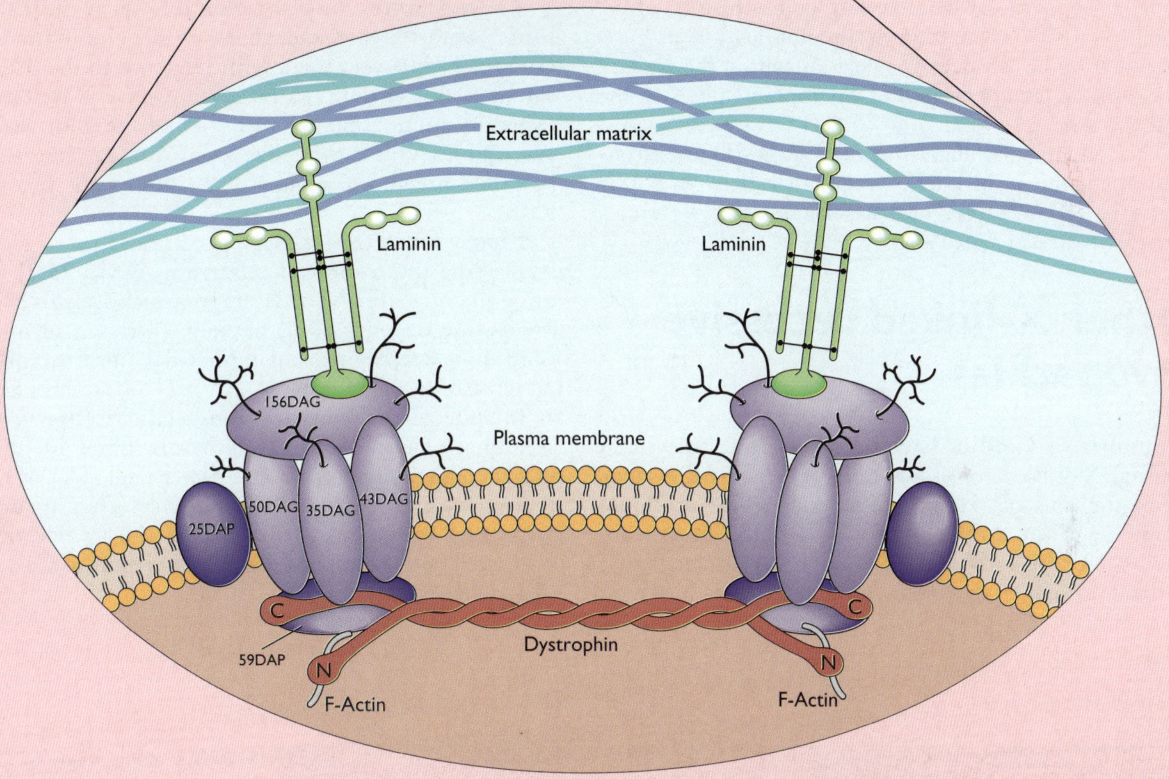

5.1b Dystrophin-glycoprotein complex within the sarcolemma (modified from Matsumura and Campbell, 1994).

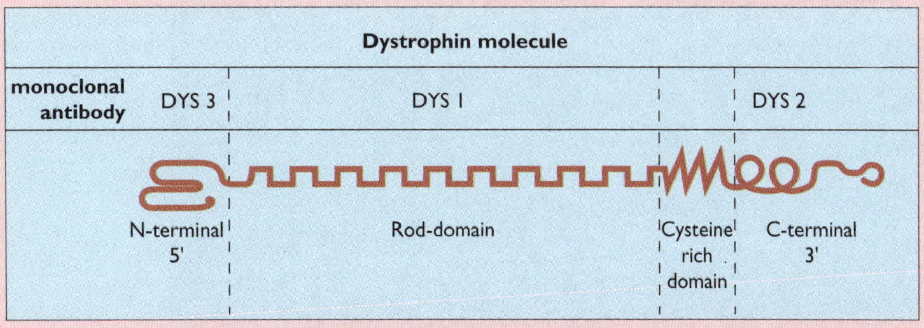

5.1c Dystrophin molecule (modified from Koenig *et al.*, 1988).

Confirmation of the diagnosis of DMD depends in the first instance on immunocytochemical and immunoblotting analyses of the muscle biopsy. Diagnosis of the **carrier** state in female relatives has depended in the past on the finding of elevated serum CK activity, but is now dependent on the demonstration of Xp21.2 deletions in the mother and her extended family. This also permits accurate antenatal diagnosis of the disease in pregnant carriers.

BMD

The natural history of late-onset X-linked myopathy varies much more than DMD, so much so, in fact, that affected subjects may remain ambulant for many years while their younger brothers become wheelchair-bound, or even die at a relatively early age.

The diagnosis of a male affected by BMD can now be made, even in the absence of any family history of X-linked recessive myopathy, by the application of gene deletional studies to identify a deletion within the Xp21.2 gene, and by the application of dystrophin studies to the muscle tissue.

The fundamental clinical features of BMD are essentially the same as in DMD, although calf hypertrophy tends to be massive, and at times asymmetrical, particularly in older patients.

Other X-linked recessive myopathies

A number of families have been described in which unusual X-linked, recessive, neuromuscular disorders occur, the musculoskeletal stigmata of which are quite different from either DMD or BMD. These are accompanied by variable A–V conduction defects, and a tendency to sudden death due to complete heart block.

These conditions have been 'lumped' into a single category, usually termed the Emery-Dreifuss syndrome. However, it is by no means certain that these disorders represent a single nosological entity.

The identification of the gene for Emery-Dreifuss myopathy (on the long arm of the X-chromosome at site q28) will assist in the clarification of these disorders, even though, to date, the gene product has not been identified.

Pathological features of Xp21.2 myopathy

The cardinal histopathological features are those of myonecrosis with decreasingly effective regeneration, and the replacement of muscle with fat and fibrous connective tissue.

These features vary considerably with the natural history of the disease, and, at any given point in time, it is essential to interpret them with reference to the patient's age and clinical state. In fact, the disease is well established in intrauterine life (well before the 28th week of gestation), and it can be diagnosed with ease during the so-called preclinical phase (up to the fourth year after birth).

Biopsy from the earliest preclinical cases gives a misleading impression of relative inactivity (**5.2**). The only changes seen at this stage are those of chronic myopathic degeneration, namely, variation in muscle fibre diameter, significant interstitial fibrosis, and the presence of hyaline fibres.

In spite of their relative nonspecificity (they can be seen in any chronic myopathy), hyaline fibres are present in such large numbers in a low power field in DMD that they are of great diagnostic significance. They are due to segmental hypercontraction within the muscle fibres, associated with excessive local accumulation of Ca^{2+} ions (**5.3, 4.28–4.31**).

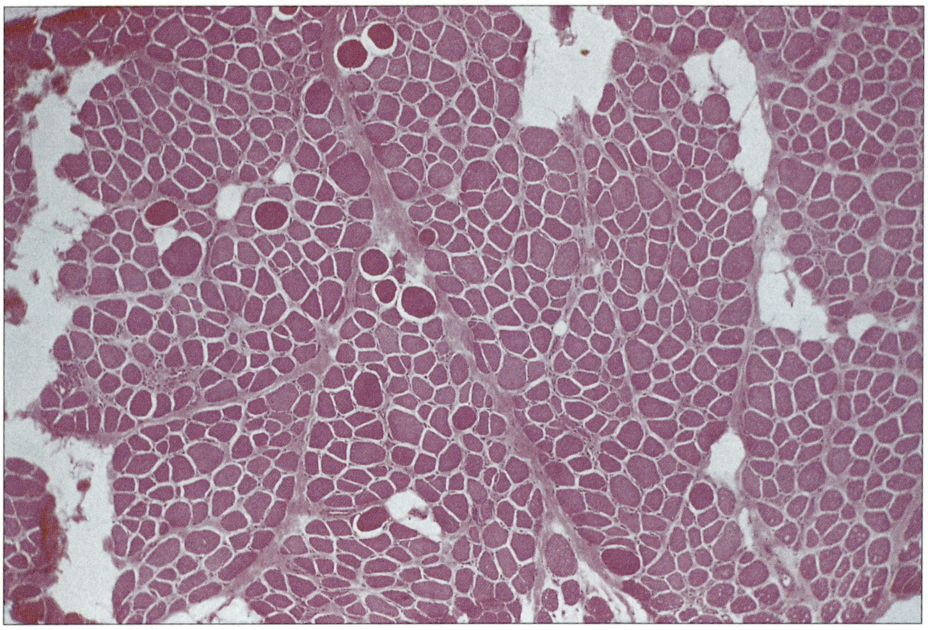

5.2 The early stage of DMD from a child aged 10 months. There is increased variation in muscle fibre size, with conspicuous, large, rounded and dense-staining hyaline fibres. Perimysial and endomysial connective tissue is increased, and adipose tissue is prominent. (H&E, ×56)

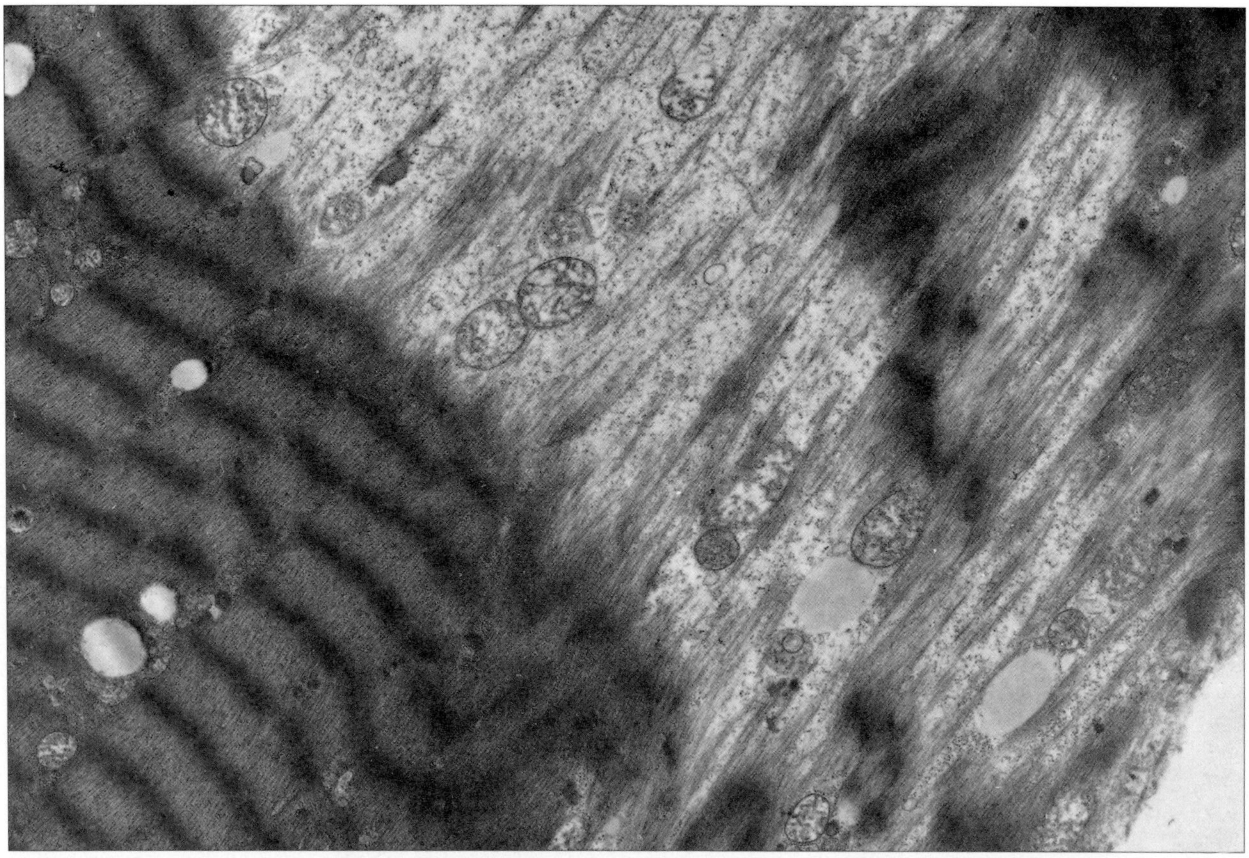

5.3a Contraction of the fibrils to form dense hyaline masses results in filaments being pulled out from adjacent regions of the fibre, causing damage to the sarcomeres. (×14440)

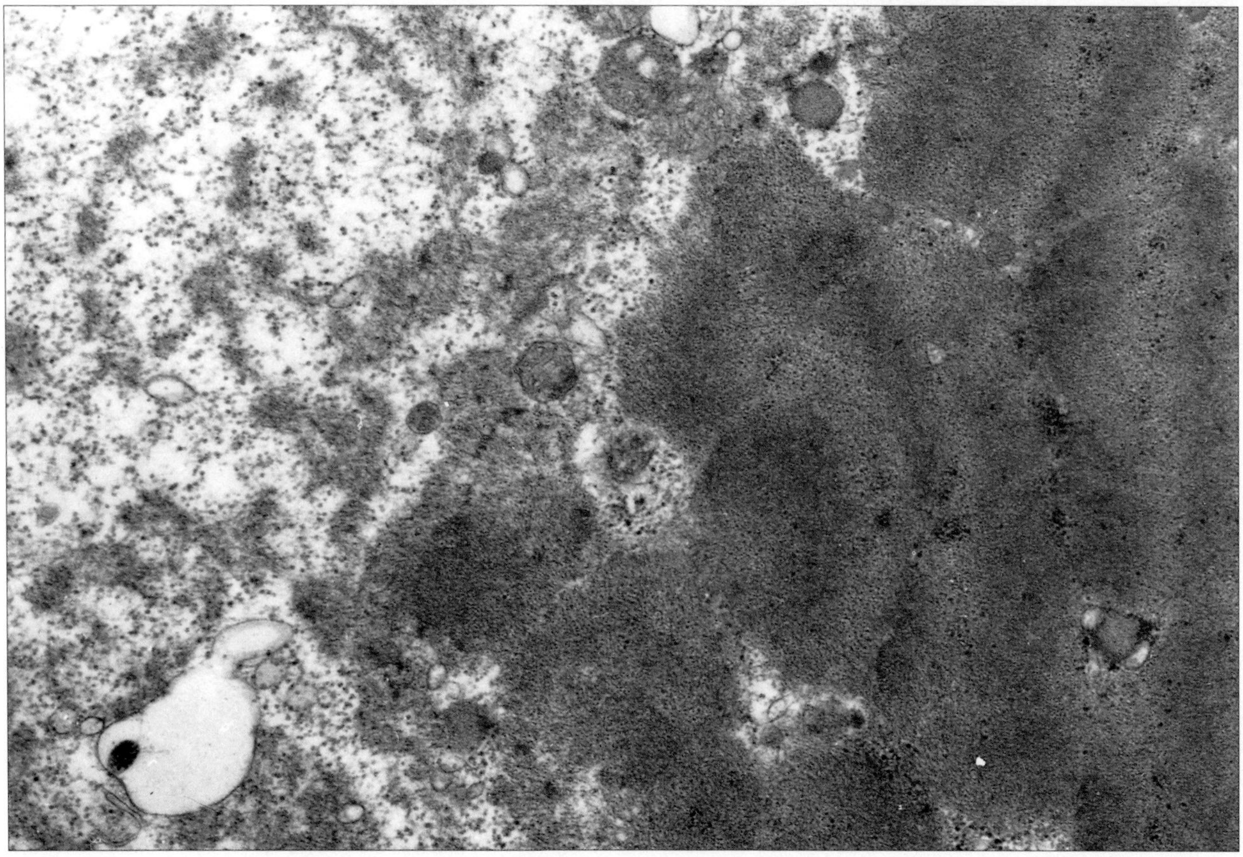

5.3b In transverse section, the hypercontracted and disorientated myofibrils are electron-dense, and the adjacent sarcoplasm is almost devoid of filaments. (×27300)

This phenomenon is believed to be due to the influx of Ca^{2+} through defects in the plasmalemma (**4.36**), leading to the development of underlying delta lesions that are the forerunners of segmental hypercontraction (**4.26**, **4.36**). These defects are likely to be related to the abnormalities of dystrophin (**5.4**).

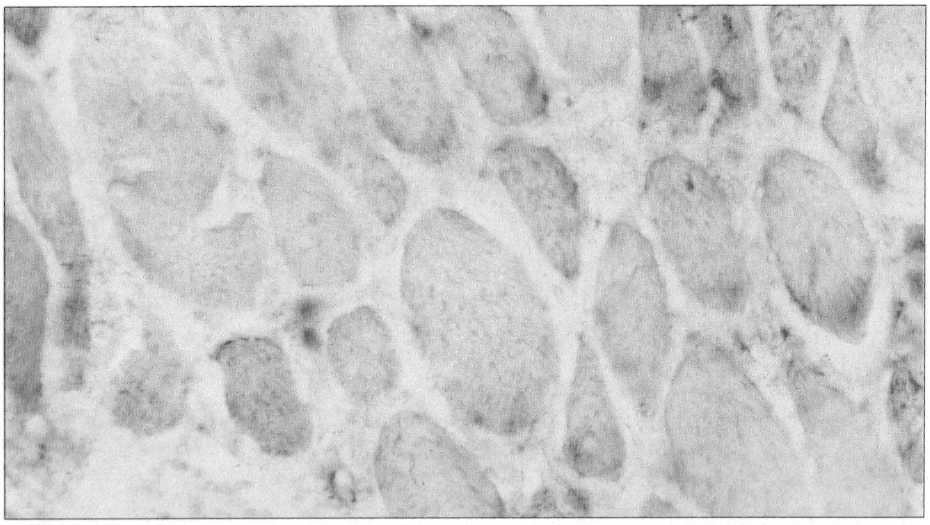

5.4 Dystrophin immunolabelling in:
a) DMD, where there is variability of labelling depending on the antibody employed. (Antidystrophin, peroxidase, ×438);

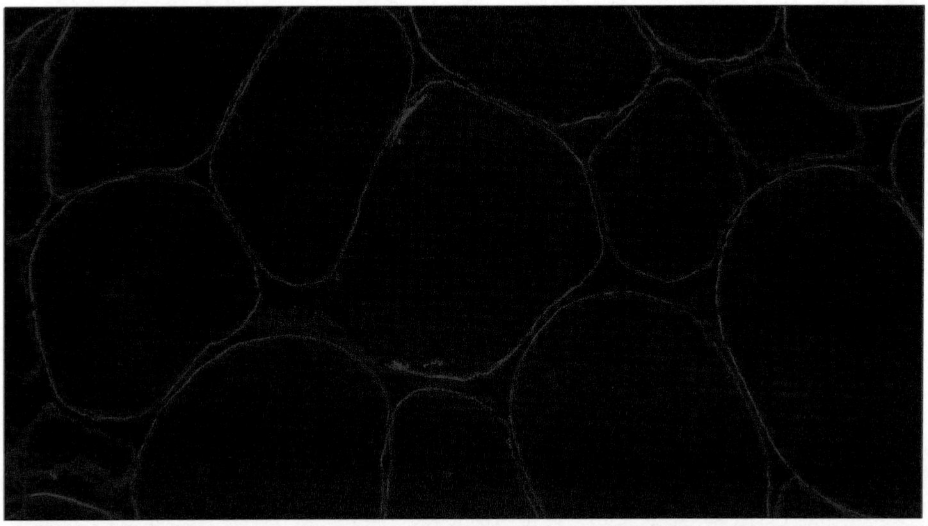

b) BMD, where labelling is reduced and shows inter-fibre and intra-fibre variation, being discontinuous in places. (Antidystrophin, Texas red fluoresence, ×210)

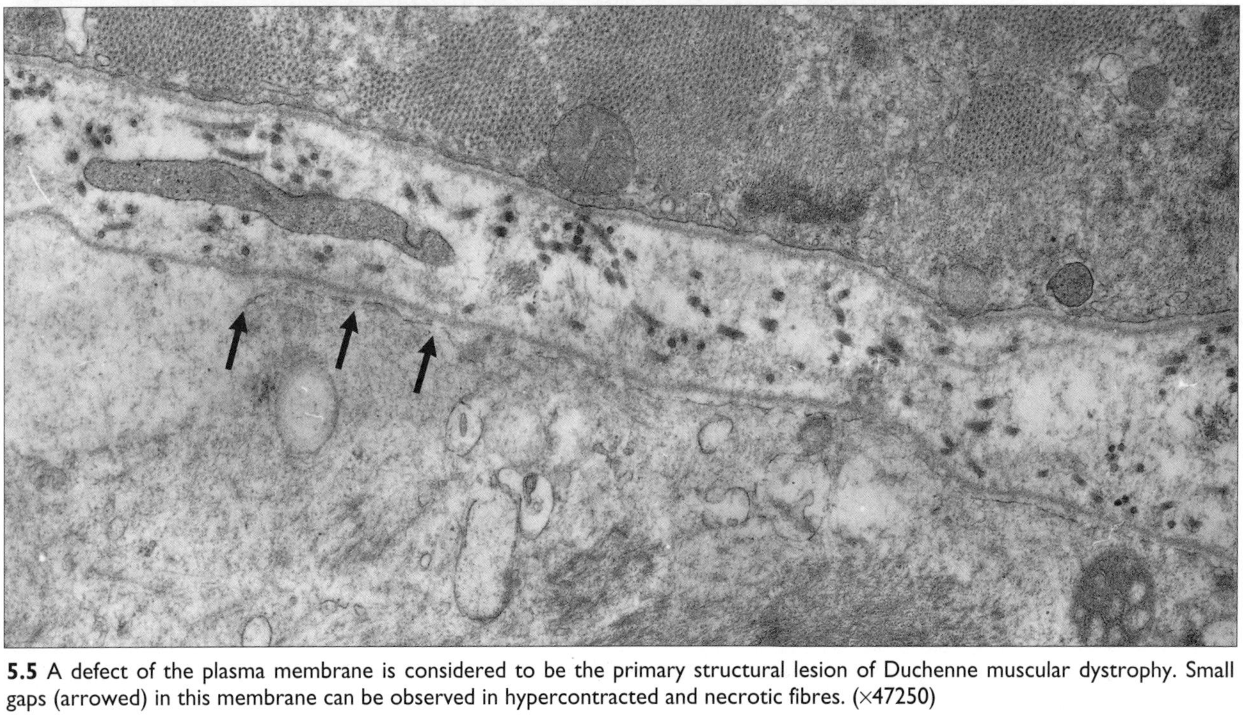

5.5 A defect of the plasma membrane is considered to be the primary structural lesion of Duchenne muscular dystrophy. Small gaps (arrowed) in this membrane can be observed in hypercontracted and necrotic fibres. (×47250)

All of the subsequent morphological abnormalities seen in muscle tissue in this disease should be regarded as direct consequences of this fundamental biological error.

The hypercontraction process described above, leads inevitably to local cell death (segmental necrosis) (**5.5–5.9**, **4.25**, **4.32**).

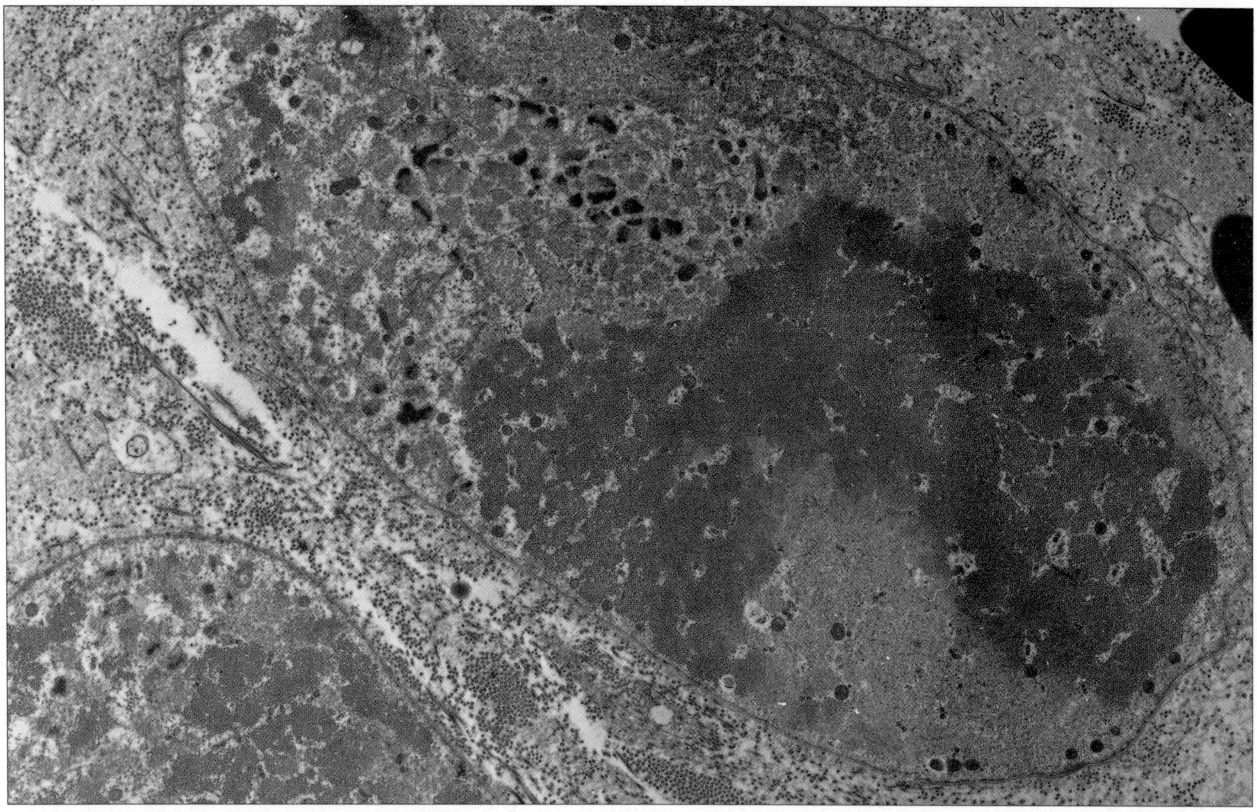

5.6 An excess of interstitial collagen surrounding each muscle fibre is a characteristic pathological feature of Duchenne muscular dystrophy. (×8820)

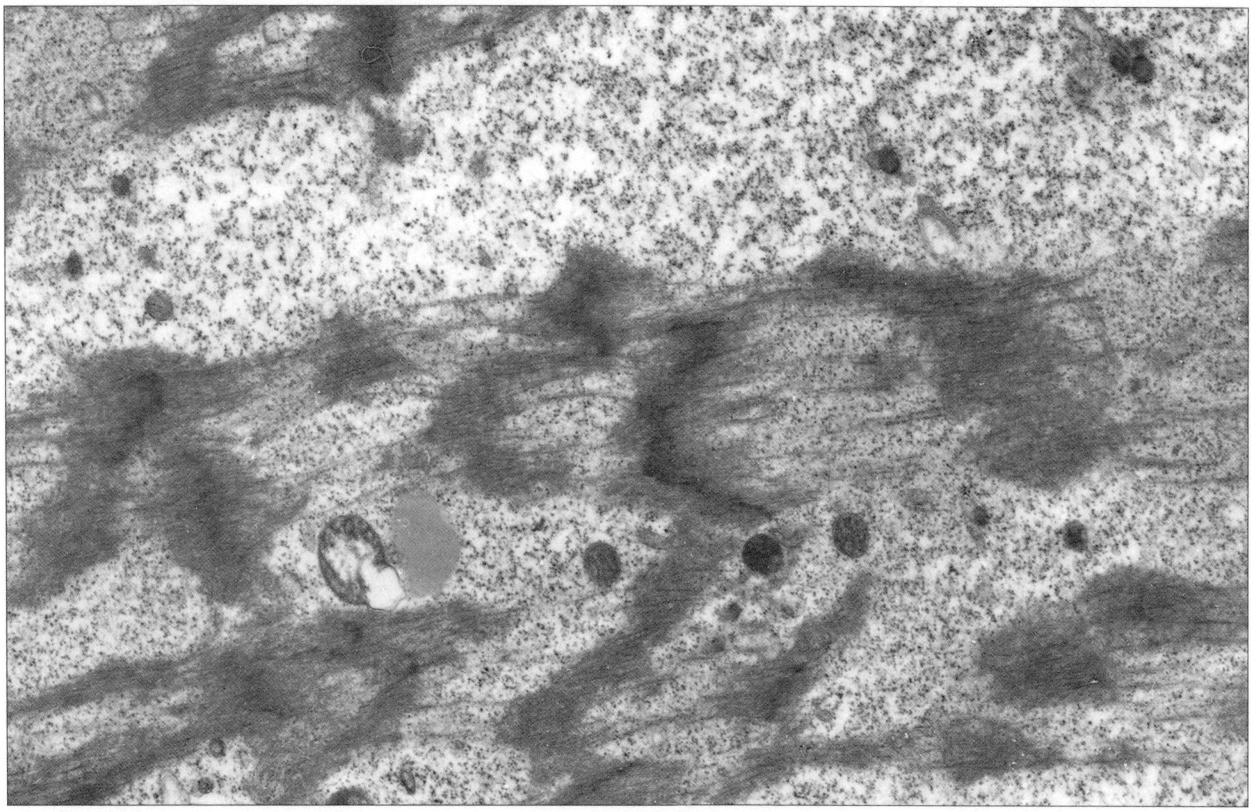

5.7 The filaments of disrupted sarcomeres appear to break down to form a granular material within the sarcoplasm. A few mitochondria are the only recognisable organelles remaining at this stage. (×18375)

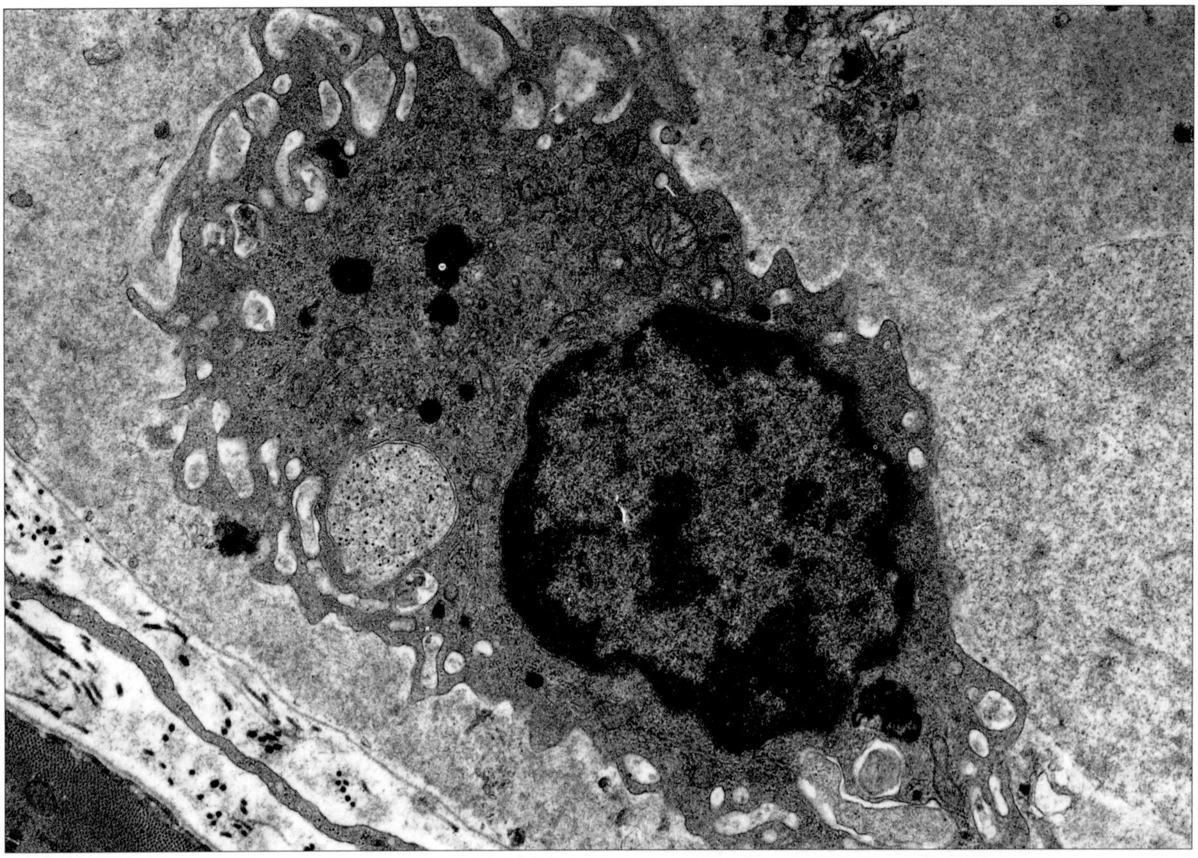

5.8 All of the myofilaments are eventually replaced by granular debris, which fills the basal lamina tube. Phagocytes can be observed within this tube, containing ingested debris and organelles. (×18000)

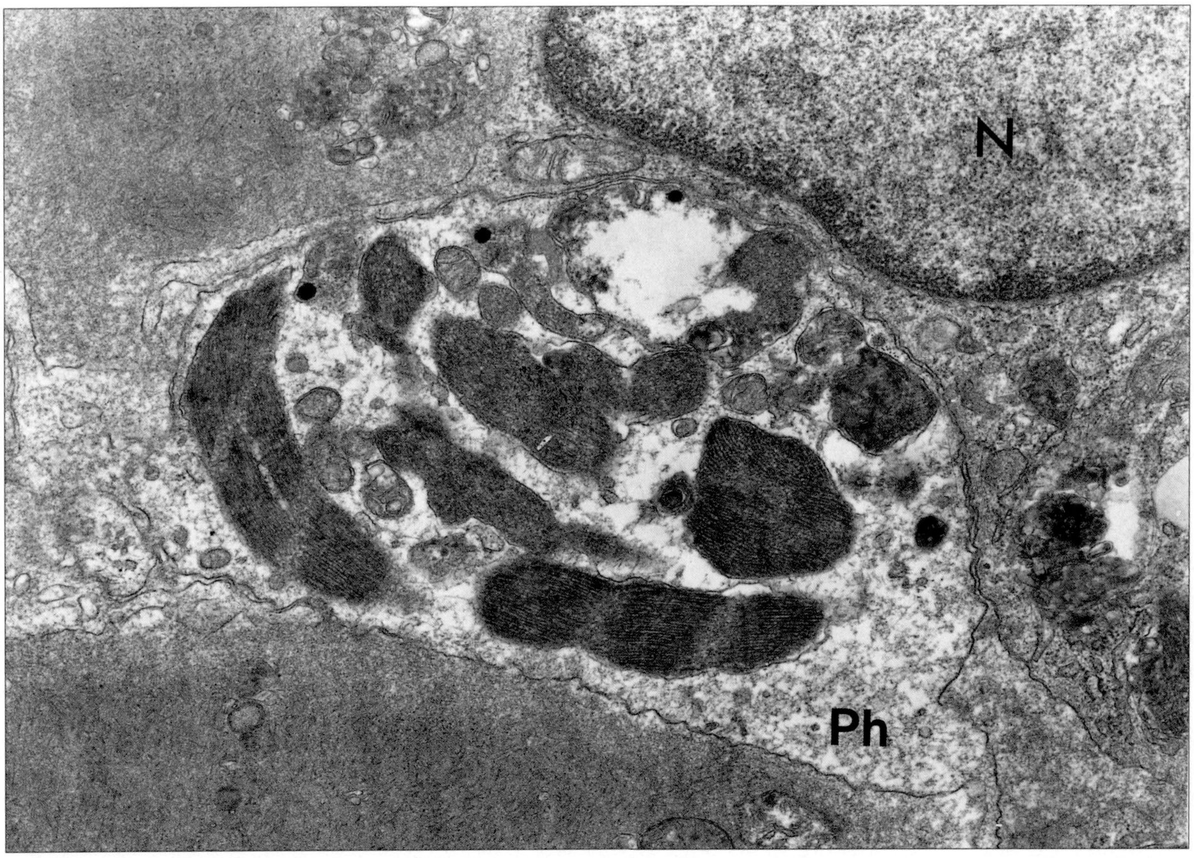

5.9 Phagocytic cells (Ph) may be observed to ingest not only the granular debris, but also segments of hypercontracted myofibrils and mitochondria of necrotic fibres, adjacent to still viable muscle nuclei (N). This may also been seen after loss of the Z-line and I-bands. (×25000)

The necrotic material is removed by invading phagocytes (**4.22–4.24**, **4.32**, **7.20–7.28**), and, simultaneously, regeneration of the necrotic segment is initiated by proliferation and subsequent fusion of 'activated' satellite cells (**5.10–5.14**, **3.31**, **4.37– 4.41**, **4.64**).

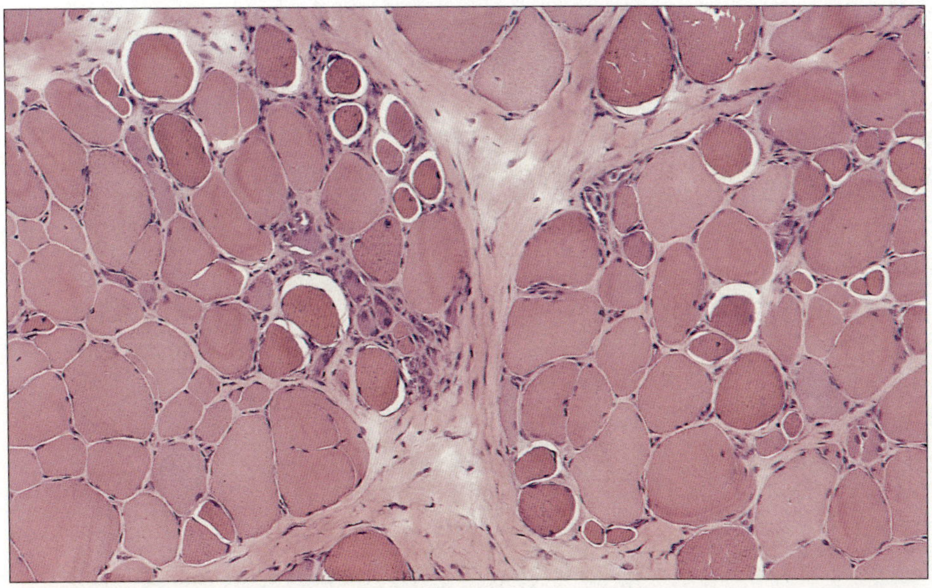

5.10 Regeneration in DMD, occurring in the presence of hyaline fibres, split fibres and an excess of connective tissue. (H&E, ×88)

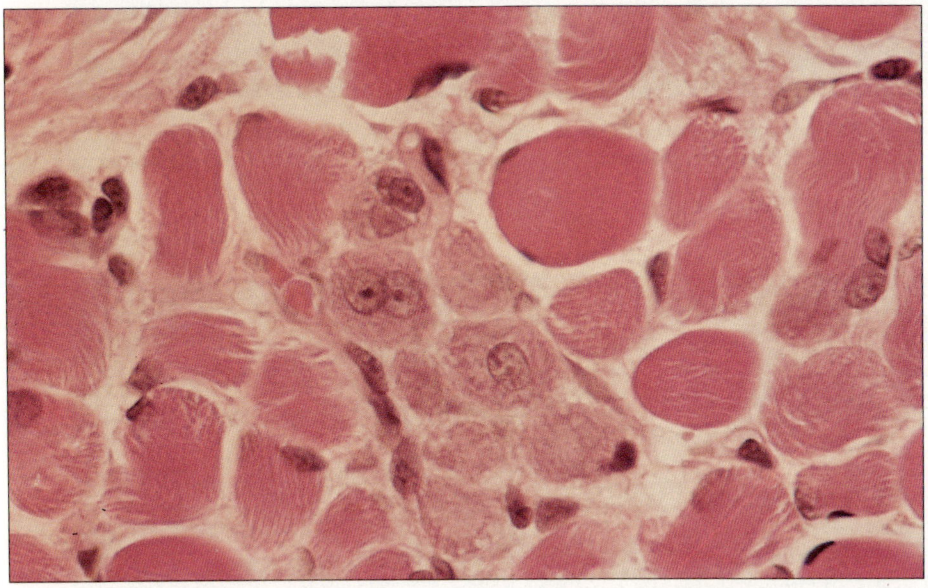

5.11 Regenerating fibres showing loose cytoplasm and large pale nuclei with prominent nucleoli. (H&E, ×350)

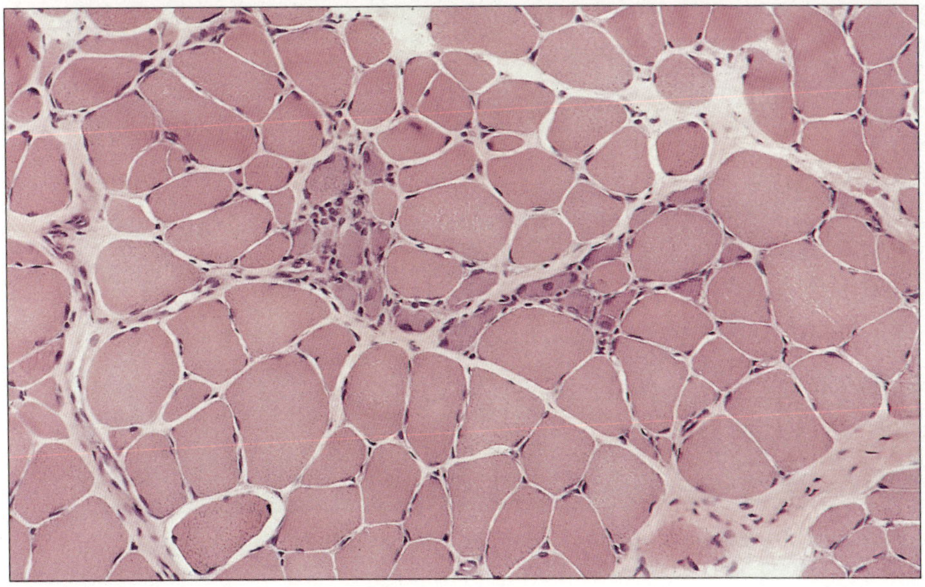

5.12 Becker muscular dystrophy showing a focus of regeneration in otherwise preserved muscle from a male, aged 16. (H&E, ×175)

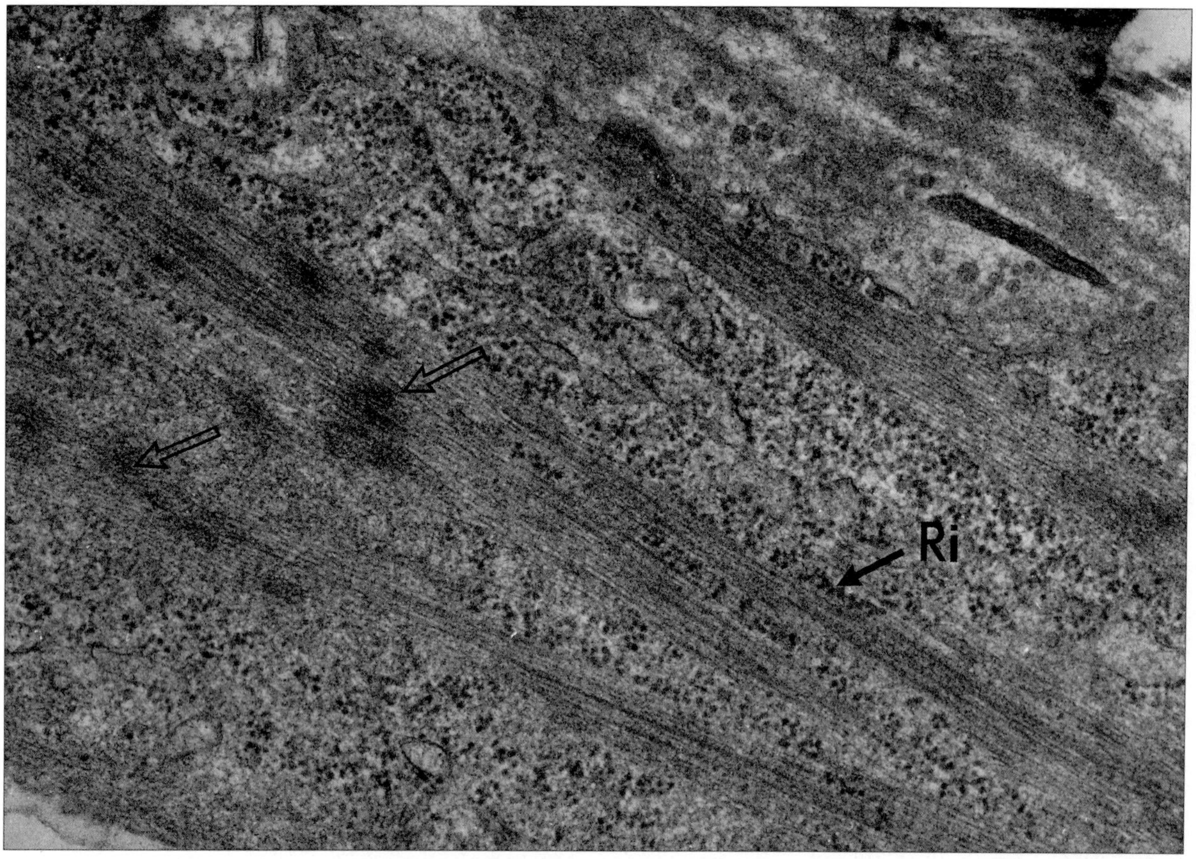

5.13 Fibrils are seen to be formed in the ribosome-rich sarcoplasm of regenerating fibres. Electron-dense regions in the filament bundles are sites of Z-line formation. Rows of ribosomes (Ri) are seen adjacent to some of the myofilaments (arrowed). ×60000)

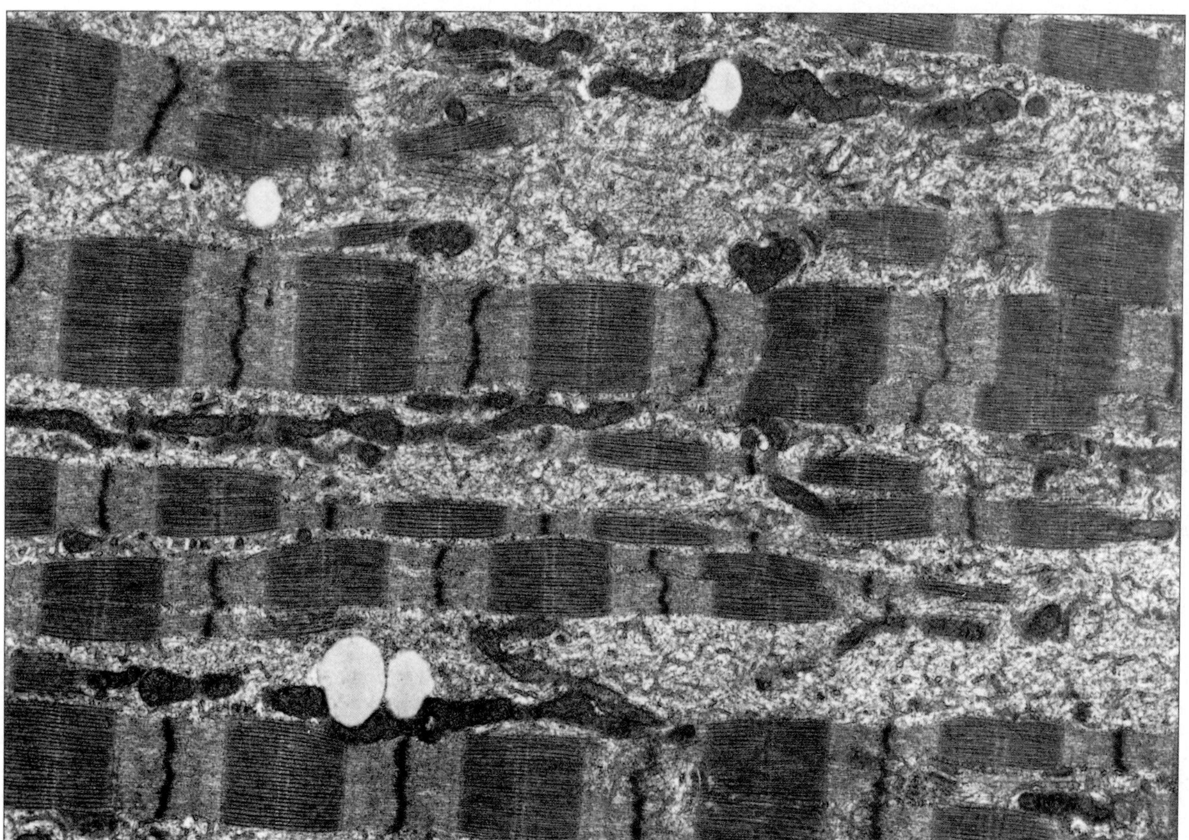

5.14 In more advanced regeneration, strands of sarcoplasmic reticulum, filaments, mitochondria and lipid droplets occupy the space between well-formed myofibrils. (×14000)

Muscle fibre splitting occurs as the result of two or more satellite cells developing within the same basement membrane. Histologically, this appears to be entirely normal in the earlier stages of the natural history of the disease, but becomes manifestly abnormal with the passage of time: the foci of what is clearly ineffective regeneration often being invaded by phagocytic cells (**4.25**).

These necrobiotic phenomena become more obvious with the passage of time, reaching a peak about the fifth year (**5.15**). Thereafter, they become gradually less prominent and, as the child reaches the stage where he will become wheelchair bound, the only evident histopathological features are, once again, those of chronic myopathic degeneration.

This progression is illustrated in **5.15–5.17**. In the later stages, the normal sharp discrimination between the major fibre types, as shown by myosin ATPase staining, is lost almost completely (**5.18**).

Although these pathological features leave little doubt as to the diagnosis of DMD, absolute confirmation rests on dystrophin analysis. Immunocytochemical studies in DMD, using antibodies to C-terminal domain, rod domain and N-terminus, reveal complete absence of immunolabelling for dystrophin in the majority of cases (**5.4**).

Isolated, clearly stained fibres (less than 1% of the total fibres present) where the labelling may be continuous around the fibre are now termed 'revertant' fibres (*Table 5.1*). It is important to look for immunolabelling in non-necrotic fibres since these fibres inevitably lose their surface staining, irrespective of the cause of the necrosis.

Ultimately, only fat and fibrous connective tissue with scattered surviving and abnormal muscle fibres will be seen in the larger proximal muscles, illustrated so graphically by Erb in his description of the autopsy pathology of the disease in 1884.

The basic features of the myopathology of BMD are essentially the same as those in DMD (**5.12**), although they tend to be less severe at the same age, and they are very much less predictable in the individual case at any given point in time. In contrast to DMD, the foci of necrosis in BMD are often seen to occur in areas of otherwise normal appearing tissue.

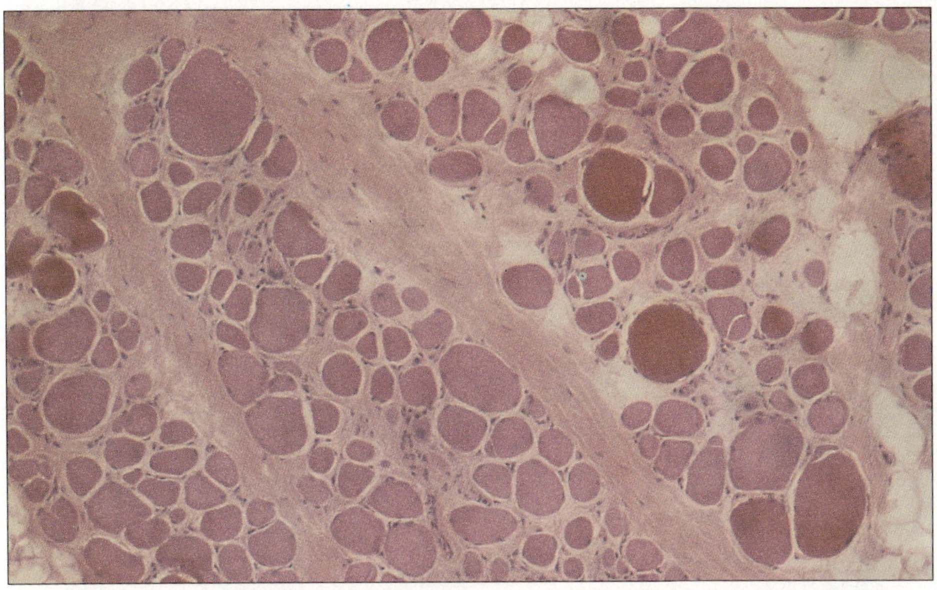

5.15 DMD in a 5 year old, showing abnormal variation of muscle fibre size, rounded fibres, fibre splitting, excess connective tissue and some regeneration. (H&E, ×300)

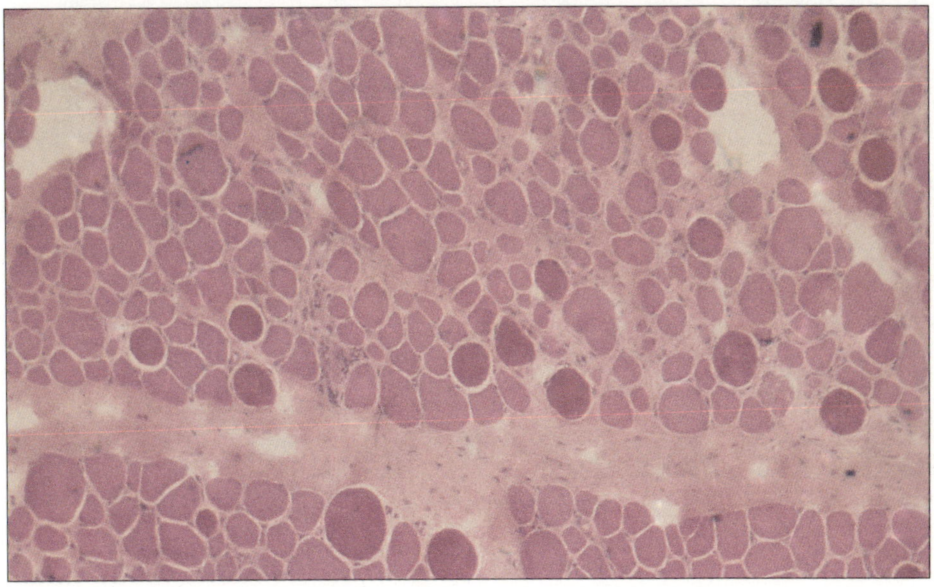

5.16 DMD in a 3 year old boy. The marked increase in the variation of fibre size, the rounding of muscle fibres and excess endomysial connective tissue is typical of this stage of the disease. Hyaline fibres and necrotic fibres are also present. (H&E, ×210)

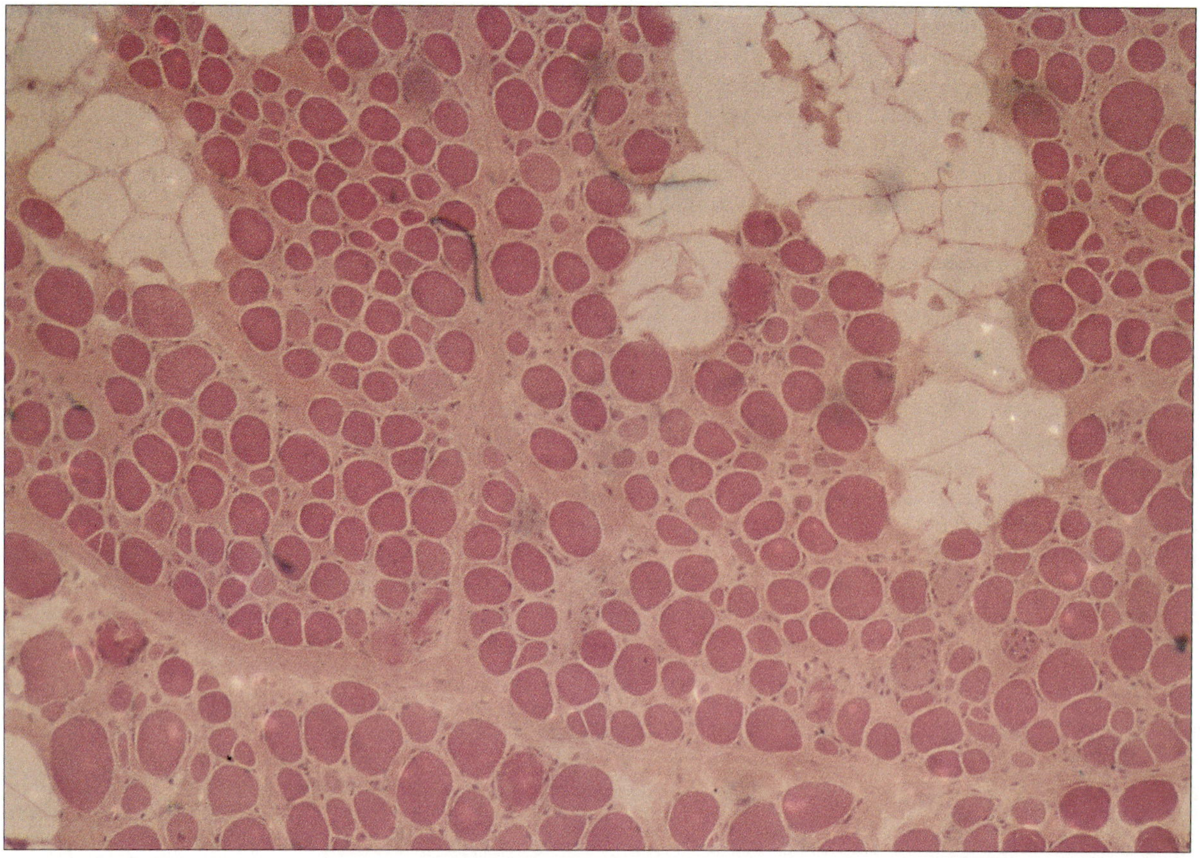

5.17 The late stage of DMD, showing increased fibrosis in the presence of adipose tissue, marked variation of fibre size, with abundant atrophic and necrotic fibres. (H&E, ×284)

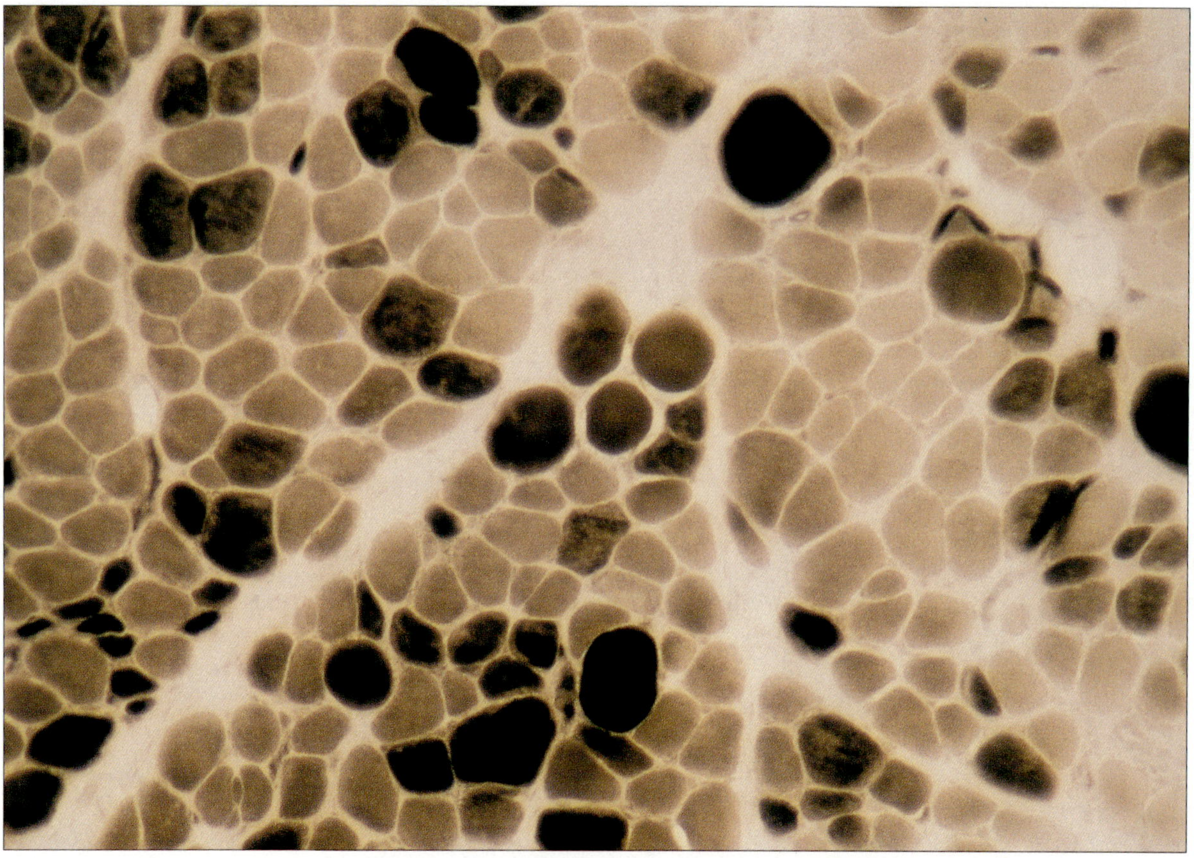

5.18 DMD, showing the variability of staining and lack of clear fibre type differentiation. (Myofibrillar ATPase, 119)

Some fibres in BMD show increased staining with NADH-Tr (the reasons for this are unclear), and fibre type differentiation is usually normal. Histochemical changes suggestive of chronic denervation have also been reported in BMD (Chapter 6), and this has been demonstrated in electrophysiological studies. However, if the clinical situation is suggestive of a Xp21.2 deletional myopathy, then dystrophin analysis is mandatory, irrespective of the pathological appearance. Dystrophin studies in BMD may require immunoblotting as well as immunocytochemistry.

A wide spectrum of changes occur in immunolabelling. Occasional cases show loss of labelling comparable to DMD. The most common change is where there is marked inter-fibre variation of labelling (**5.4**). Intrafibre variation from continuous to discontinuous labelling may occur, as may a virtually normal pattern, indistinguishable from controls. In this situation, immunoblotting is required to reveal quantitative differences in the size and abundance of dystrophin (*Table 5.1*).

In the **carrier** female, histological analysis of the biopsy is unrewarding unless the female is 'manifesting'. However, even in that situation, the changes may well be confined to a mild variation in fibre size with an excess of internal nuclei. Immunolabelling for dystrophin will reveal a mosaic pattern of dystrophin positive and dystrophin negative fibres, although the number of dystrophin negative fibres may be small.

Many of the histopathological changes observable with the light microscope may be examined in greater detail with the electron microscope. Hyaline, necrotic and regenerating fibres (**5.6–5.8**, **5.13–5.14**, **5.19**, **4.27**, **4.29–4.30**), and the increase in endomysial connective and adipose tissue (**5.5**) are all easily identified.

Changes that are only detectable with the electron microscope include early changes of Z-line streaming (**4.35**), Z-Line loss (**4.33**), and defects of the sarcolemma (**5.5**, **4.36**) and capillaries (**5.20**).

Immunogold labelling (**5.21**) of cytoskeletal and structural proteins serves only to confirm that seen with the light microscope using other markers (**5.4**).

Table 5.1 Differential diagnosis of DMD/BMD.

	DMD	BMD	Manifesting carrier of DMD/BMD	Normal muscle
Immunostaining pattern	Negative	Faint and/or patchy	Mosaic	Normal (continuous)
Amount of dystrophin	< 3%	Normal or decreased	Normal or decreased	Normal
Molecular weight of dystrophin	Undetectable	Larger or smaller	Normal	Normal (427 kD)
Gene abnormality	Yes	Yes	Heterozygote	No

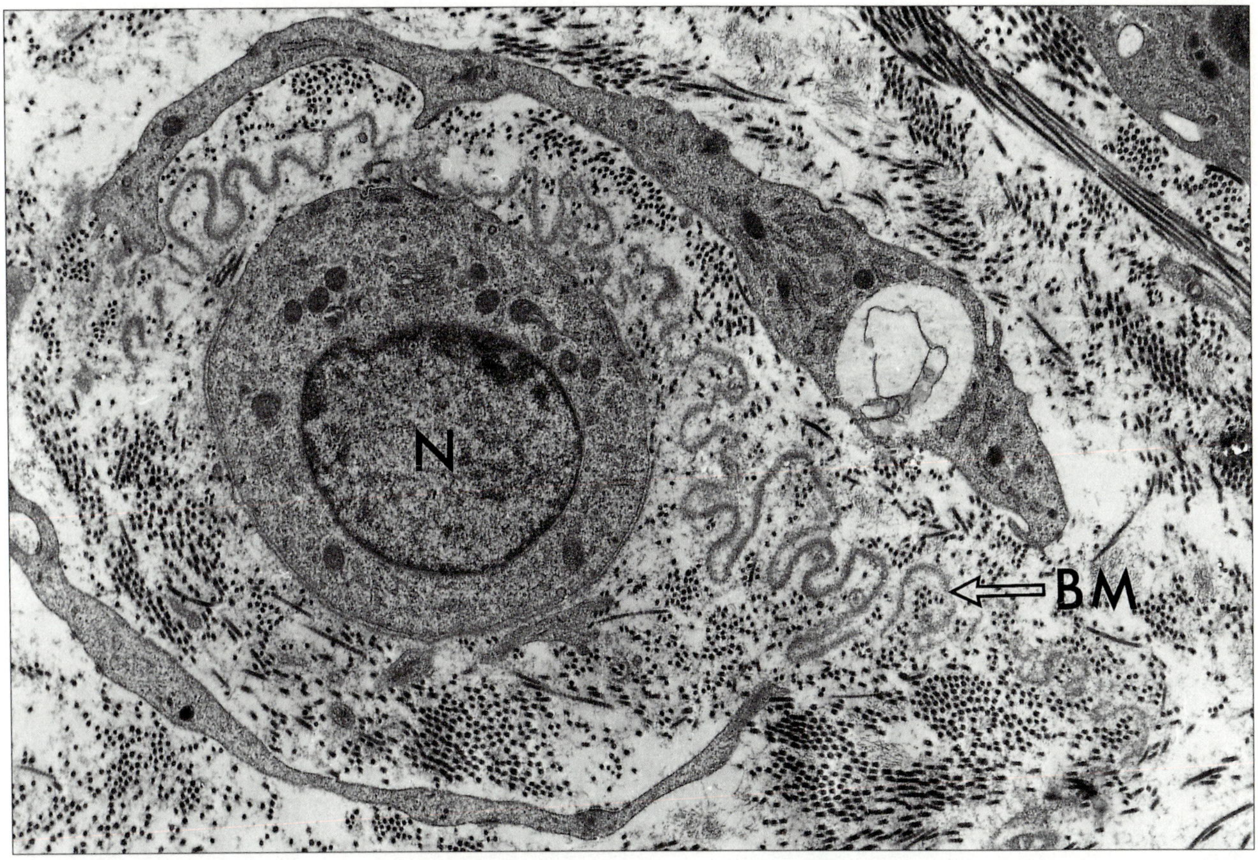

5.19 A small regenerating fibre, in transverse section, characterised by the basal lamina, central nucleus (N) and Golgi apparatus. The surrounding interstitium is seen to contain many collagen fibrils and a segment of redundant basement membrane (BM). (×10500)

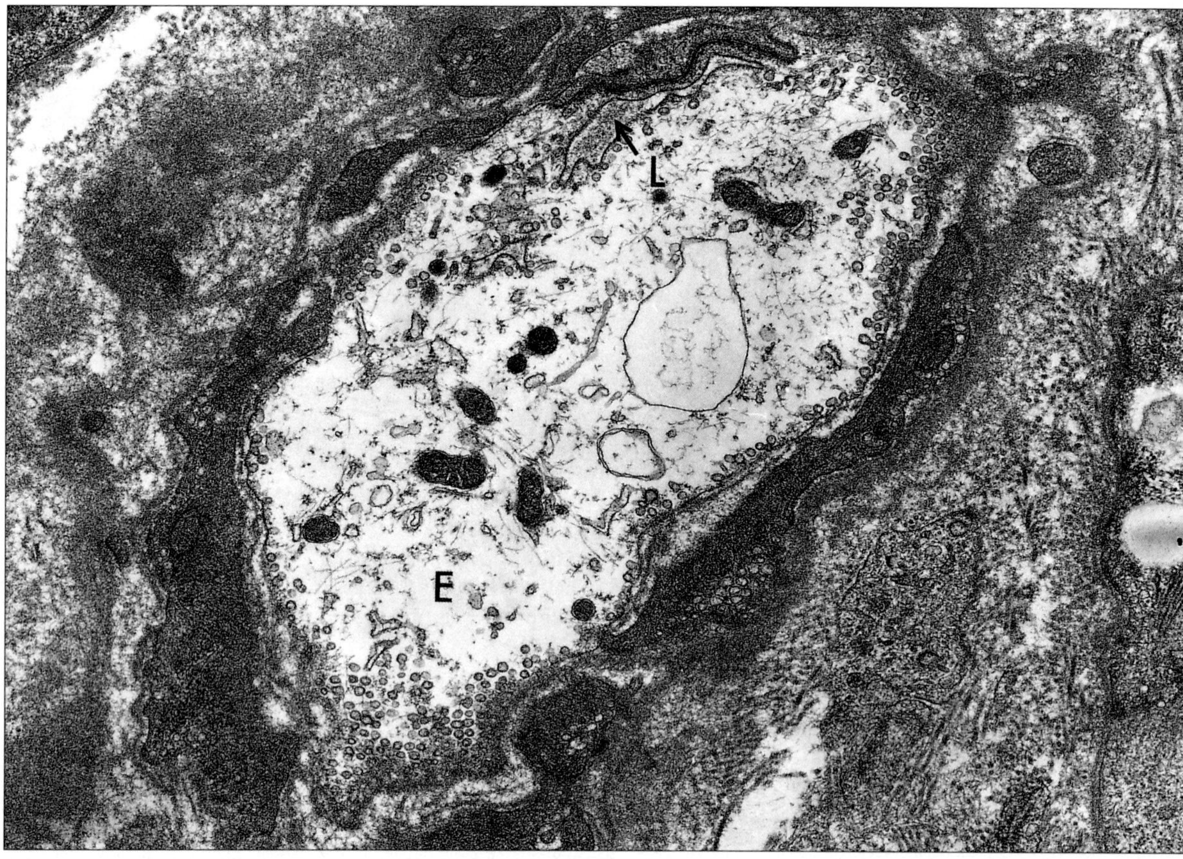

5.20 The endothelial cell cytoplasm (E) of degenerating capillaries is greatly distended, and contains small pinocytotic vesicles and scattered electron-dense mitochondria. The capillary lumen (L) is difficult to identify. (×22500)

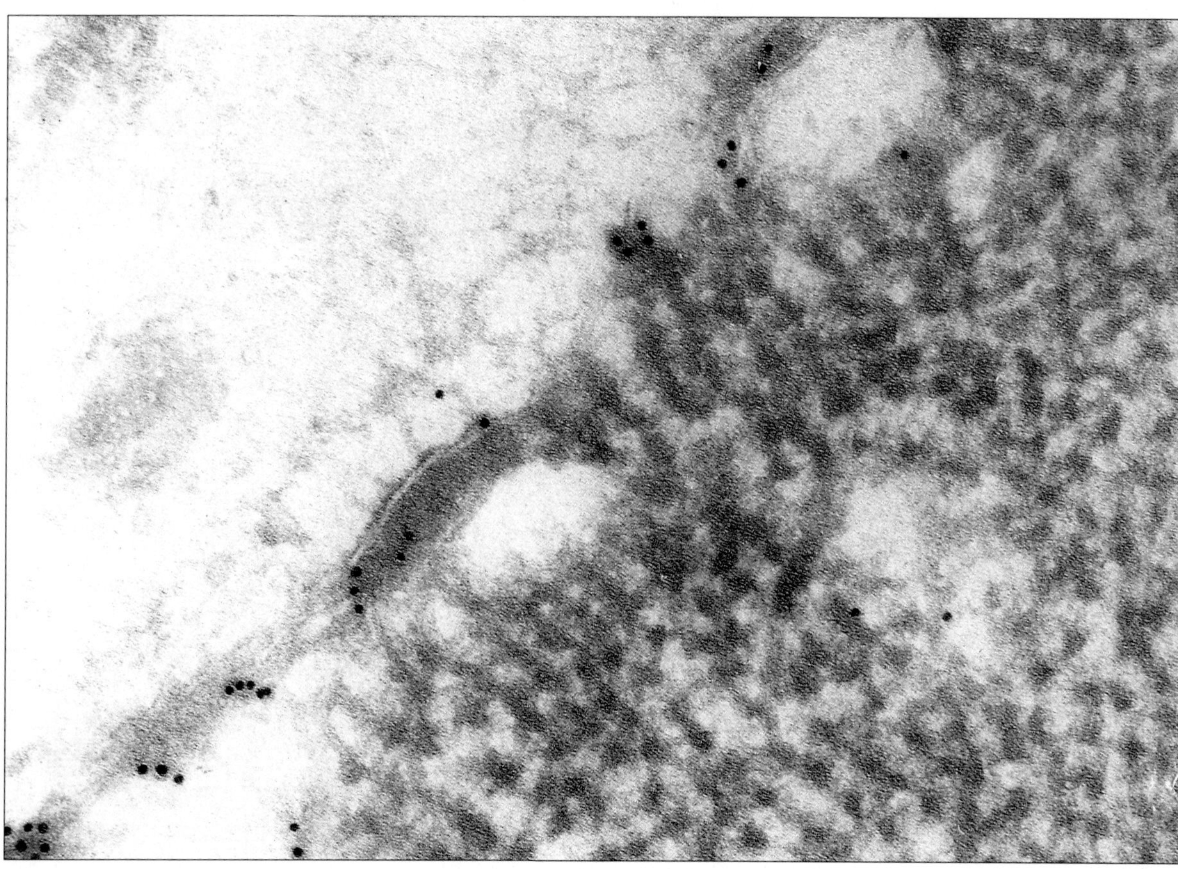

5.21 Antidystrophin labelling is revealed by localisation of gold particles over the sarcolemmal membrane in normal muscle. (×75000)

6

Spinal muscular atrophies

Whereas perturbation of any element of the motor unit will lead to neurogenic disturbance of the muscle fibre, this chapter will deal only with disorders of the anterior horn cell. As muscle biopsy is rarely indicated in a peripheral neuropathy, we have not considered this topic, and readers are referred to the standard authority (Dyck *et al.,* 1993). End-plate disorders are considered briefly in Chapter 10.

The genetically determined anterior horn cell disorders collectively constitute the second commonest neuromuscular problem seen in clinical practice.

Recent advances in molecular genetics have confirmed genetic homogeneity in autosomal recessive SMA with phenotypic heterogeneity (Brzustowicz *et al.,* 1990; Munsat *et al.,* 1990): the gene for AR-SMA maps to the long arm of chromosome 5 at site q11-13.

The genetic locus for the autosomal *dominant* forms of SMA (which have common pathological features with AR-SMA), have yet to be elucidated.

X-linked *recessive* SMA (Kennedy syndrome) has been localised to Xq11.12 (the androgen receptor). The pathological features are those of any other form of SMA, with the distinction being made on clinical grounds (i.e. buccal wasting and fasciculation, gynaecomastia, testicular atrophy and peripheral neuropathy).

The modes of presentation of SMA are as follows:

- Acute infantile SMA (AR-Type I SMA, Werdnig-Hoffman disease).
- Chronic infantile SMA (AR-Type II SMA).
- Late-onset forms of SMA (AR, AD, X-linked Type III SMA or Kugelberg-Welander disease, adult-onset SMA).
- Adult motor neurone disease (MND).

We should make it clear at this stage that we do not consider adult MND to be aetio-pathogenetically related to the other three types of SMA listed above, although we acknowledge that genetic influences play a part in the development of some forms.

Acute infantile SMA

Acute infantile SMA (Type I SMA) is the commonest form of SMA seen clinically, and is one of the most severe forms of infantile hypotonia. In virtually all of the cases recorded in the literature, inheritance appears to be determined by an autosomal recessive factor.

Tragically, its course is inexorably downhill, with affected children succumbing within 2–3 years at the most, and usually after much shorter periods. The mode of death in these cases is without exception respiratory, and the domestic and psychosocial morbidity engendered by the experience of caring for them is considerable. To date, there are no reports of the recent finding of the gene locus being applied to at-risk foetuses, although it is to be hoped that this situation will not be long delayed.

Chronic infantile SMA

In chronic infantile SMA (Type II SMA) the affected children are usually, but not always, severely hypotonic at birth, and their body habitus is essentially the same as that in Werdnig-Hoffman disease. However, their condition appears to be relatively non-progressive, and the development of their heads and higher cerebral functions are normal to the extent that many can go on to tertiary education, and can live semi-independent lives, given the requisite mechanical aids.

Late-onset forms of SMA

The later-onset forms of SMA appear to fall into two major groups:

- Type III SMA, also known as Kugelberg-Welander disease (so-called 'pseudo-myopathic' SMA, which is usually misdiagnosed as limb girdle dystrophy).
- Late-life SMA (including the Kennedy syndrome).

It goes without saying that the individual morbidity and mortality in these cases is considerably better than in Types I and II. However, many of these patients will be incapacitated to a degree in middle life, with some being wheelchair dependent and requiring the same sort of day-to-day support as children with advanced DMD. Types I, II and III SMA may occur within the same family or even in the same sibship, although this is uncommon.

As indicated above, SMA Types I and II present with severe infantile hypotonia. Patients with Type II survive into adult life with pitifully thin, wasted limbs and trunk, and a normal sized head.

Type III SMA presents with mainly proximal weakness and wasting, which may be strikingly asymmetrical, and which is usually accompanied by either reduced or absent deep tendon reflexes (this can be a useful distinguishing feature from primary muscle disease). In many cases, proximal weakness and wasting will be accompanied by hypertrophy of other muscle groups. In others, muscle weakness and wasting may have a scapuloperoneal distribution, or limb-girdle weakness will be accompanied by involvement of the facial and/or bulbar musculature, although involvement of the extraocular muscles is extremely uncommon in our experience.

In patients with either SP or FSH distributions, inheritance tends to be determined by autosomal dominant or X-linked recessive genes. As discussed in Chapter 5, it is important to remember that isolated males may present with what appears clinically, electrophysiologically and pathologically to be SMA, but on immunolabelling for dystrophin is in fact a Becker Xp21.2 myopathy. Unless this is borne in mind, the female relatives of the affected male will not receive the necessary genetic advice, with what could be disastrous results.

The diagnosis of the SMAs depends upon electrophysiological and biopsy confirmation of the presence of a chronic denervating process in the affected muscles (*see below*). This is, of course, crucial for prognostic and genetic counselling purposes, as well as simply confirming the diagnosis in the individual case.

Motor neurone disease

Motor neurone disease is in most cases a rapidly progressive disorder affecting the anterior horn cells in a random, patchy way, resulting in incapacitating weakness in the limbs and often the bulbar and ventilatory musculature as well.

In most cases, anterior horn cell involvement is accompanied by clear-cut evidence of corticospinal tract degeneration (so-called amyotrophic lateral sclerosis, ALS). In practice, the clinical distinction between ALS and the other modes of presentation of MND is artificial, as virtually all cases progress to a common endpoint in which anterior horn cells, bulbar motor neurones and the long descending tracts are all involved.

Diagnosis depends on recognition of the clinical presentation and the electrophysiological confirmation of a rapidly progressive denervating process, the exclusion of multifocal conduction block, and evidence of a peripheral neuropathy as in the Kennedy syndrome.

Biopsy confirmation of the diagnosis is rarely if ever necessary, although it may be helpful in excluding other disorders, particularly acid maltase deficiency (Chapter 10).

Anterior horn cell pathology

Our understanding of the pathogenesis of anterior horn cell degeneration in either the genetically determined or acquired motor unit disorders is fragmentary, to say the least.

Basic histopathologic data from autopsy studies is limited, and there do not appear to be any major differences between the two groups at either the light microscopic or ultrastructural levels. It is of interest that recent studies have shown deposition of the abnormal cytoskeletal protein, ubiquitin, in degenerate/dying anterior horn cells in adult motor neurone disease, but this does not appear to occur in the other motor unit disorders. However, this still does not shed much light on the fundamental processes underlying what appears to be a pure abiotrophy and, for this reason, much attention has been centred on the way in which the motor unit itself disintegrates.

With particular reference to the diagnosis of adult MND, we have already made it clear that we do not consider muscle biopsy to be essential in the vast majority of cases. Indeed, we feel that the use of invasive diagnostic techniques should be avoided wherever possible, to minimise discomfort and inconvenience to patients who face an extremely unpleasant and foreshortened existence. However, the essential features are shown in **10.1**.

This most certainly does not apply to genetically determined anterior horn cell degeneration, irrespective of the age at onset or the rate of progress. Precise pathological diagnosis is essential in these cases, particularly in the clinical limb girdle syndrome, in the hope of excluding cases that may have a potentially treatable cause and to give the patient's family an accurate prognosis. Disintegration of the motor unit, as a consequence of gradual denervation, produces a mosaic of atrophying muscle fibres because of the normal anatomy of the motor unit itself (**3.6–3.9**).

In very slowly progressive disorders, the morphological consequences may be difficult to detect at first glance, as

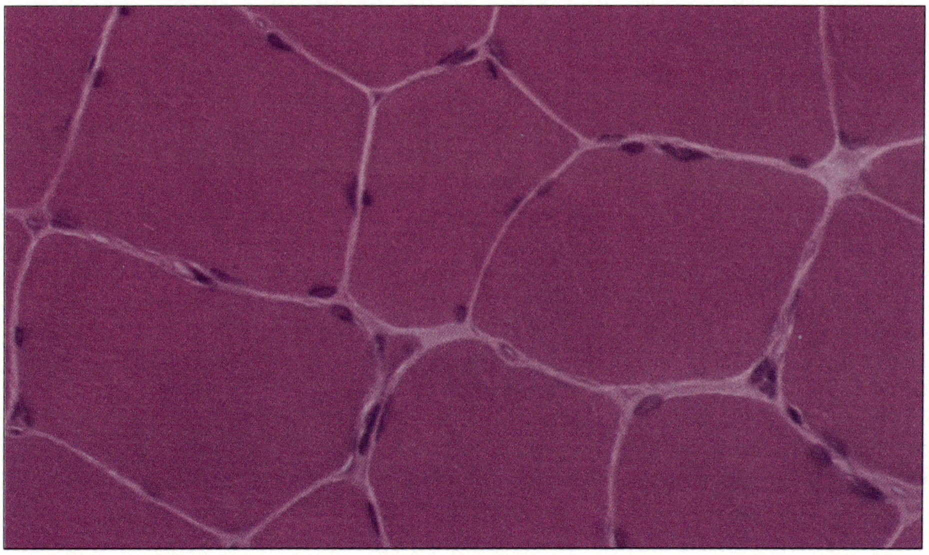

6.1 A severely atrophic muscle fibre, with angular outline, amongst fibres of normal size and architecture is suggestive of mild denervation. (H&E, ×336)

there may be only a few scattered tiny atrophic fibres or nuclear clumps (**4.1–4.5**). Indeed, it may be impossible to see any definitely atrophic fibres, although comparison of the dimensions of the fibres in adjacent fascicle may indicate that fibres in one of two overlapping motor units are undergoing atrophy (**6.1, 6.2a**). These subtle changes may be detected in samples of muscle taken from muscles of normal (MRC 5) strength on clinical examination. Any uncertainty created by finding such fibres in H&E stained sections will usually be resolved by studying material stained for myofibrillar ATPase activity across the pH range described in Chapter 2 (**6.2**).

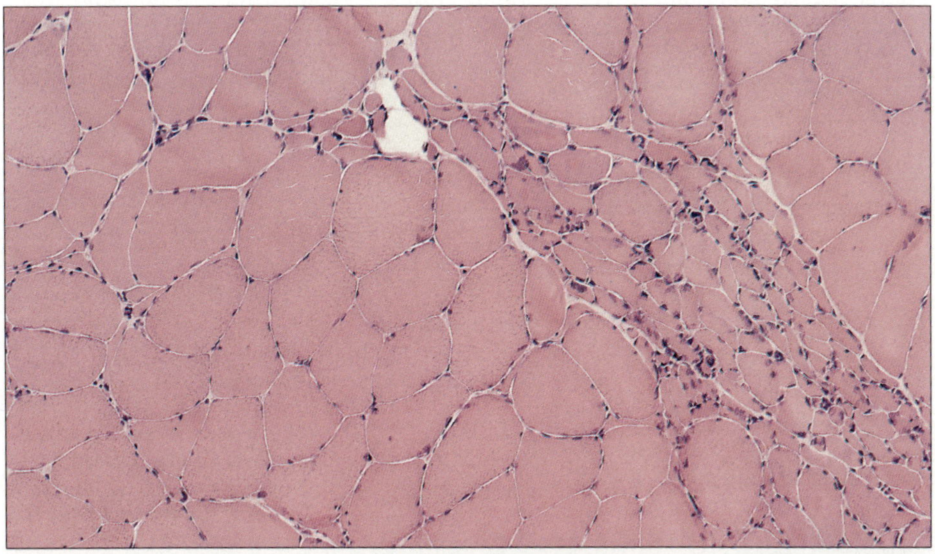

6.2a A fascicle consisting mainly of grouped atrophic fibres from a patient in the early stages of SMA.
There is increased variation in fibre size, frequent occurrence of small angular fibres and relatively good preservation of structure in other areas.
(H&E, ×88)

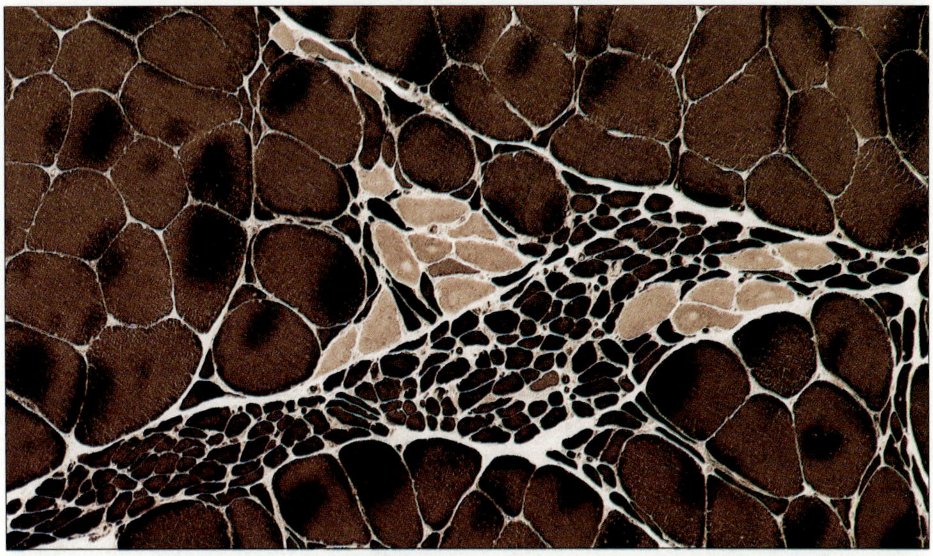

6.2b Section showing the grouping of fibre types.
(Myofibrillar ATPase, pH 10.2, ×88)

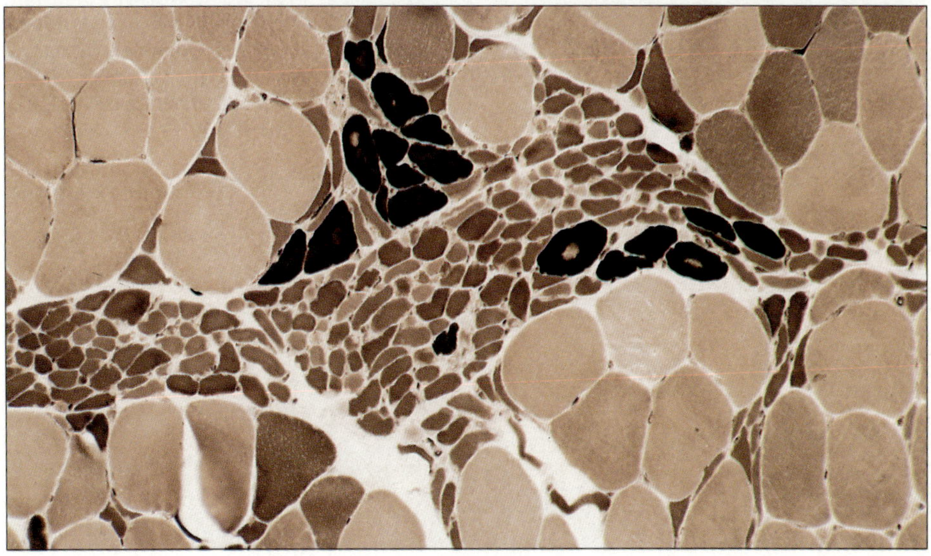

c) Serial section of **6.2b**, showing the grouping of fibre types.
(Myofibrillar ATPase, pH 4.3, ×88)

In more rapidly progressive or more advanced cases, small, medium sized, or even large groups of atrophic fibres may be seen, the size of the fibres in the individual groups being determined by the stage of degeneration reached by the innervating anterior horn cell. The appearance of these groups (**6.3–6.4**) is of considerable diagnostic help, but has given rise to incorrect assumptions about the normal anatomy of the human motor unit in the past. It was assumed that all of these fibres belonged to the same motor unit, but the formation of the 'group' in this instance is due to simultaneous denervation of adjacent/overlapping motor units at the same rate.

Further diagnostic assistance may be provided by the NADH-Tr preparation, which may demonstrate the presence of disturbed distribution of reaction product, including target and targetoid fibres (**6.5–6.6**, **4.8**), fibre type grouping (**6.7**), lobulated fibres (**4.54**) and ringbinden (**4.19–4.20**).

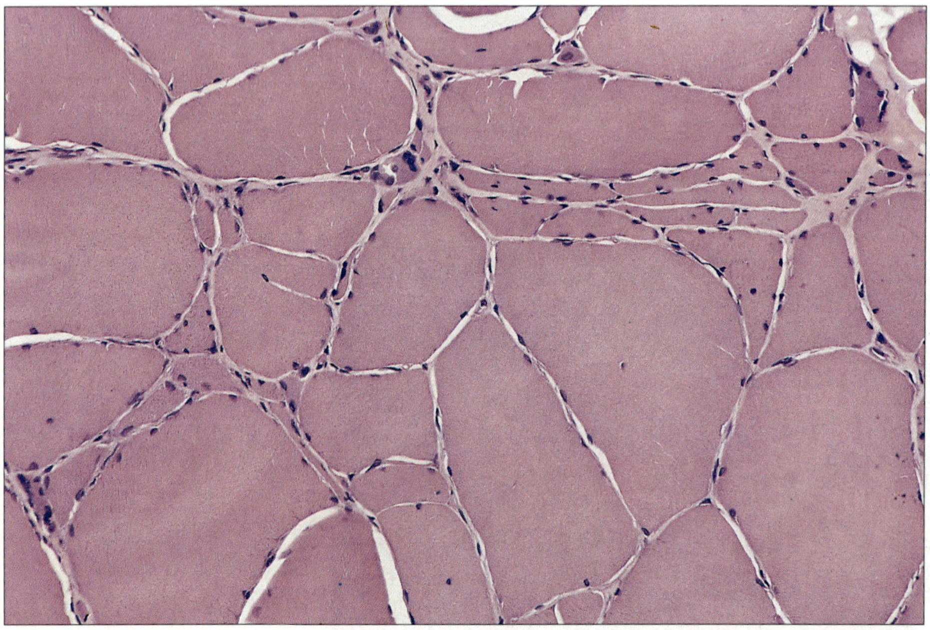

6.3 Groups of severely atrophic fibres and scattered fibres showing angular outlines are present amongst fibres of normal appearance. (H&E, ×175)

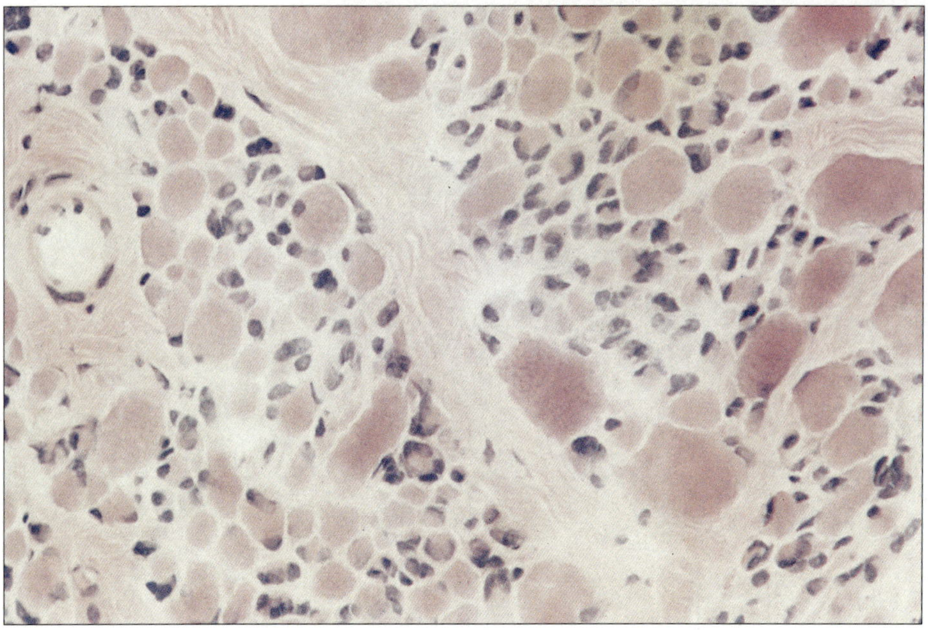

6.4 High power light micrograph of a biopsy from a patient with SMA, showing marked variation in fibre size. Most fibres are extremely atrophic, containing nuclear clumps within the endomysium. (H&E, ×336)

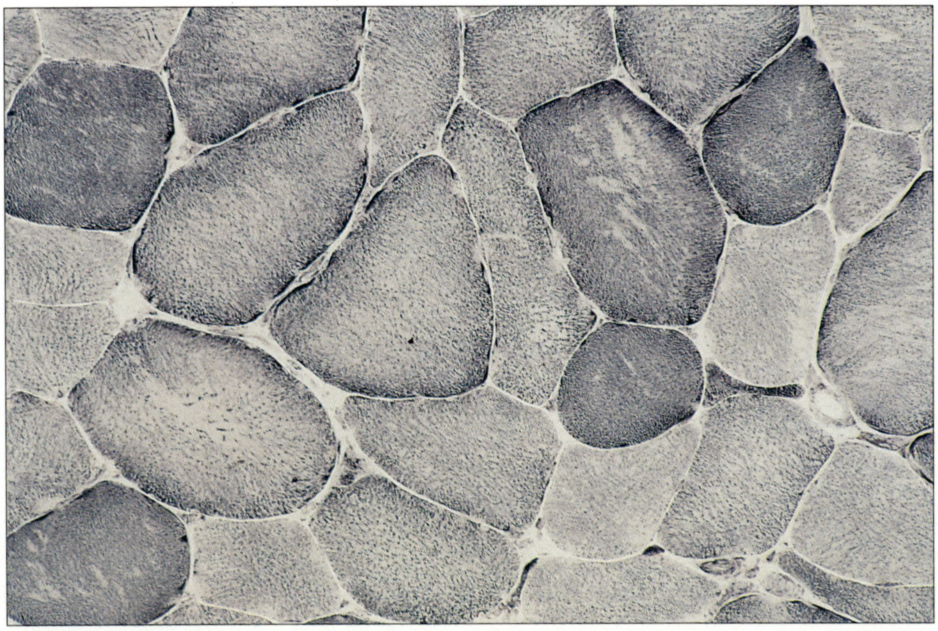

6.5 This illustrates the presence of target and motheaten fibres.
(NADH-Tr, ×175)

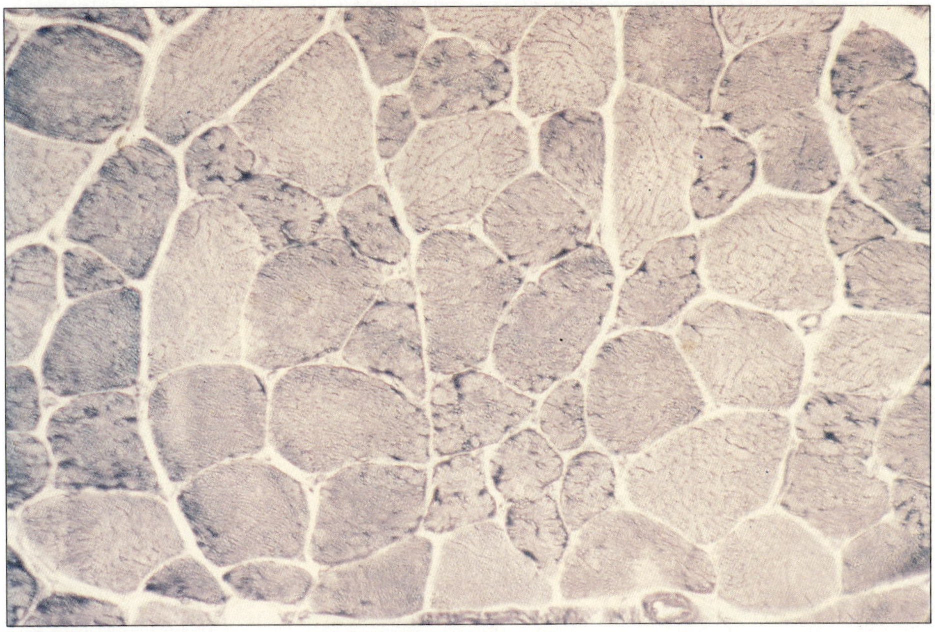

6.6 In addition to the increased variation in fibre size, the irregular or lobulated appearance of the majority of the smaller fibres (mainly Type I oxidative fibres) is striking.
(NADH-Tr, ×140)

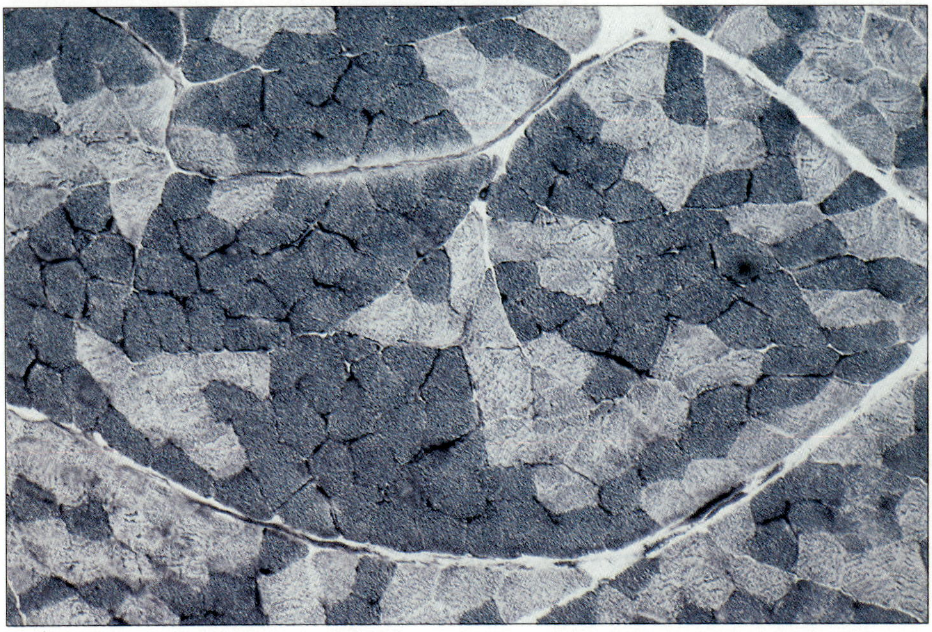

6.7 The grouping of fibre types following reinnervation can also be seen using oxidative stains.
(NADH-Tr, ×105)

However, the definitive diagnosis of a chronic denervating state depends ultimately upon the appearance of the myofibrillar ATPase preparations; this technique will demonstrate the phenomenon of grouped reinnervation. The grouping is due to the compensatory reinnervation of denervated motor units by collateral sprouting from terminal motor nerves originating from still functioning anterior horn cells (**6.2b**, **6.2c**, **6.8–6.9**, **3.6**). Grouped reinnervation is best seen in the more slowly progressive forms of spinal muscular atrophy, and, while there is no hard and fast rule, the larger groups tend to be Type I in profile. The explanation for this observation is not yet forthcoming.

In the more rapidly progressive forms of SMA, one sees what can best be described as generalised and gross muscular atrophy, with scattered fibres of normal or increased diameter (**6.10–6.12**). Once again, the larger fibres tend to be Type I in profile.

More precise identification of the nature of the anterior horn cells innervating these surviving fibres or motor units, may be provided by the application of antibody staining for fast and slow myosin chains (**3.12**).

Diagnostic uncertainty may arise in extremely chronic forms of SMA, in which prolonged motor unit denervation has been complicated by the development of secondary myopathic degeneration. This usually takes the form of random variation in muscle fibre diameter, the presence of internal nuclei, splits (particularly in hypertrophied fibres), and interstitial fibrosis (**4.9**, **4.12**, **4.14**, **4.17–4.18**). In a very few instances, evidence of more active muscle fibre degeneration, in terms of necrosis and phagocytosis, and even regeneration may be found (**4.9**).

It should be remembered that the extent and severity of these secondary myopathic features reflects, in general terms, the severity and rate of progress of the clinical disorder. However, they tend to be patchy and have to be viewed against the background of what is clearly a chronic, partial and often quite well compensated denervating process.

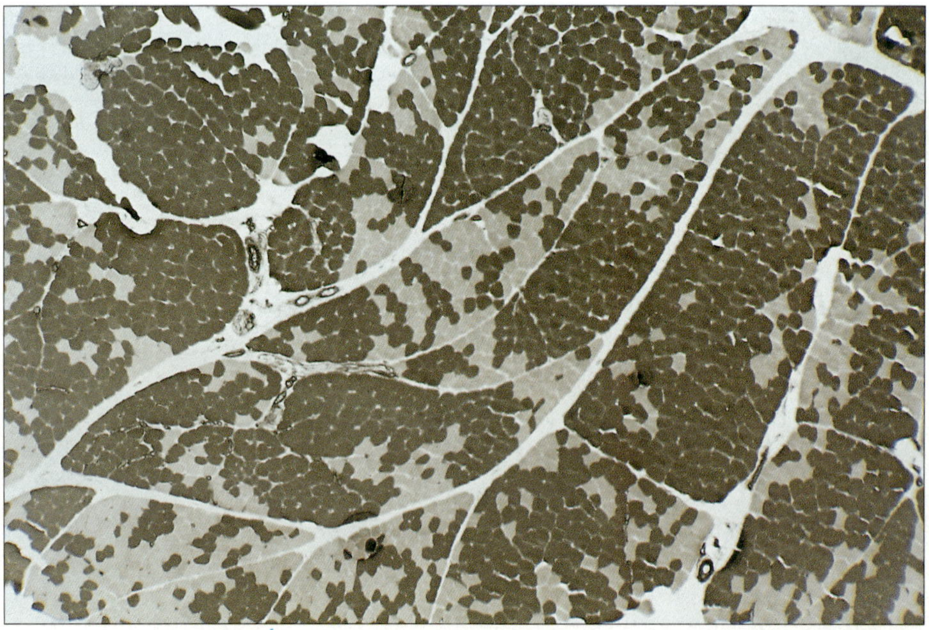

6.8 Marked grouping of fibre types of both types of muscle fibre, together with relatively normal size and architecture, is a feature of reinnervation in a well-compensated denervating condition of central origin. (Myofibrillar ATPase, pH 4.35, ×56)

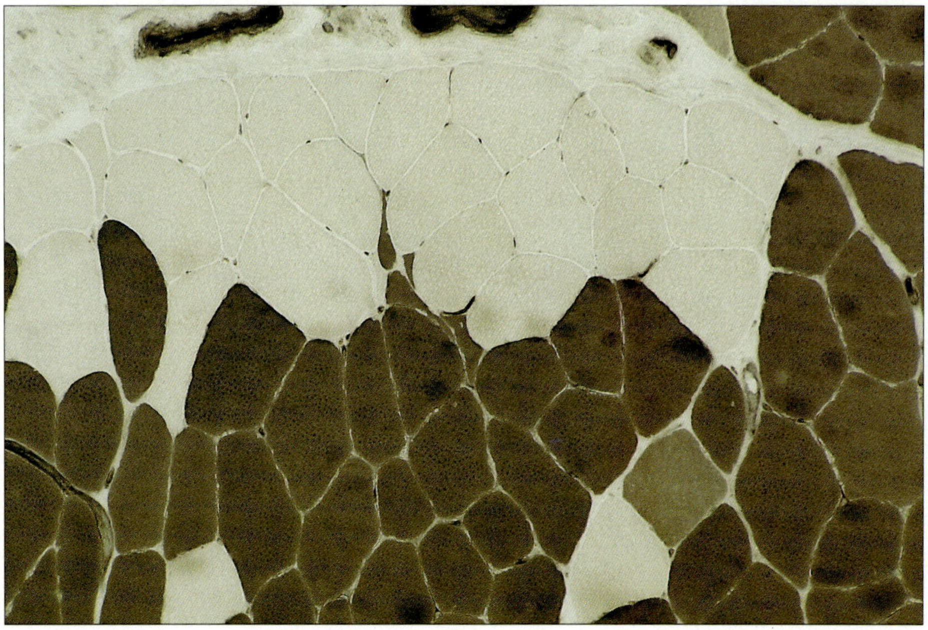

6.9 Higher magnification showing large groups of Type I and Type II muscle fibres in a long-term case of SMA. (Myofibrillar ATPase, pH 4.6, ×140)

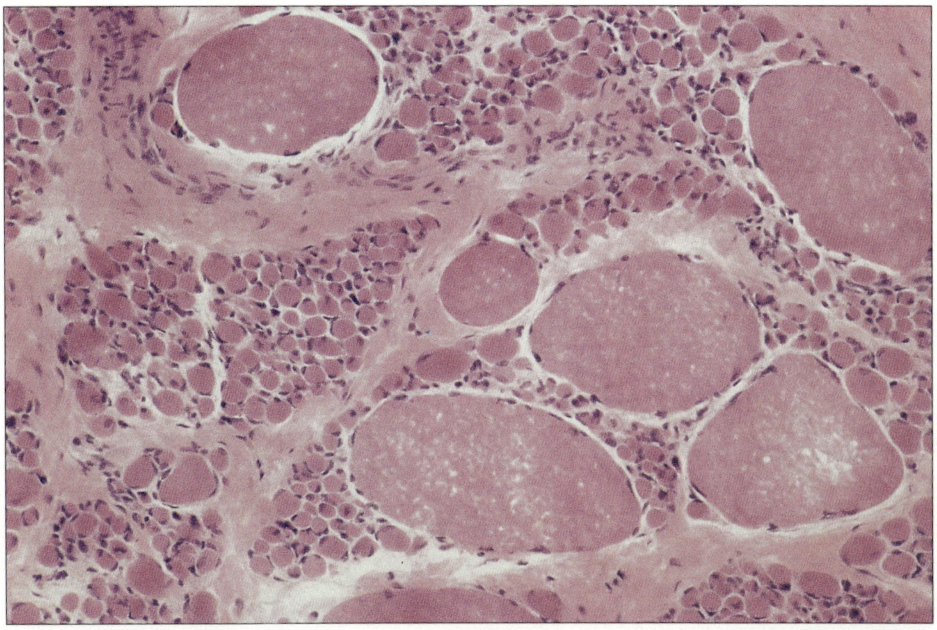

6.10 Biopsy from an infant with the Werdnig-Hoffman form of SMA. The majority of muscle fibres are much smaller than normal, although a few extremely hypertrophic fibres are also visible.
(H&E, ×217)

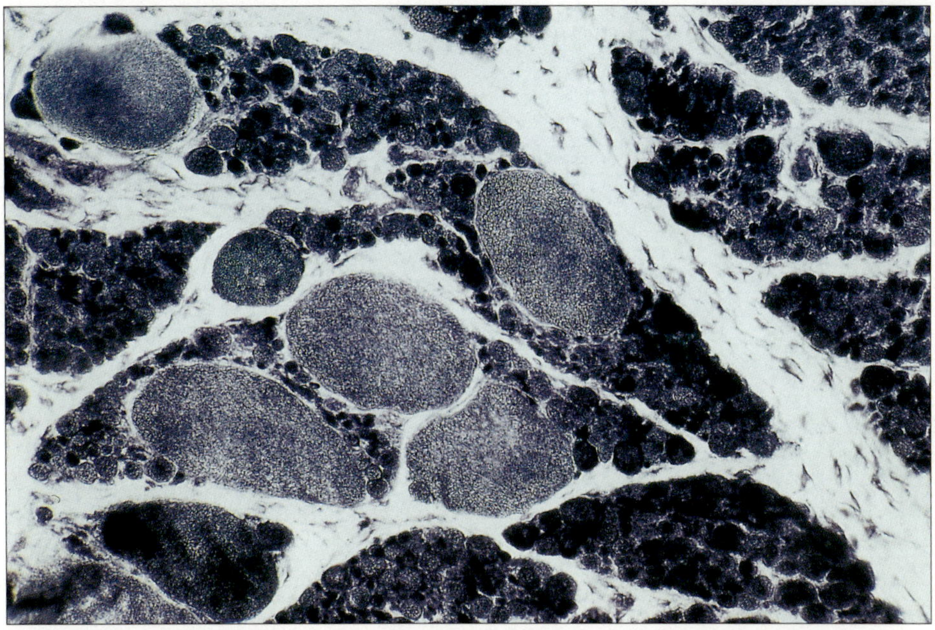

6.11 A serial section to **6.10**, showing the dense staining of the severely atrophic fibres.
(NADH-Tr, ×140)

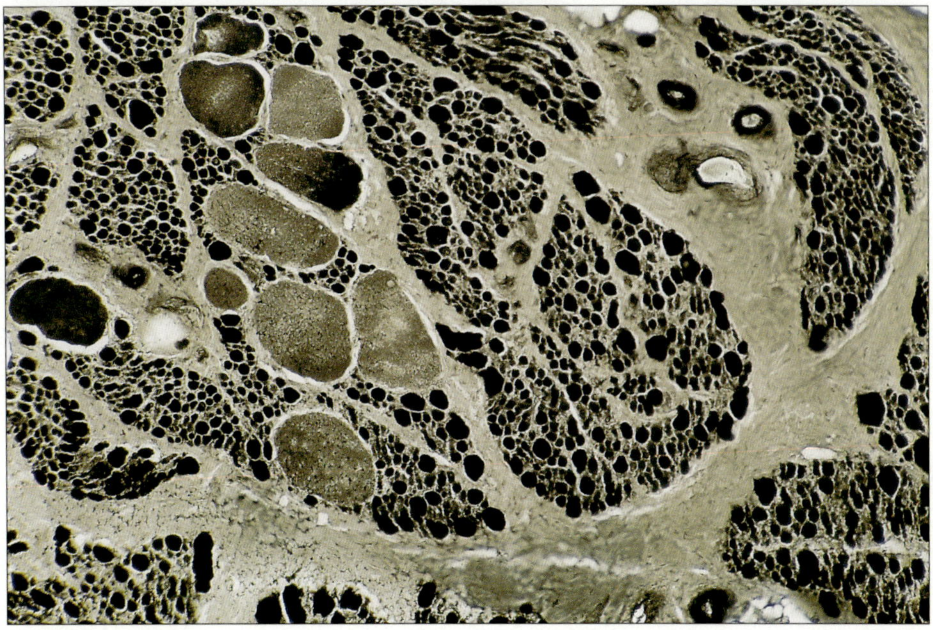

6.12 A serial section of **6.11**.
(Myofibrillar ATPase, pH 10.2, ×88)

7

Inflammatory myopathies

Although infective myositis is the commonest disorder seen worldwide (particularly in HIV positive and other immunocompromised patients), muscle biopsy is rarely indicated, and will therefore not be considered further in this chapter.

The non-infective inflammatory myopathies are the commonest, acquired, primary muscle cell disorders seen in clinical practice in both children and adults. From the clinician's point of view, they constitute a group of muscle disorders that are amenable to treatment, and thus have a potentially satisfactory outcome.

The patients with polymyositis (PM) or dermatomyositis (DM) usually present with a slowly progressive, painful, limb-girdle syndrome, with, in the DM type, a 'heliotrope' rash around the eyes, the malar region, and the extensor aspects of the limbs. In childhood, the commonest condition seen is DM.

In inclusion body myositis (IBM), the patient is typically male (the male:female ratio being 3:1). IBM presents as a slowly progressive, predominantly lower limb weakness with distal involvement, without pain (*Table 7.1*). (For review, see Dalakas, 1992.)

The diagnosis of PM and DM, which is by far the most important group, depends upon fulfilling two out of the following four criteria:

- A characteristic clinical presentation, with a symmetrical, proximal myopathy which may be painful, and may be accompanied by a skin rash.
- Elevated serum CK activity.
- An abnormal EMG, showing evidence of acute denervating as well as 'myopathic' features.
- Evidence in the biopsy of some, or all, of the following features: necrosis, phagocytosis, regeneration in the muscle fibres, and perivascular or endomysial mononuclear inflammatory cell infiltration.

Treatment of the idiopathic inflammatory myopathies is based on the assumption that these conditions are examples of disordered-cell mediated and/or humoral immunity, with the myovascular lesions being due to the cytotoxic effects of specifically-sensitised mononuclear cells

Idiopathic inflammatory myopathy (PM/DM/IBM)

PM/DM

In 'pure' PM in children or adults, the clinical presentation may be acute with rhabdomyolysis (rare), subacute, or chronic (in middle-aged females). The subacute type is commonest in young to early middle-aged adults. PM, and to a lesser extent DM, are seen in association with other auto-immune disorders. Both PM and DM are seen in association with neoplastic disorders.

IBM was initially considered to be a rare condition, but is now seen with increasing frequency, particularly in patients who either fail to show a response to conventional immunosuppressive therapy for their presumed PM, or who initially respond, but then relapse following therapy. This is in addition to the patients described above, with the more typical presentation of the disease.

The cardinal pathological features is the presence of necrosis with inflammatory cell infiltrates (**7.1–7.3**).

Table 7.1 Clinical–pathological correlation

	Frequency Child	Frequency Adult	Clinical	Pathology	Immunology
PM	+/–	+++	LG	Inflammatory infiltrate, predominantly endomysial. Scattered necrosis.	T-cells (CD8+) MHC class I expression
DM	++	+++	LG	Inflammatory infiltrate, predominantly perifasicular. Necrosis. Grouped and single perifasicular atrophy and microvascular changes commoner in child.	CD4+ common CD8+ rare B-cells (CD22) MAC expressed (C5b-9)
IBM	–	+++	Distal and LG age > 50	Vacuolar change and filamentous inclusions. Inflammatory variable.	Sometimes appears to be CD8+

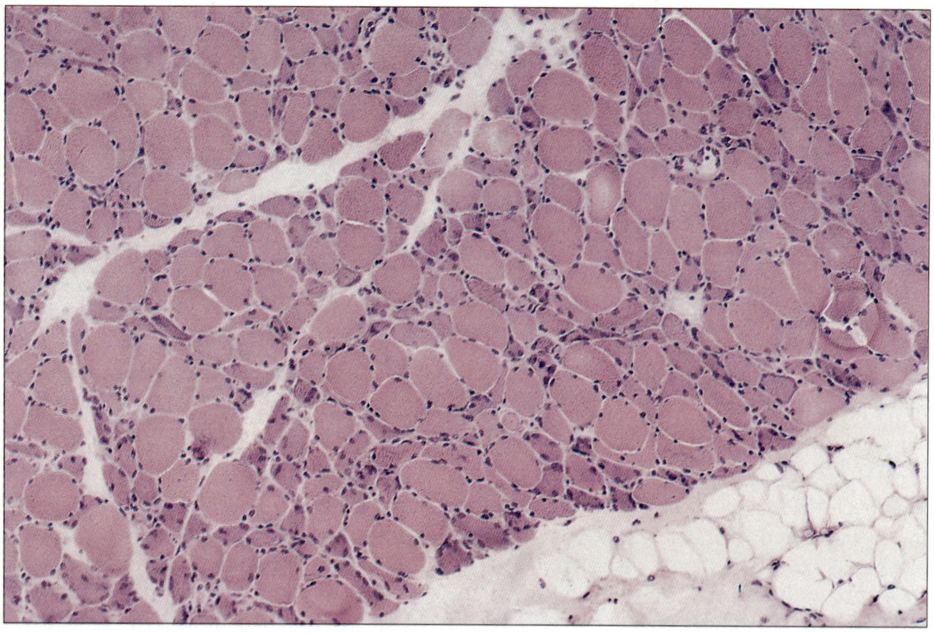

7.1 Diffuse, endomysial infiltrate, showing necrotic and regenerating muscle fibres. (H&E, ×88)

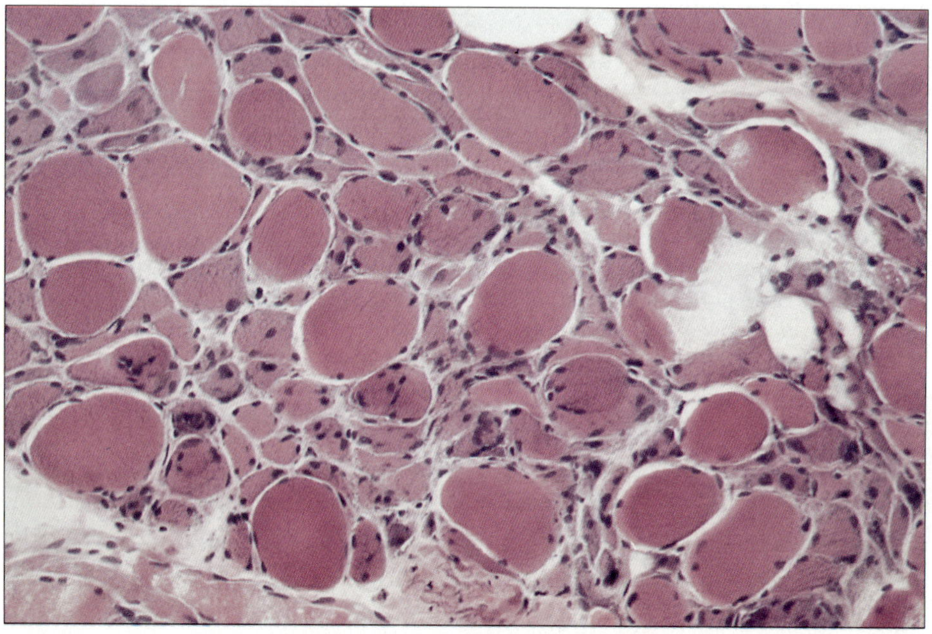

7.2 Diffuse, endomysial, inflammatory infiltrate, with muscle fibre atrophy, internal nucleation and necrosis. (H&E, ×140)

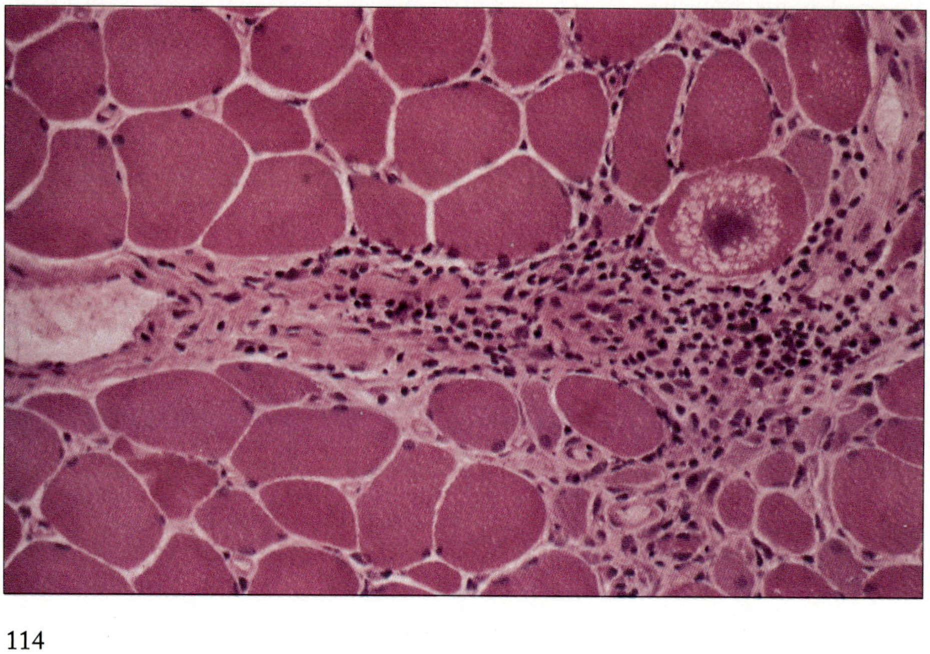

7.3 Infiltrate around a small blood vessel, with associated myofibre atrophy and degeneration. (H&E, ×140)

Granulomatous myositis

The term 'granulomatous myositis' encompasses a variety of relatively uncommon disorders, and muscle involvement in such cases is usually only part of a generalised or systemic disorder. Indeed, there may be no clinical evidence of myopathy in such cases.

However, muscle biopsy is of diagnostic value when other procedures such as renal or liver biopsy have been unsuccessful.

- Sarcoidosis – Cases of granulomatous myositis, without systemic involvement at the time of diagnosis, almost always turn out to have sarcoidosis. Sarcoid myopathy is probably the commonest variety of this genre, and the histopathology of the muscle lesion in these cases reflects the appearance of non-caseous granulomas elsewhere in the body tubercle (**7.4–7.5**). Occasionally, typical Langhans-type giant cells will be seen in these tubercles.
- Giant-cell myositis, in association with giant-cell myocarditis, myasthenia gravis (MyG), auto-immune thyroiditis and thymic neoplasms (benign or malignant).
- Wegener's granulomatosis
- Eosinophilic myositis as part of the hypereosinophilic syndrome (HES).

Some of the more aggressive vasculitides may be accompanied by 'myositis'. Polyarteritis nodosa and its Churg-Strauss variant, and Wegener's granulomatosis being the best examples (**7.6–7.7**).

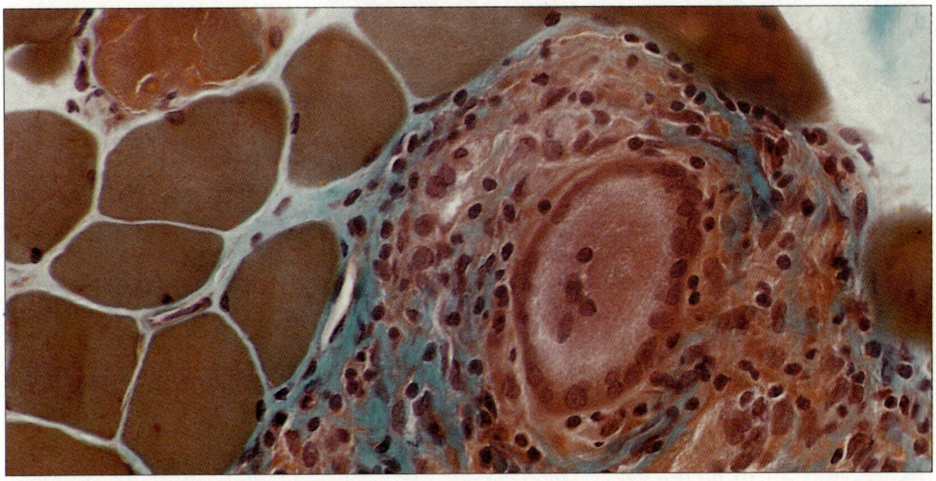

7.4 Granuloma, showing extensive destruction and phagocytosis in the interstitium of the muscle fibres.
(Masson's trichrome, ×350)

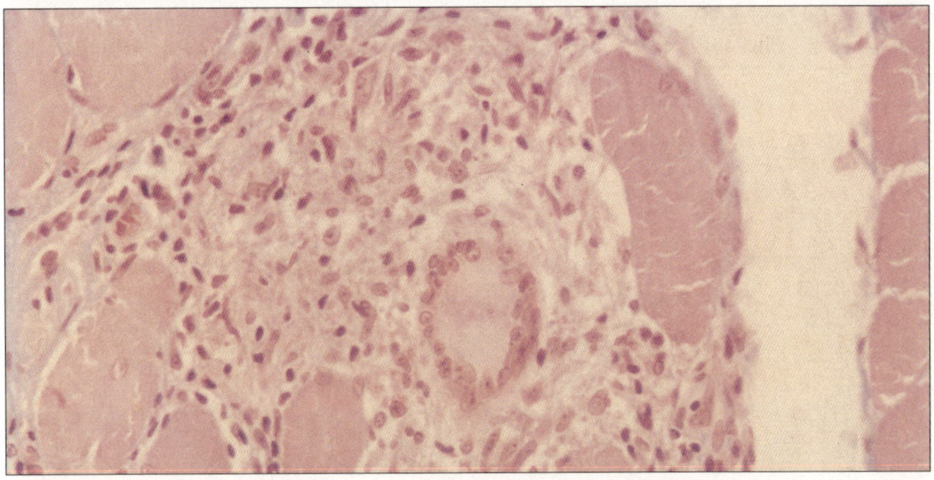

7.5 Granuloma showing extensive destruction and phagocytosis in the interstitium of the muscle fibres.
(H&E, ×280)

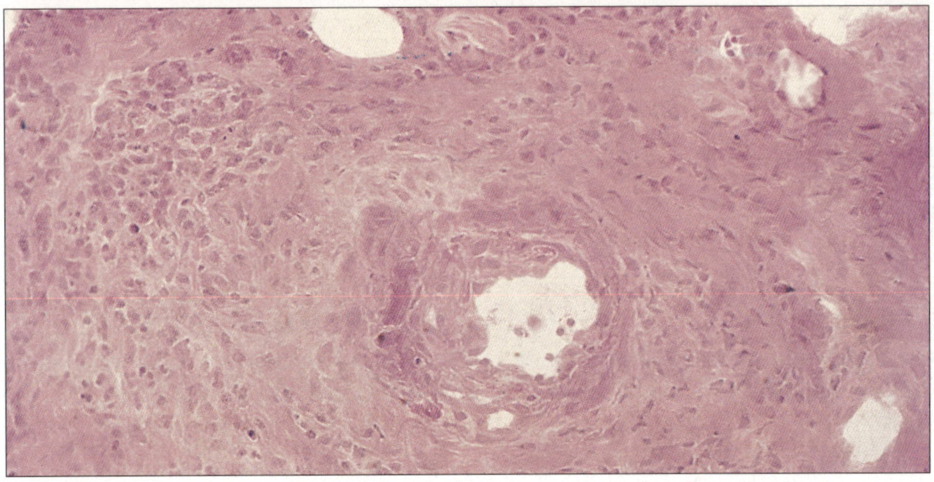

7.6 Polyarteritis nodosa, showing vascular related inflammation
(H&E, ×280)

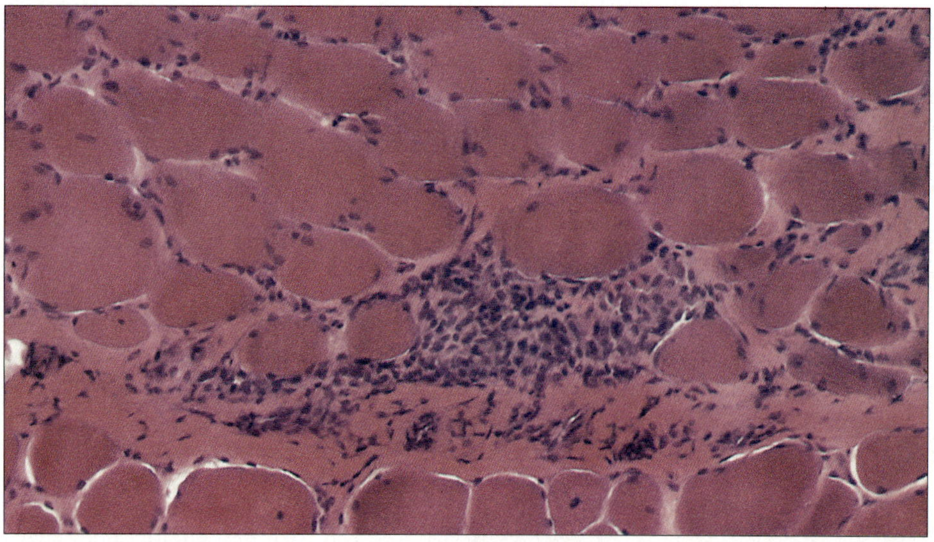

7.7 Wegener's granulomatous reaction. (H&E, ×210)

Pathological reactions in the inflammatory myopathies

The basic histopathological features of the inflammatory myopathies have been recognised since the late 19th century.

The lesions in these disorders are distributed in a patchy fashion, so, if the needle biopsy technique is being employed, multiple tissue samples are required. Similarly, when employing the open biopsy technique, it is essential to examine multiple blocks.

In assessing biopsy material from a patient with a suspected inflammatory myopathy, evidence of inflammatory cell infiltrate, perifascicular atrophy, and necrosis, phagocytosis and regeneration may be observed to varying degrees, as indicated in *Table 7.1*.

The relative proportions and distribution of the necrobiotic and inflammatory lesions vary considerably from one patient to another, although the clinical profiles for these patients may be virtually indistinguishable. This underlines the importance of multiple samples.

Inflammatory cell infiltrates

The infiltrates are predominantly mononuclear, with the majority of the cells being lymphocytes and macrophages (**7.8–7.9**, **4.43**). Plasma cells, neutrophils, eosinophils and mast cells are occasionally encountered in small numbers (**7.10–7.15**). In the particular case of eosinophilic myositis, eosinophils predominate in the infiltrate. The inflammatory lesions may be focal (**7.9**), diffuse endomysial (**7.1–7.2**), or concentrated around vessels (**7.3**).

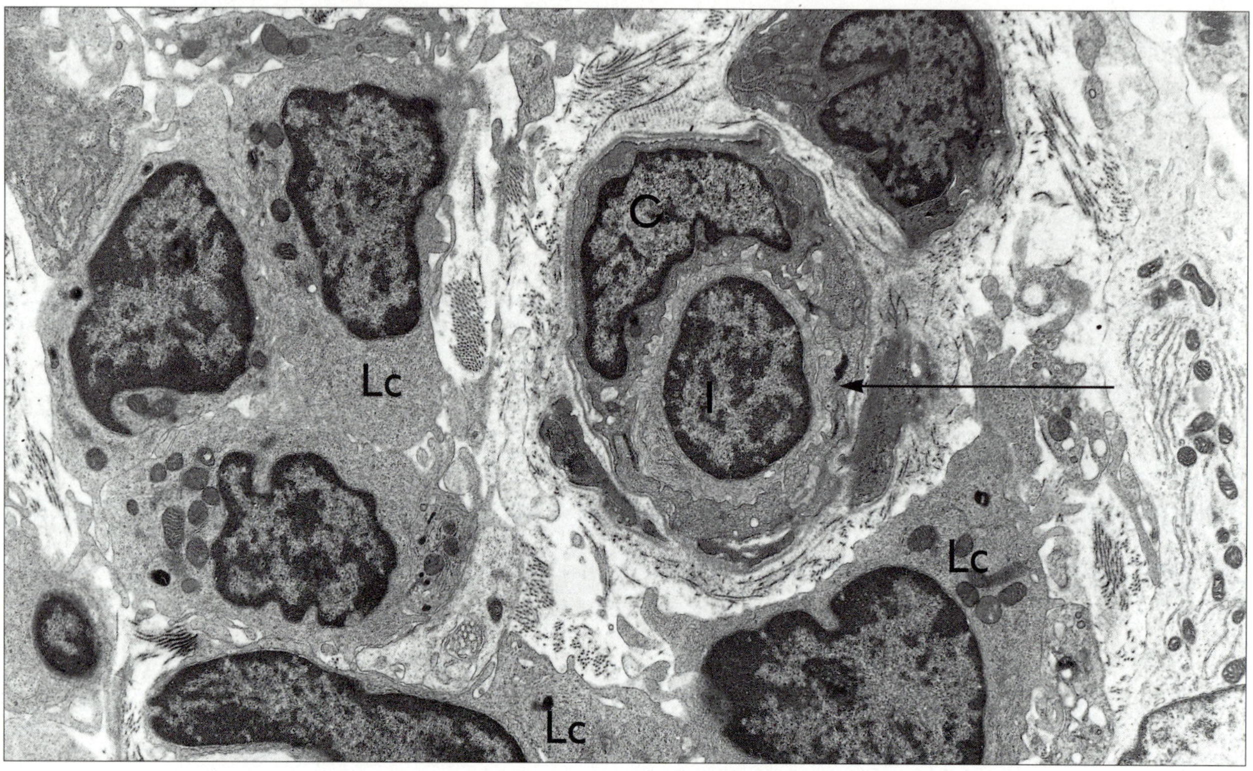

7.8 A group of mononuclear inflammatory cells form a perivascular infiltrate. The most common cell type is the lymphocyte (Lc). Cells are occasionally seen within the lumen of capillaries (arrowed). (×7875)

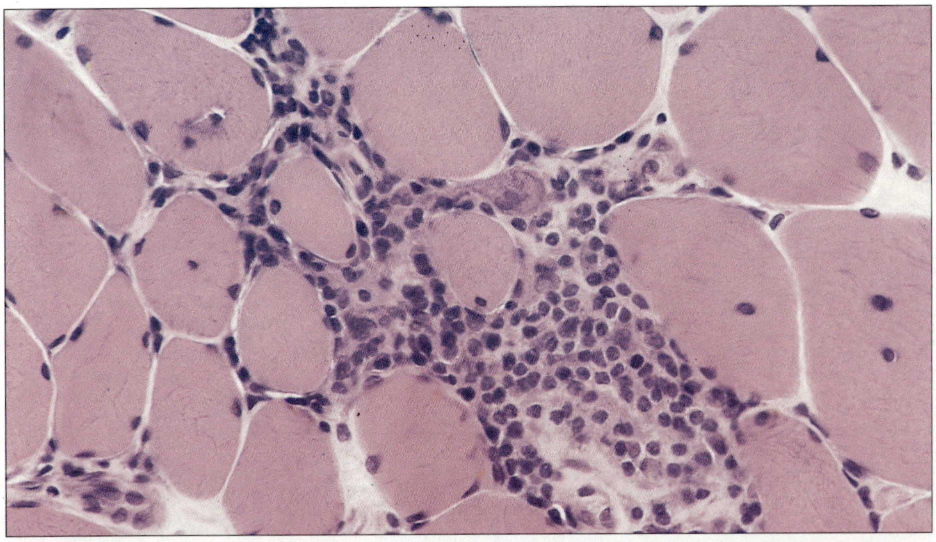

7.9 Focal mononuclear cell infiltrate. (H&E, ×350)

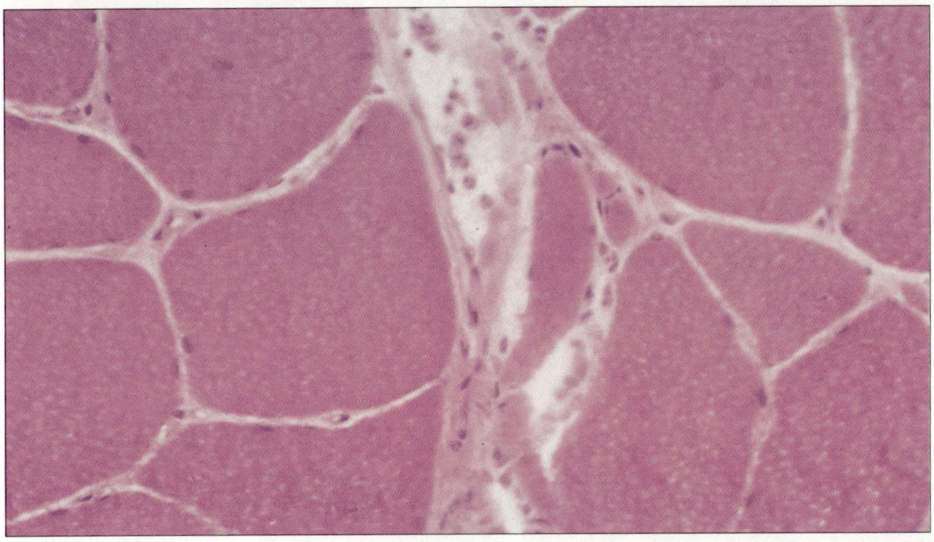

7.10 Inflammatory infiltrate showing the presence of neutrophils. (H&E, ×280)

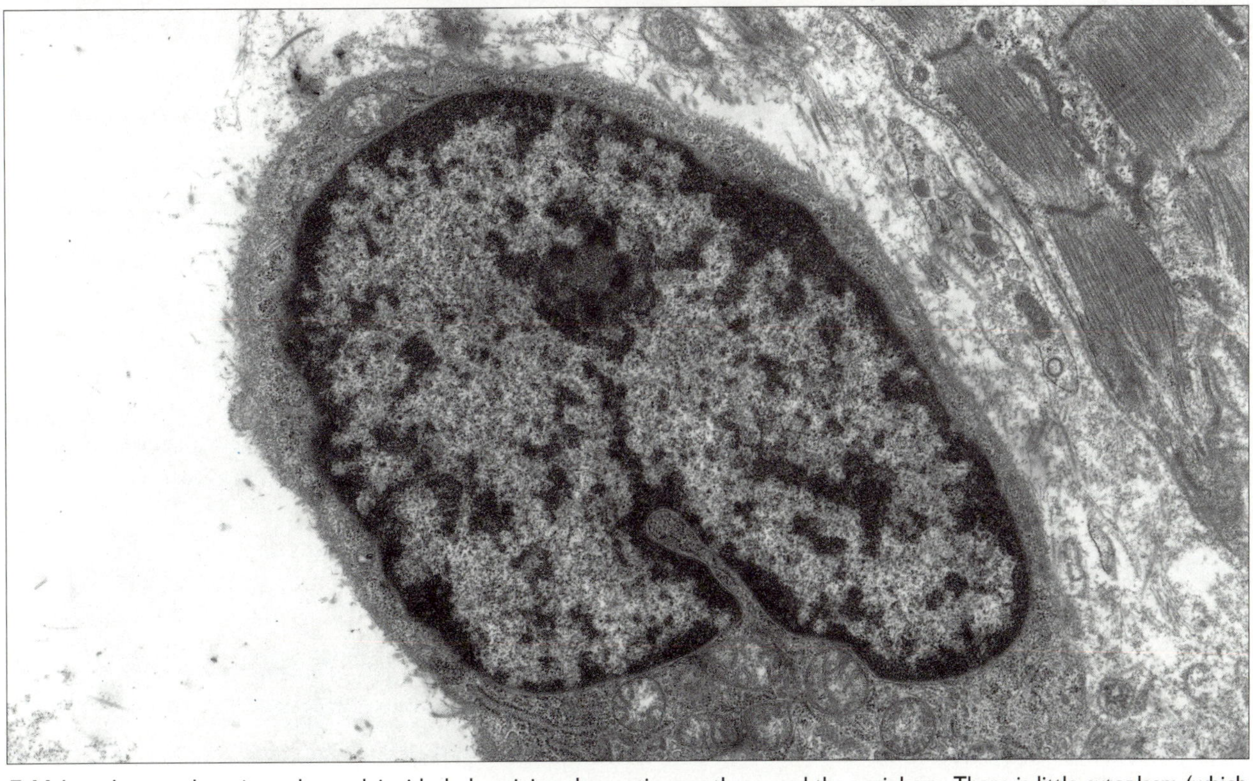

7.11 Lymphocytes have irregular nuclei with dark staining chromatin, mostly around the periphery. There is little cytoplasm (which tends to stain darkly), due to the presence of many ribosomes. (×17325)

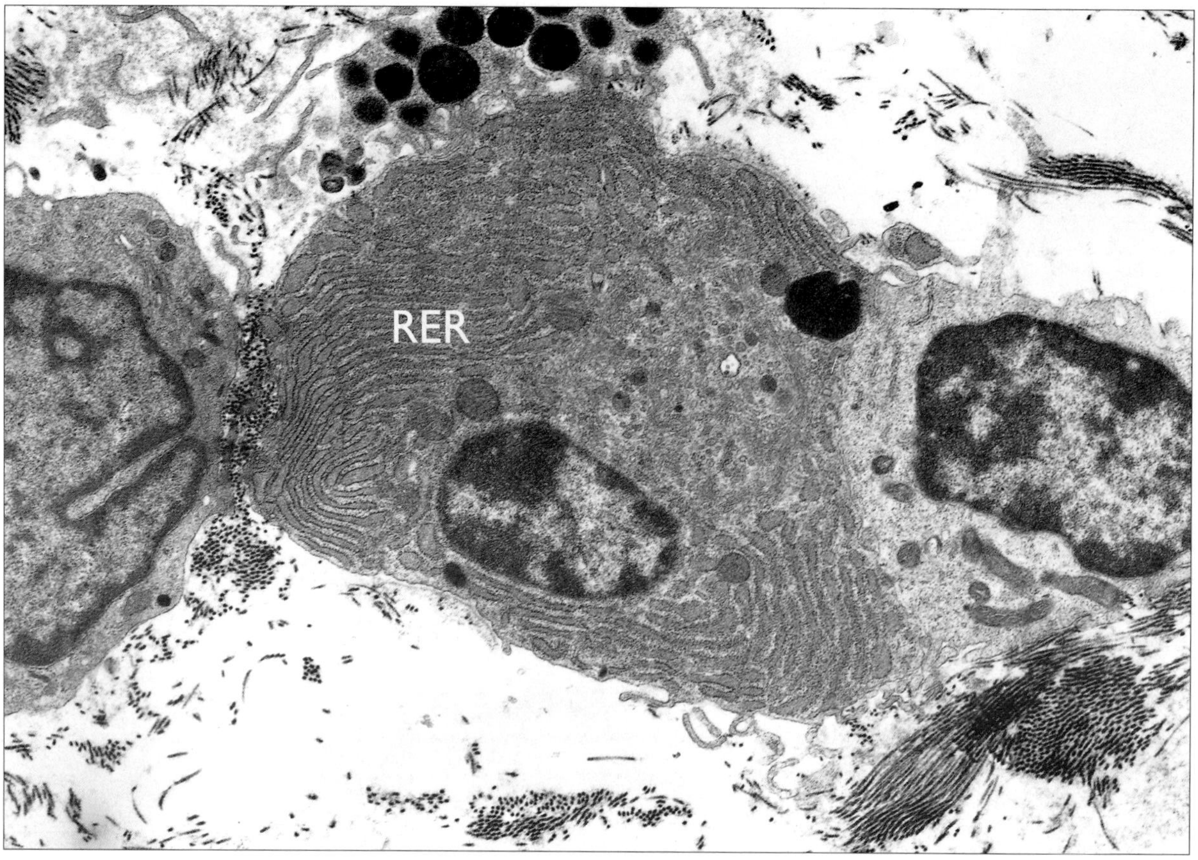

7.12 Plasma cells are characterised by the large amount of rough endoplasmic reticulum (RER), which is arranged in the cytoplasm in layers. (×12000)

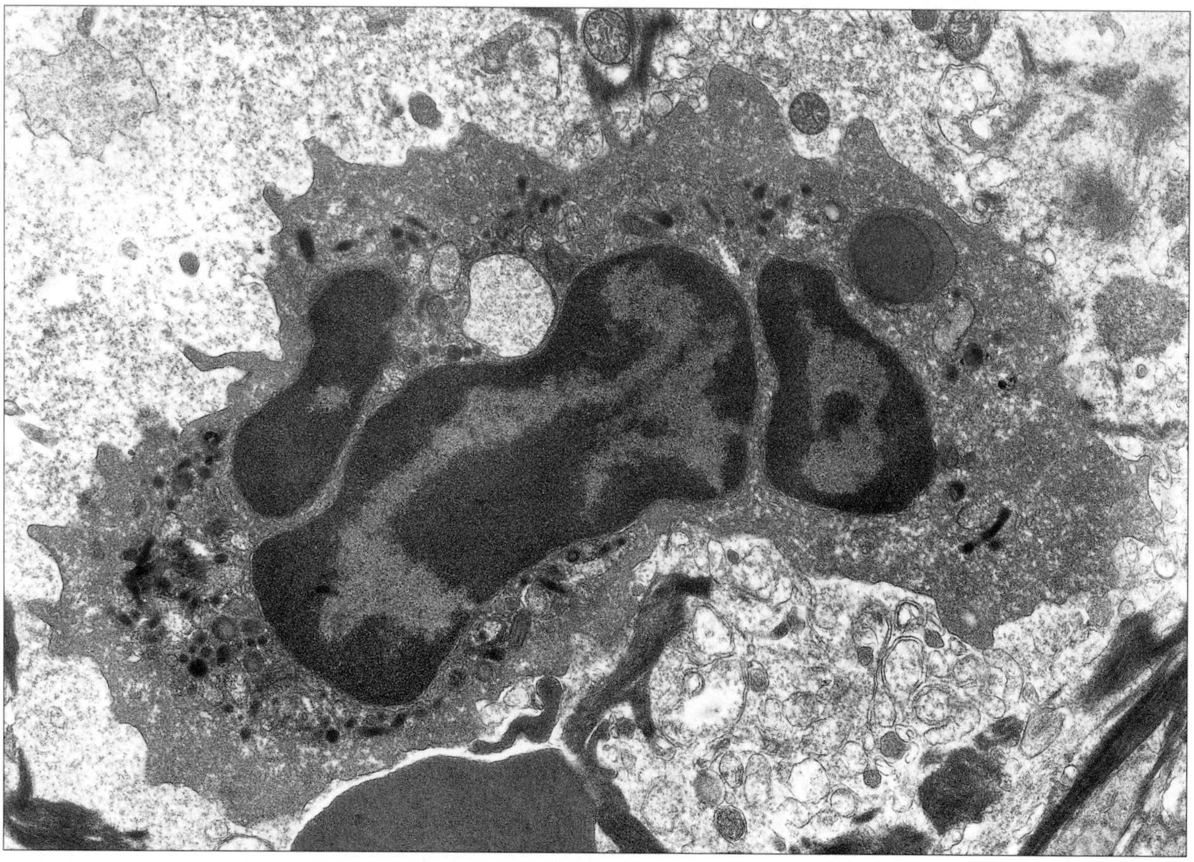

7.13 Polymorphonuclear leucocytes have lobed nuclei with darkly staining chromatin. Within the cytoplasm, there are numerous small, dense, membrane-bound vesicles. (×14000)

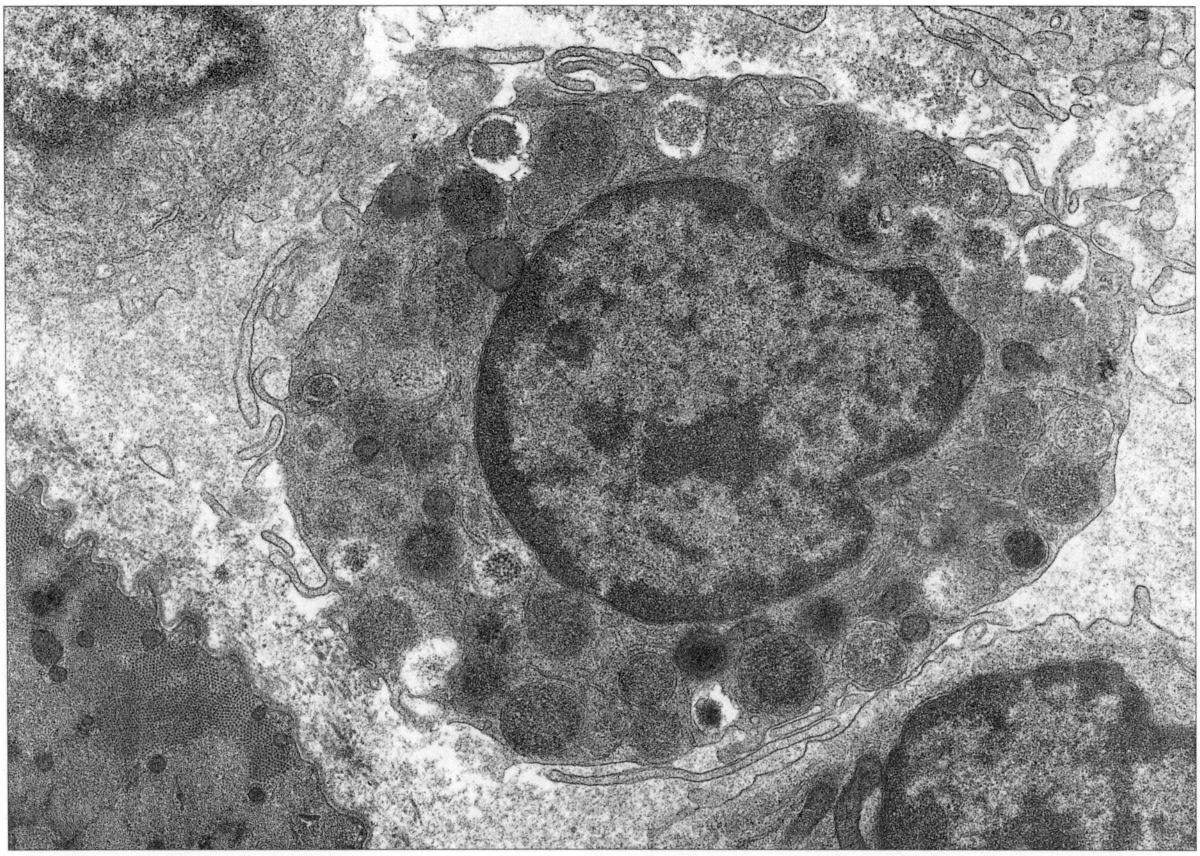

7.14 Mast cells may be found in the interstitium of both normal and diseased muscle. The cytoplasm contains many vesicles, which in turn contain small granules. The number of granules and the density of staining are both variable. (×16500)

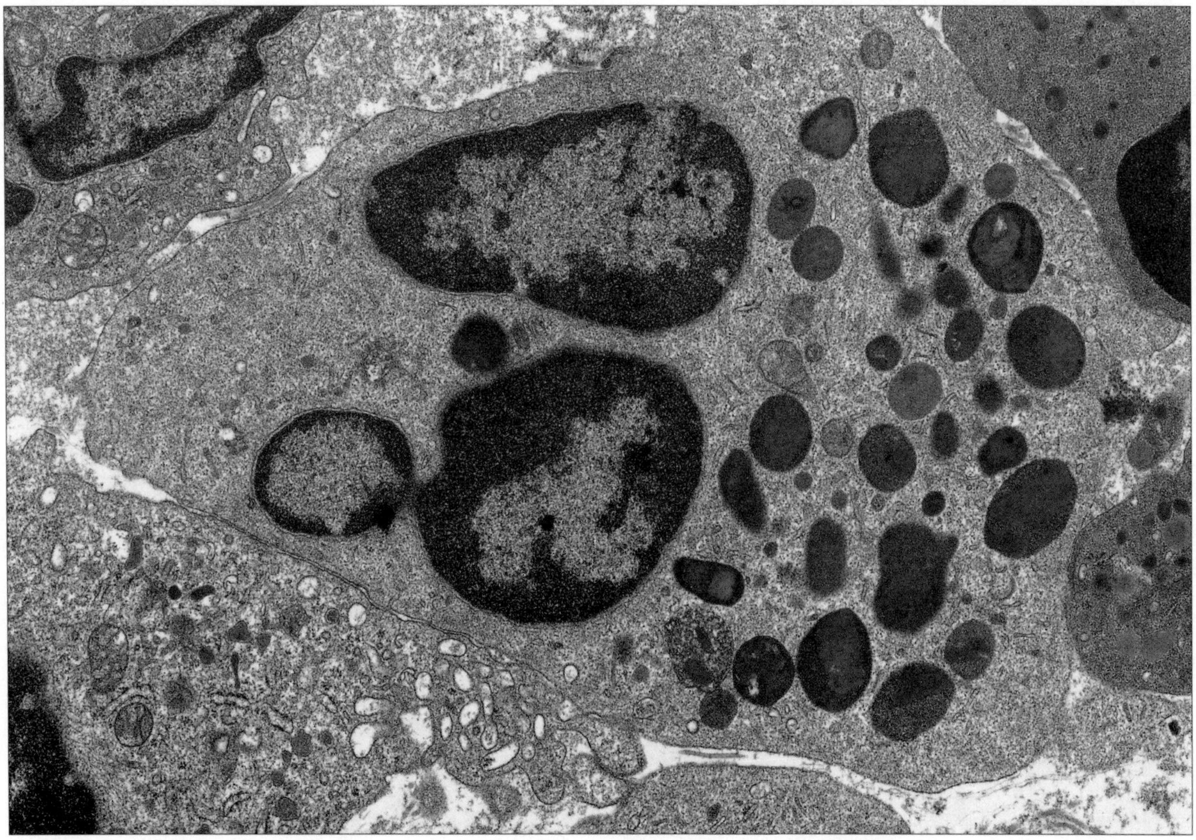

7.15 The cytoplasm of eosinophils is packed with small, darkly staining structures, some of which contain paler, angular inclusions. (×16000)

Perifascicular atrophy

Perifascicular atrophy is a feature of DM (**7.16–7.18**), being more common in the childhood form of the disease than the adult.

The perifascicular distribution of these lesions is due to proximity to the blood vessels, from which sensitised mononuclear cells migrate to initiate the myovascular lesions. These lesions consist of atrophy and necrosis, probably due to ischaemia.

The muscle cell shows evidence of acellular structural alteration (**7.19**), and cell-mediated damage, where, in addition to structural alteration, there is invasion of the fibres by mononuclear cells (**7.20–7.30**).

Necrosis

All the changes with respect to necrosis, as described in Chapter 4, may be seen in inflammatory myopathy (**7.17, 7.20–7.21, 7.26–7.33, 4.25–4.26, 4.32**).

In acute viral myositis and rhabdomyolysis, all the muscle fibres will show the same stage of response. In the inflammatory myopathies however, different stages of necrosis will be seen.

In general terms, the distribution varies according to the aetiology of the myopathy (*Table 7.1*).

Necrosis of the muscle fibres may be a consequence of cell-mediated cytotoxicity, or of ischaemia resulting from the damage to the microvasculature. Cells may be seen invading the muscle fibre and beneath the basement membrane in intimate contact with the muscle fibre (**7.22–7.25**).

In PM, the distribution of the necrosis tends to be diffuse (**7.2**), whereas in DM, the necrotic fibres are grouped (**7.9**).

In some instances, the necrosis may be so severe that the muscle appears to have undergone ischaemic change with oedema or even frank infarction (**7.19**). However, this is an exceptional circumstance; it should suggest the diagnosis of localised nodular myositis (**7.34**), which is a rare variant of PM, and one which often antedates the clinical expression of the generalised disorder. Recovery from this curious condition may leave areas of scarring in the affected muscle, which may have been misdiagnosed on biopsy as venous haemangiomas (**7.35**).

Phagocytosis

Phagocytosis (Chapter 4), which is indicated by the presence of macrophages within the fibre (**7.22–7.24, 7.26–7.28, 4.22–4.23**), is a prominent feature in the inflammatory myopathies. At the ultrastructural level, macrophages have been observed invading the muscle fibres by penetrating the sarcolemma (**7.22–7.24**).

Regeneration

Fibres in varying stages of regeneration (Chapter 4) are a feature of the inflammatory myopathies, being more prominent in the established disease (**7.17, 7.36, 4.24**).

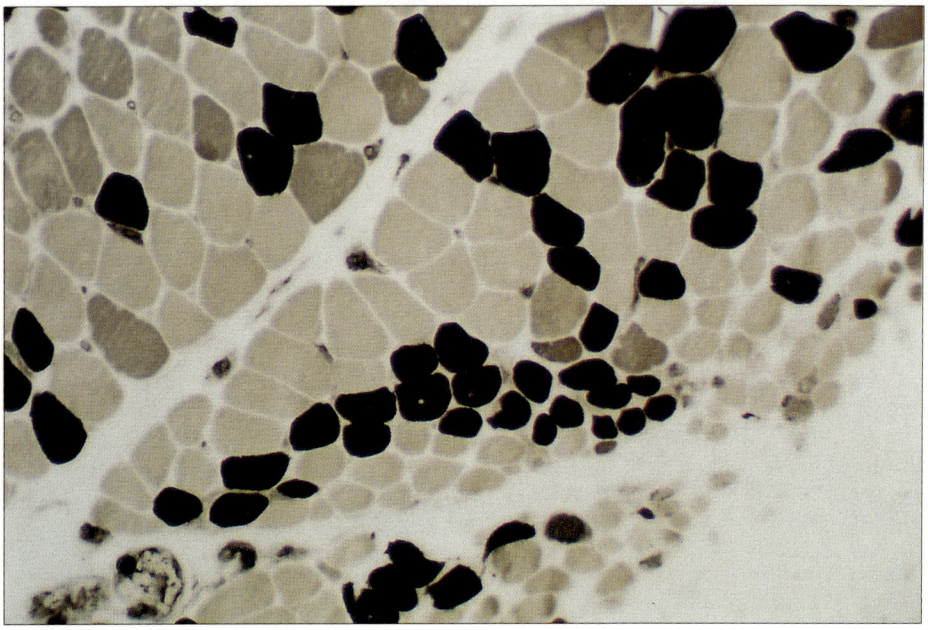

7.16 Perifascicular atrophy. (Myosin ATPase, pH 4.6, ×140)

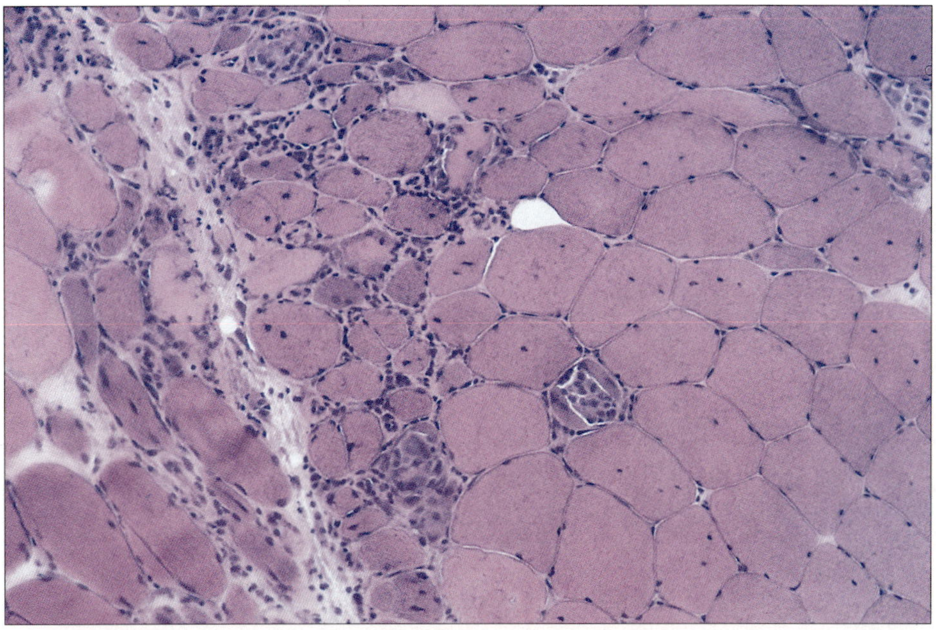

7.17 Focal perifascicular inflammation, necrosis, and regeneration of muscle fibres. The remaining portion of the fascicle is relatively normal, except for the presence of internally located nuclei. (H&E, ×140)

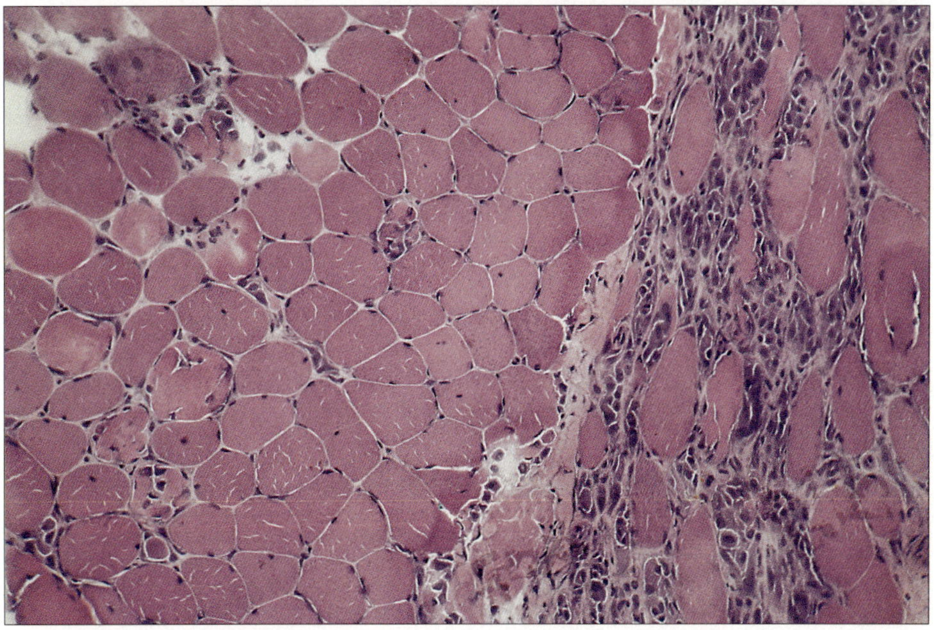

7.18 Perifascicular inflammatory infiltrate, necrosis and regeneration. (H&E, ×140)

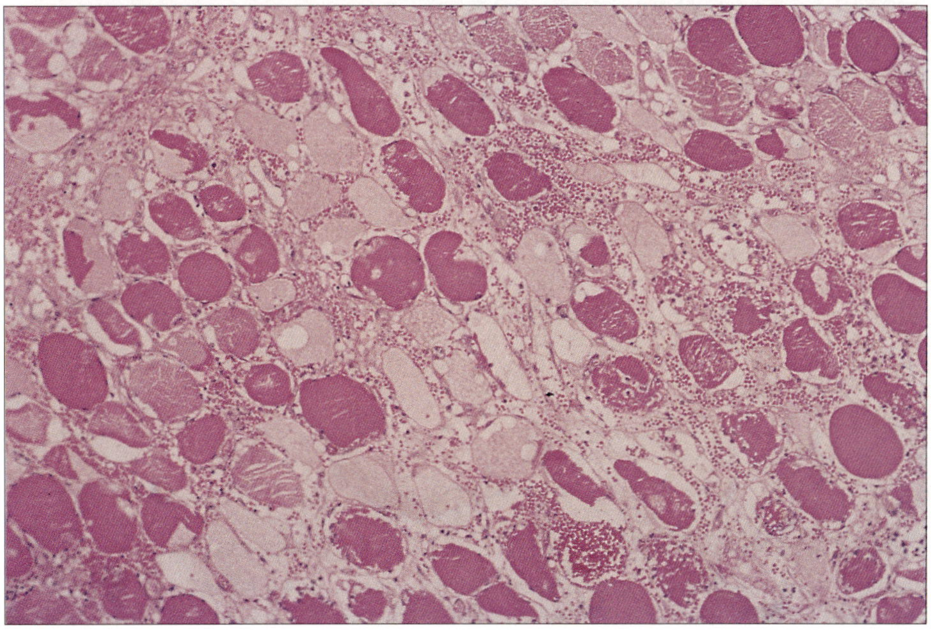

7.19 Large area of ischaemic necrosis, resembling a muscle infarct. (H&E, ×140)

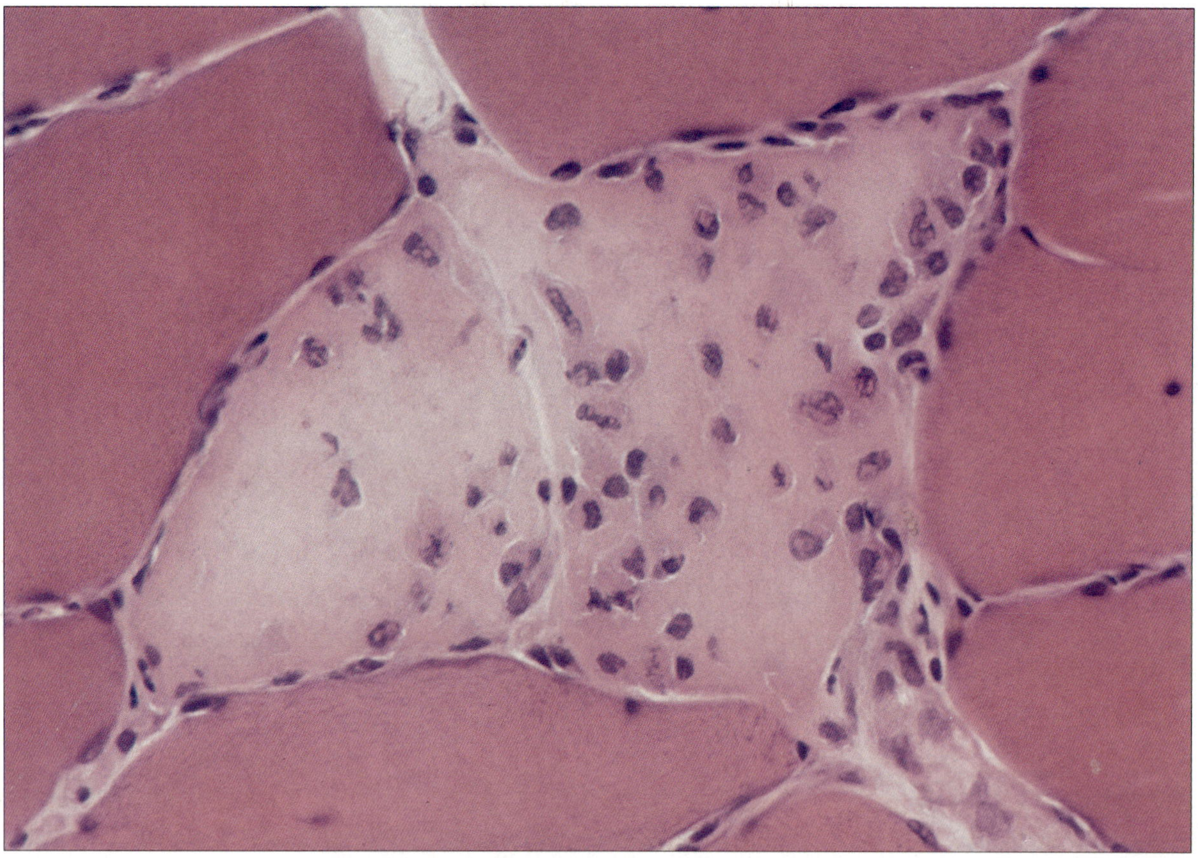

7.20 Necrotic muscle fibres, showing almost complete loss of myofibrillar material in the presence of macrophages. (H&E, ×473)

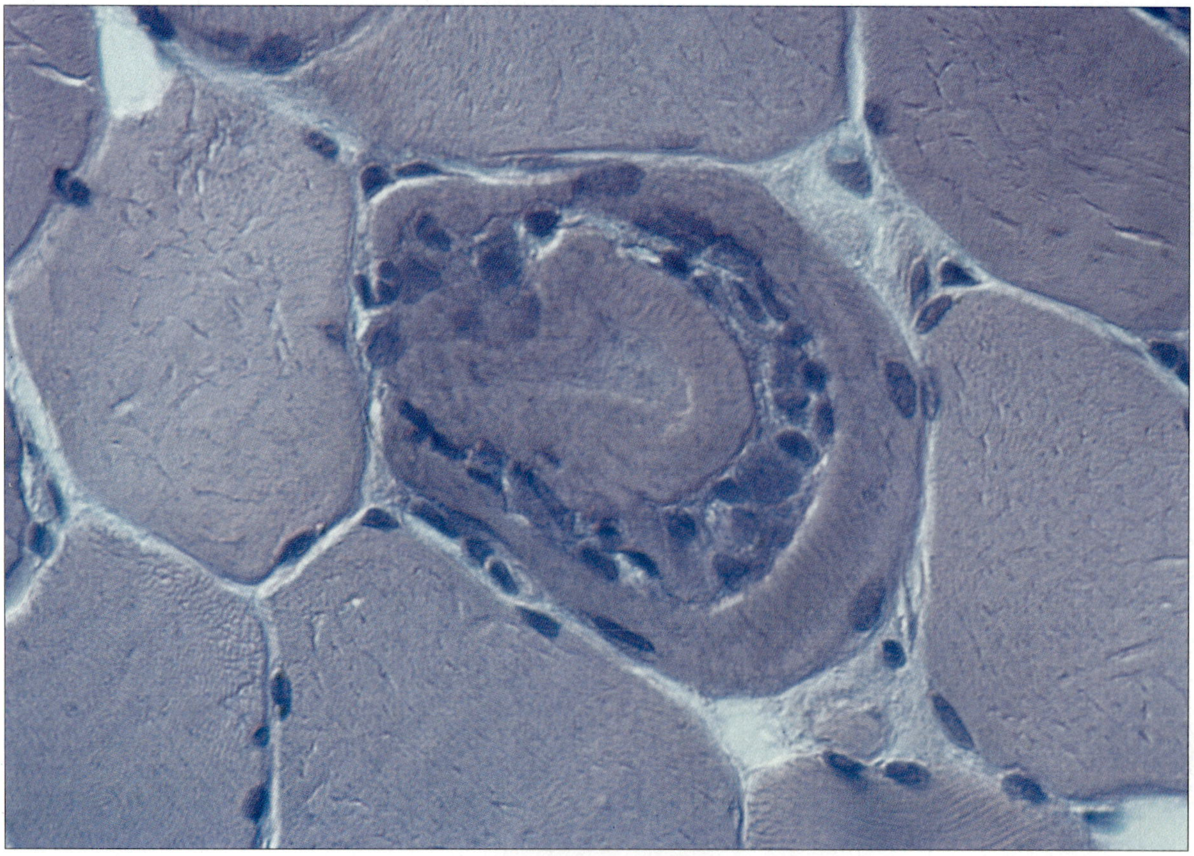

7.21 Muscle fibre showing invasion of macrophages, in the context of intact myofibrillar material. This emphasises the patchy and focal nature of the changes at the single fibre level. (H&E differential interference contrast, ×605)

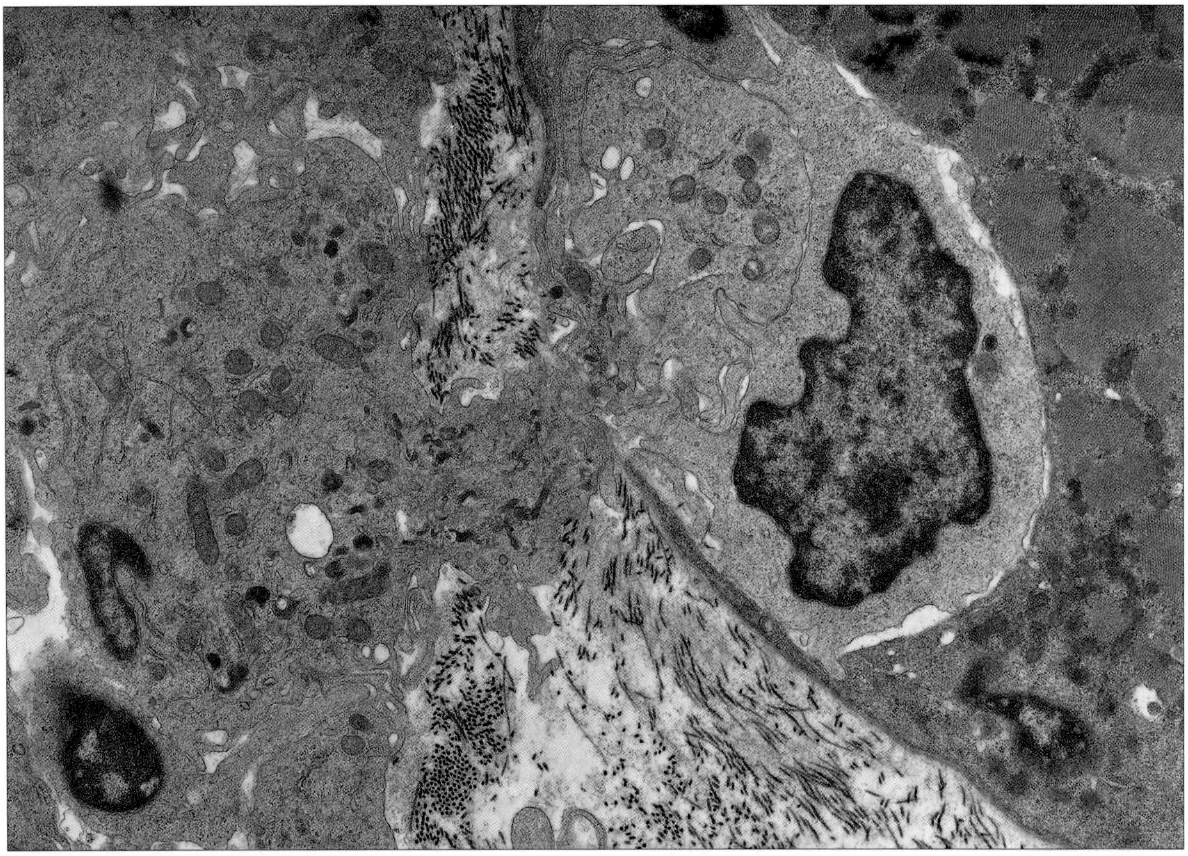

7.22 Phagocytic cells can be observed to invade muscle fibres by passing through the basement membrane and plasma membrane into the sarcoplasm. (×11000)

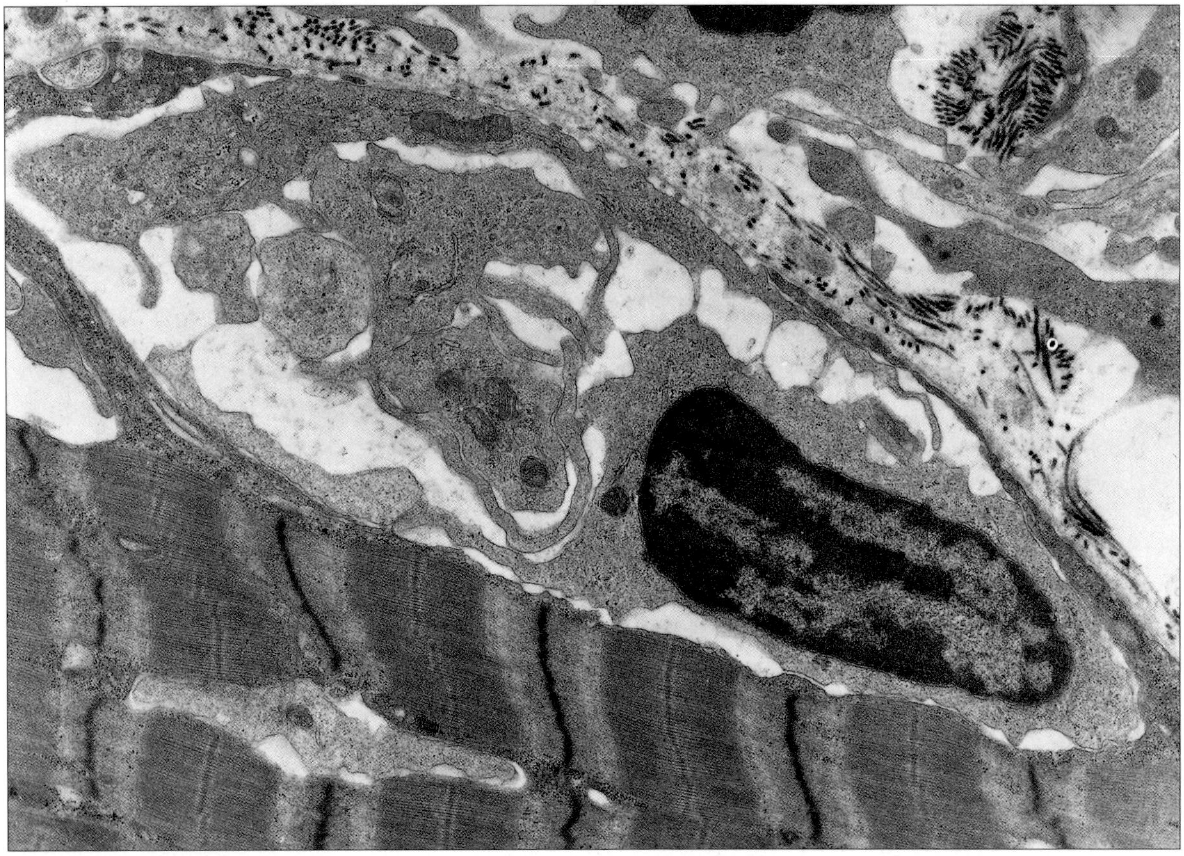

7.23 Mononuclear cells can be found lying beneath the basal lamina of muscle fibres, with cytoplasmic extensions penetrating between the myofibrils. (×15000)

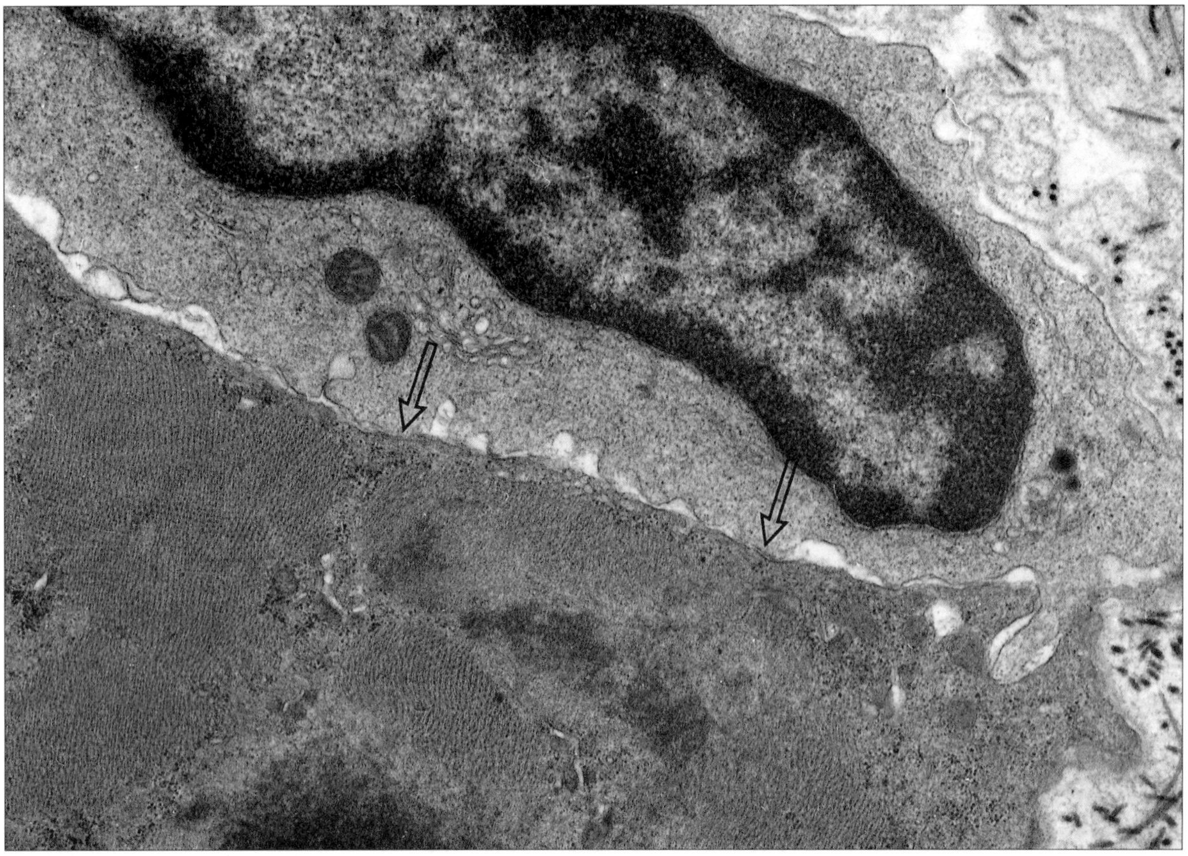

7.24 The cytoplasm of mononuclear cells that have invaded a fibre can be seen to make contact with the sarcoplasmic membrane of the muscle cell (arrowed). (×22500)

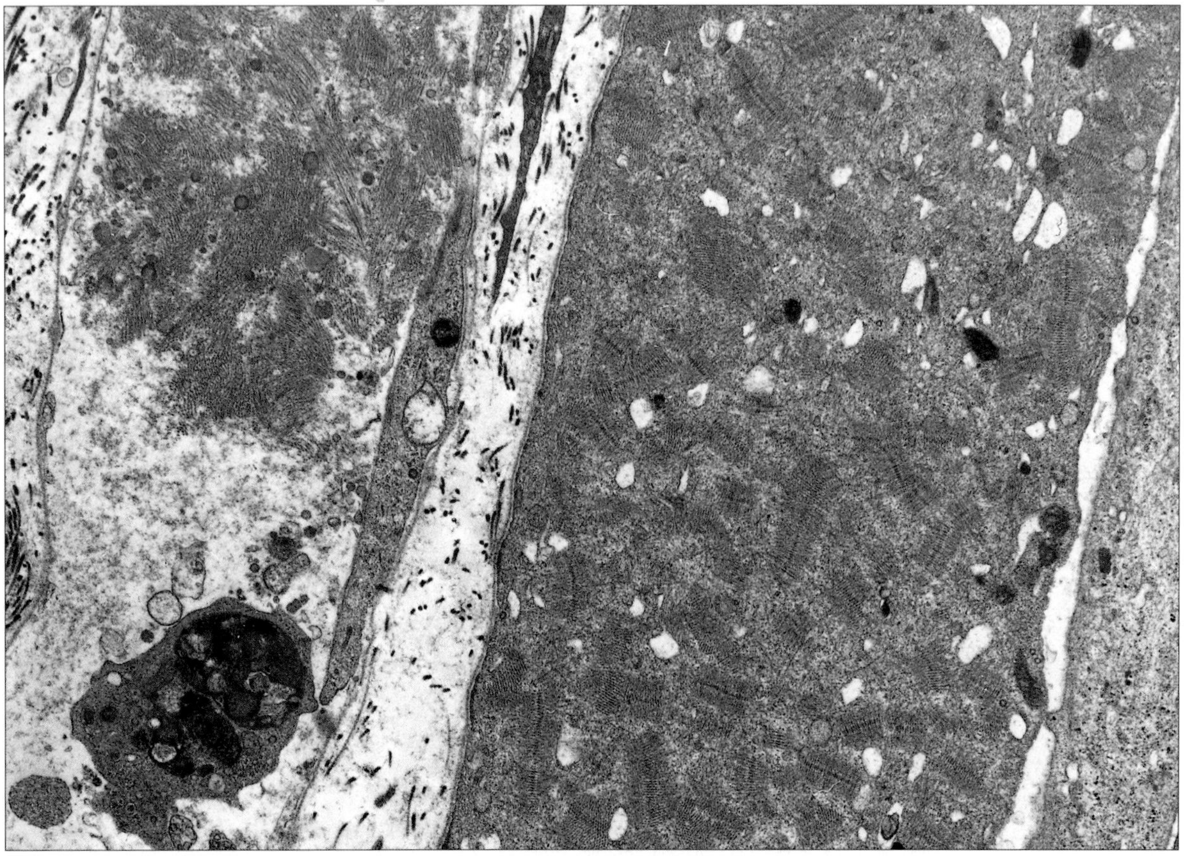

7.25 Fibre breakdown may result as a consequence of Z-line loss and I-band dissolution. The remaining A-bands become disorientated and surrounded by granular debris. (×12000)

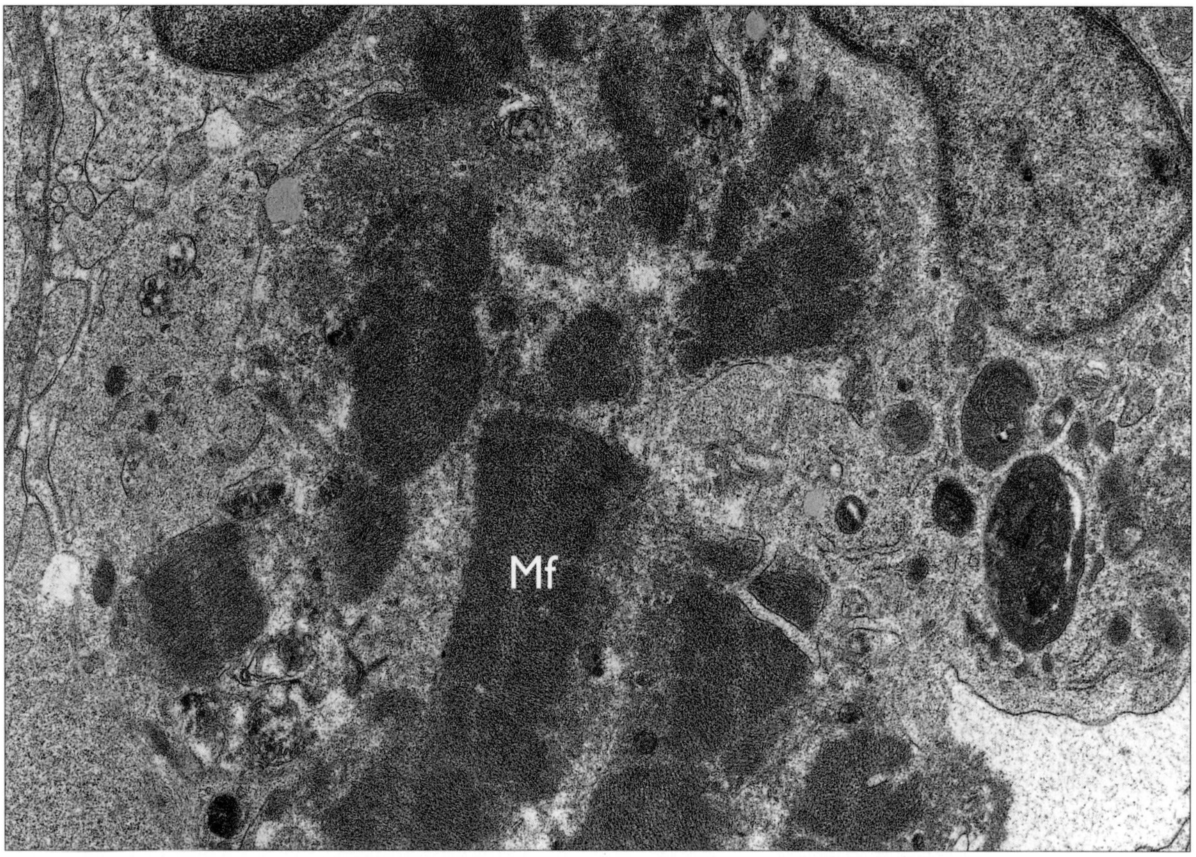

7.26 Phagocytic cells can be observed to engulf the isolated A-band segments and other filament debris (Mf). (×18000)

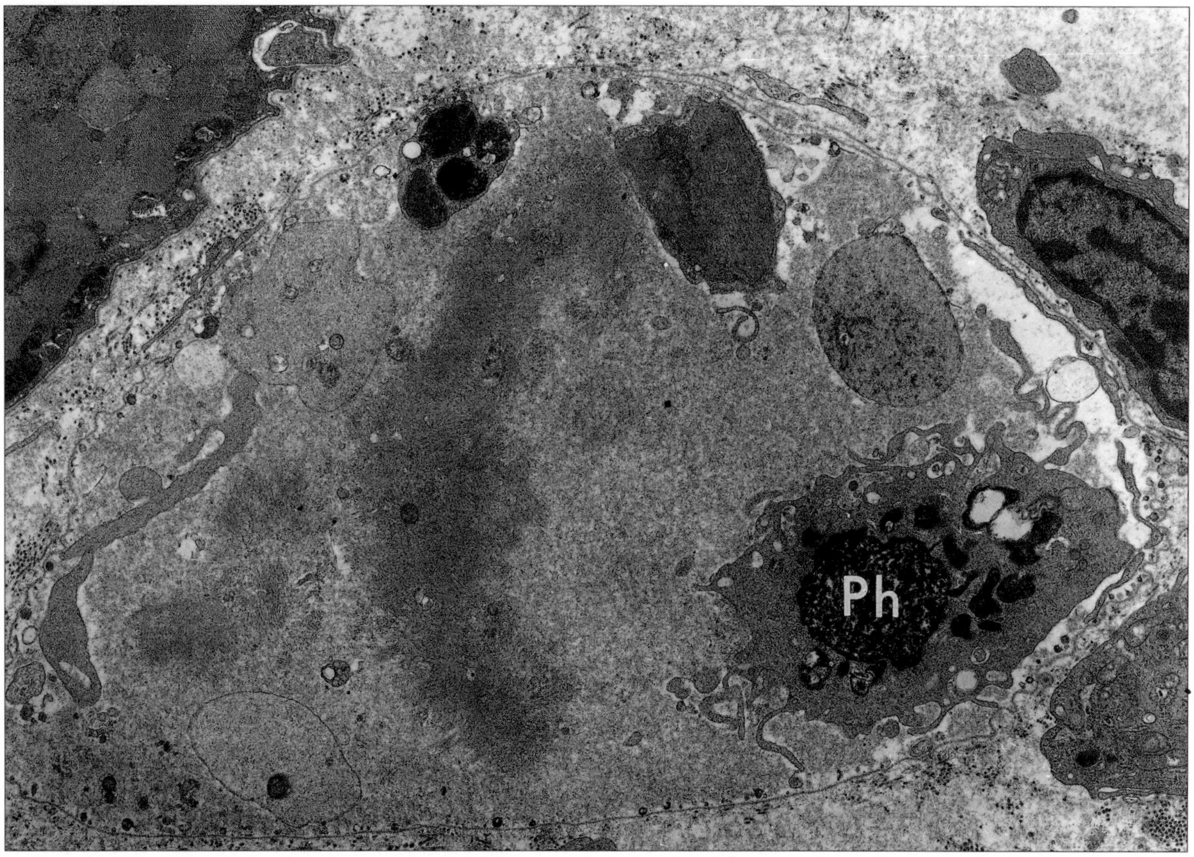

7.27 Necrotic fibres lack a plasma membrane but have an intact basal lamina. Phagocytic cells (Ph) invade the necrotic sarcoplasm and ingest the filament debris. (×10000)

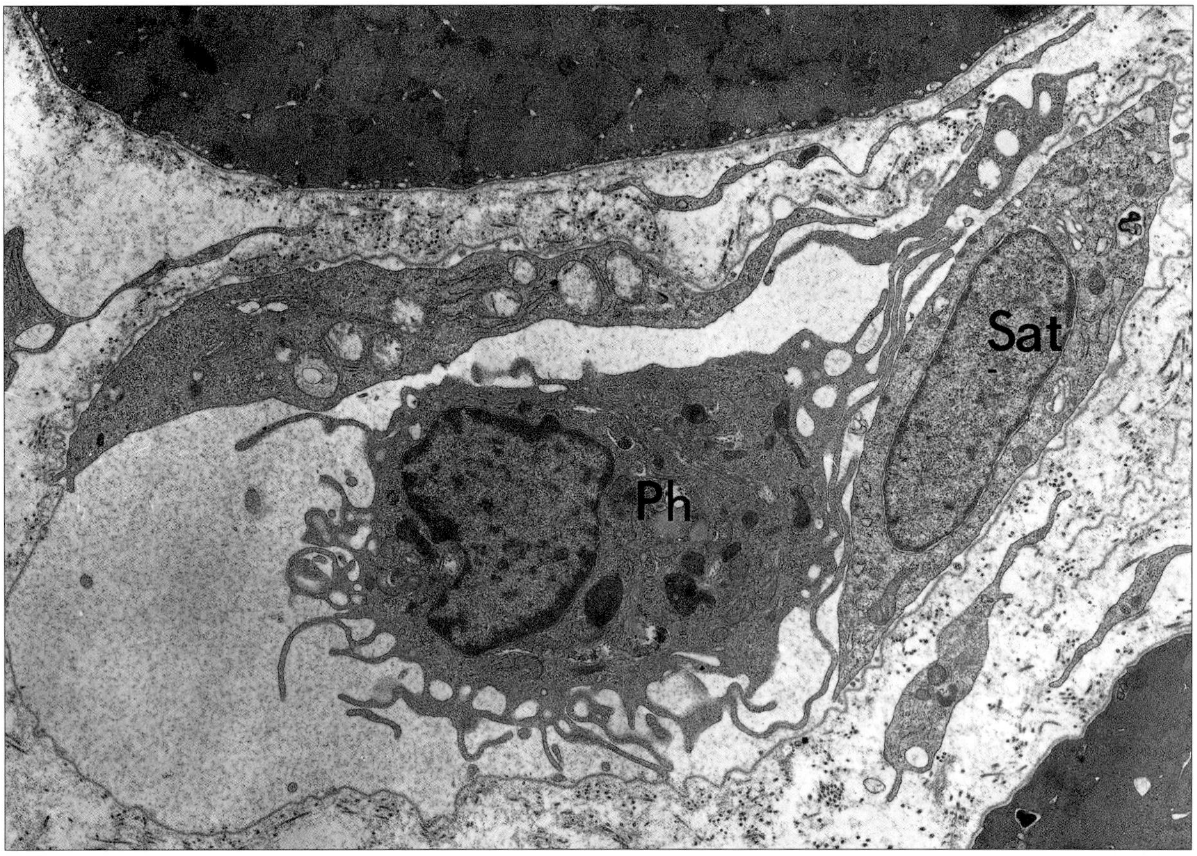

7.28 Monocytic cells with a smooth outline, which are possibly satellite cells (Sat), appear to survive the breakdown of the myofibrils and other organelles. They can be found within basal lamina tubes in which phagocytes (Ph) are still present. ×11250)

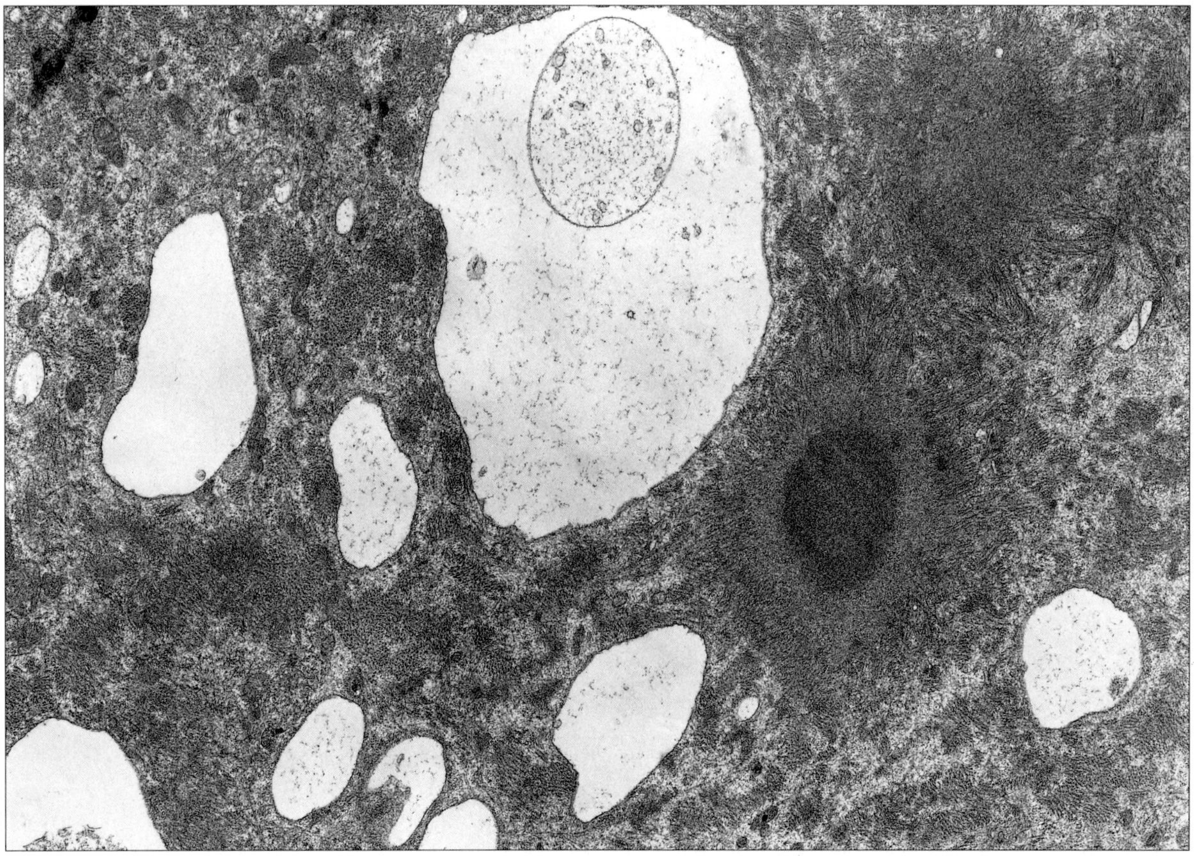

7.29 Vacuolation is a common feature of myositis. The origin of the vacuoles is unidentified. Some may contain stored material. (×12500)

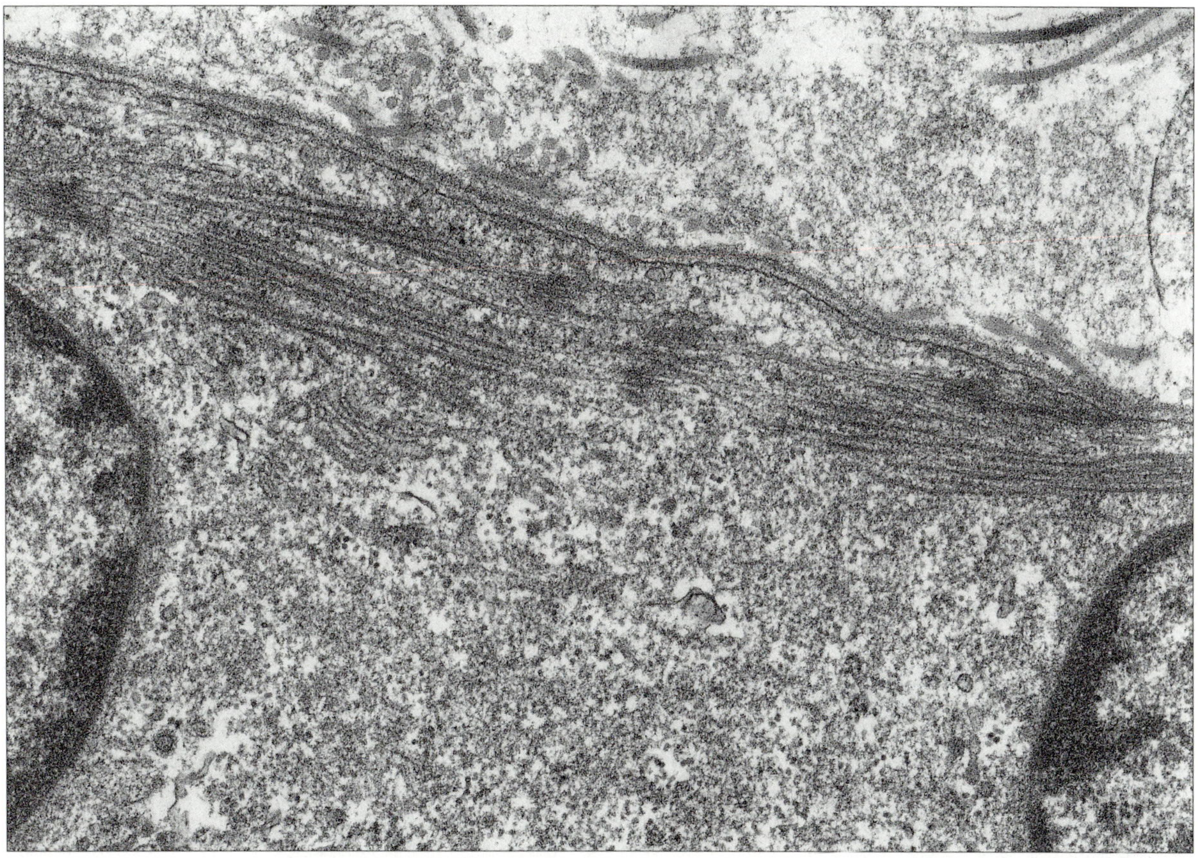

7.30 In regenerating fibres, filaments form in the cytoplasm at the periphery of the myocytes, amongst the ribosomes and sarcoplasmic reticulum (see also **5.12**). (×37500)

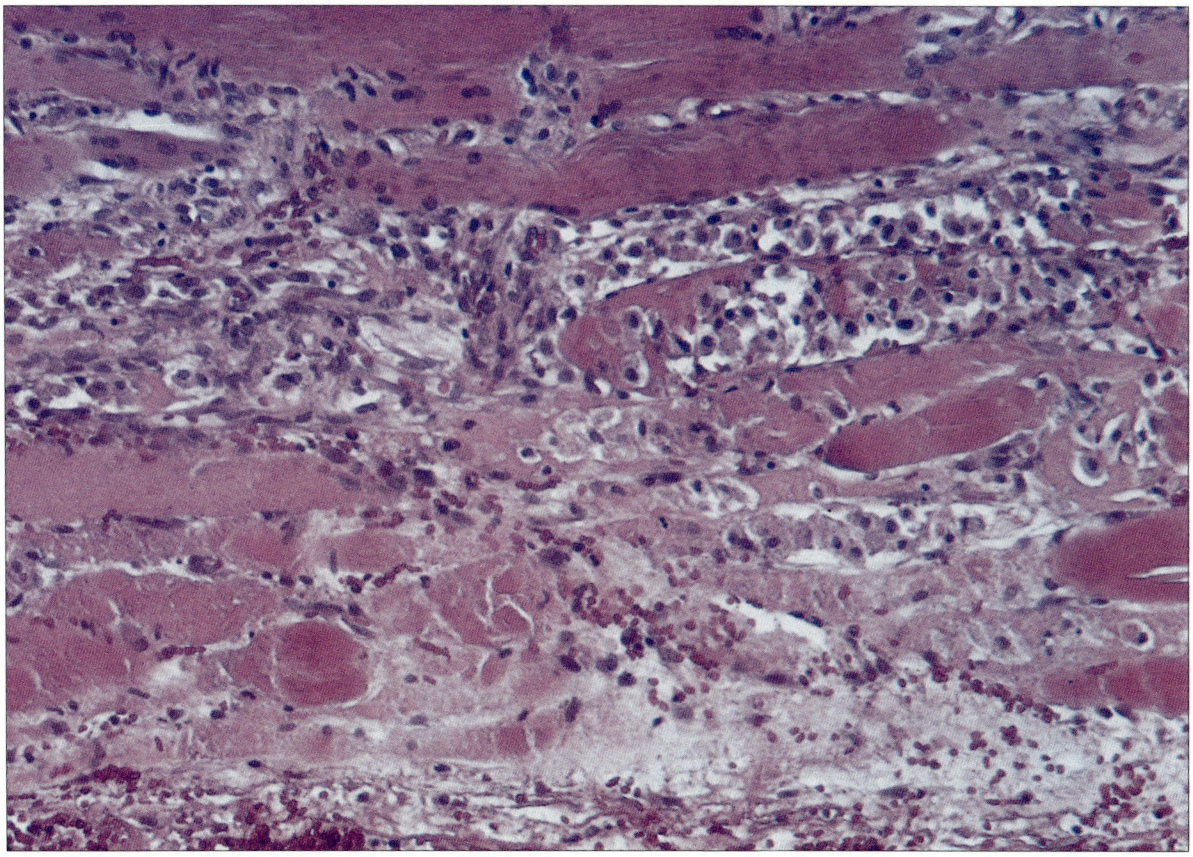

7.31 Longitudinal section, showing inflammatory infiltrate and focal fibre necrosis. (H&E, ×236)

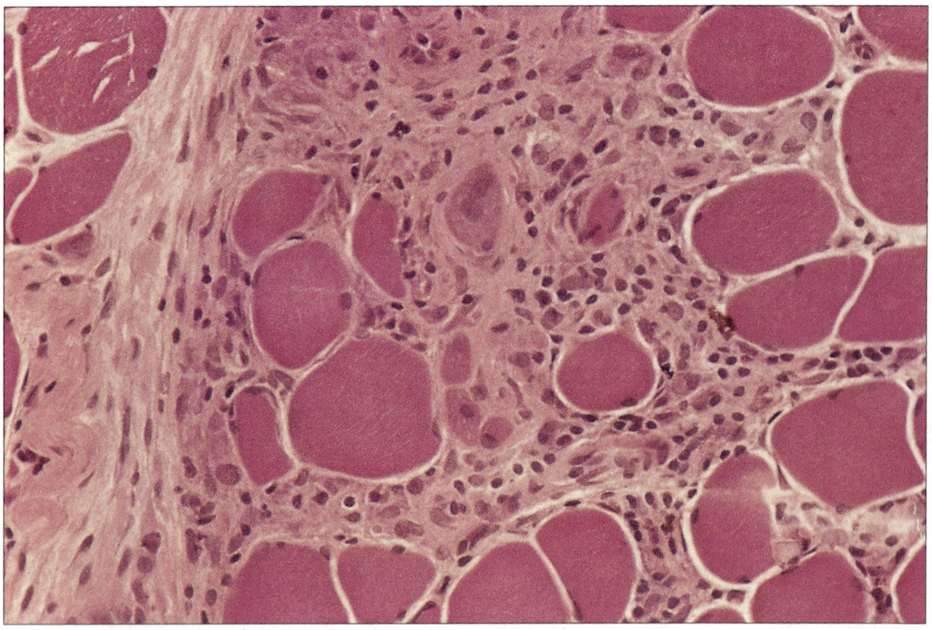

7.32 Regenerating fibres indicated by vesicular nuclei amongst an area of inflammatory infiltrate and necrosis.
(H&E, ×280)

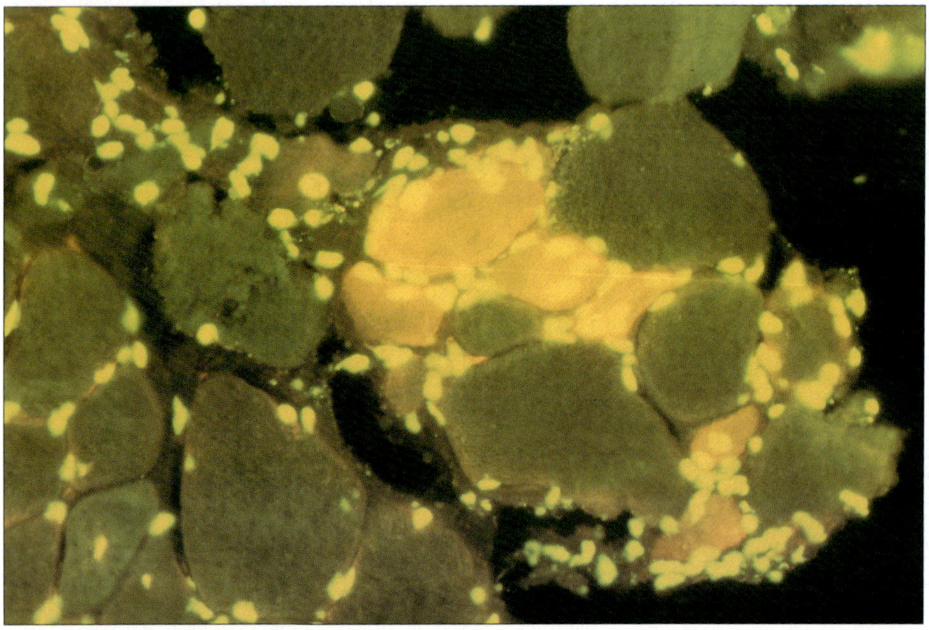

7.33 Regenerating muscle fibres are indicated by their high content of RNA, as shown by their orange staining cytoplasm. The nuclear DNA of inflammatory cells appears bright yellow.
(Acridine orange fluorescence, ×263)

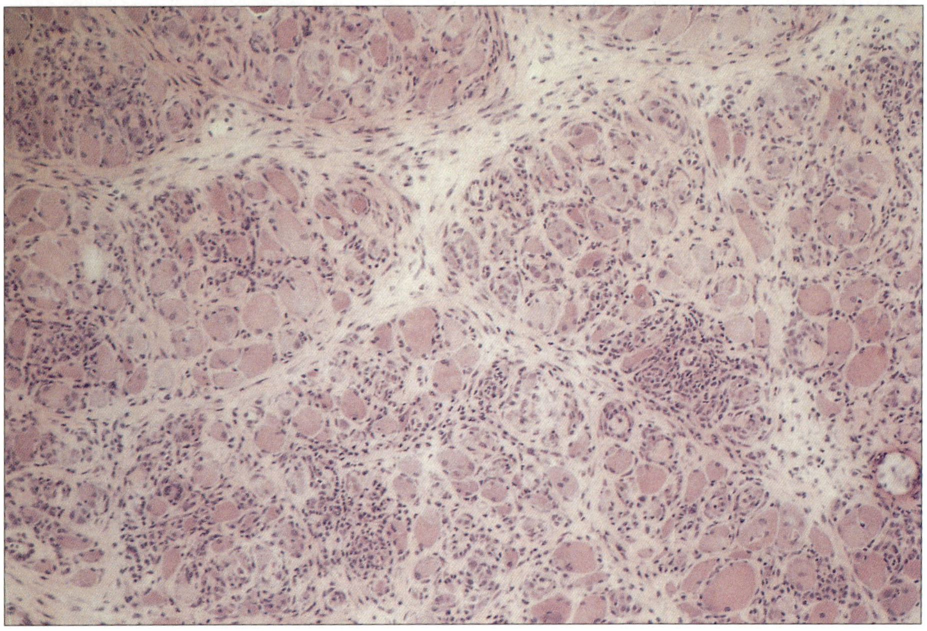

7.34 Localised nodular myositis, showing a massive inflammatory infiltrate, myofibre necrosis, phagocytosis and regeneration.
(H&E ×88)

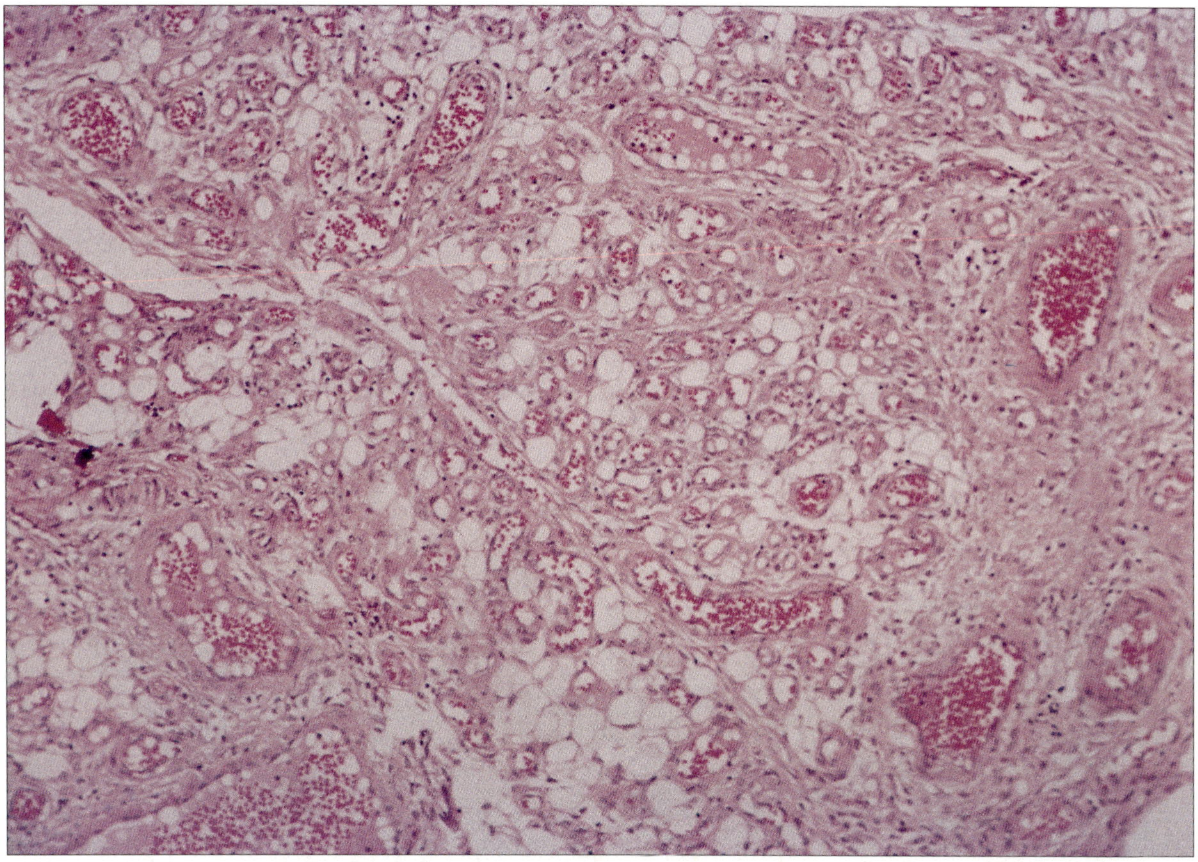

7.35 It is difficult to identify myofibrillar material in this section of end-stage, localised, nodular myositis. Hence, the previous erroneous description as 'venous haemangioma of muscle'. (H&E, ×189)

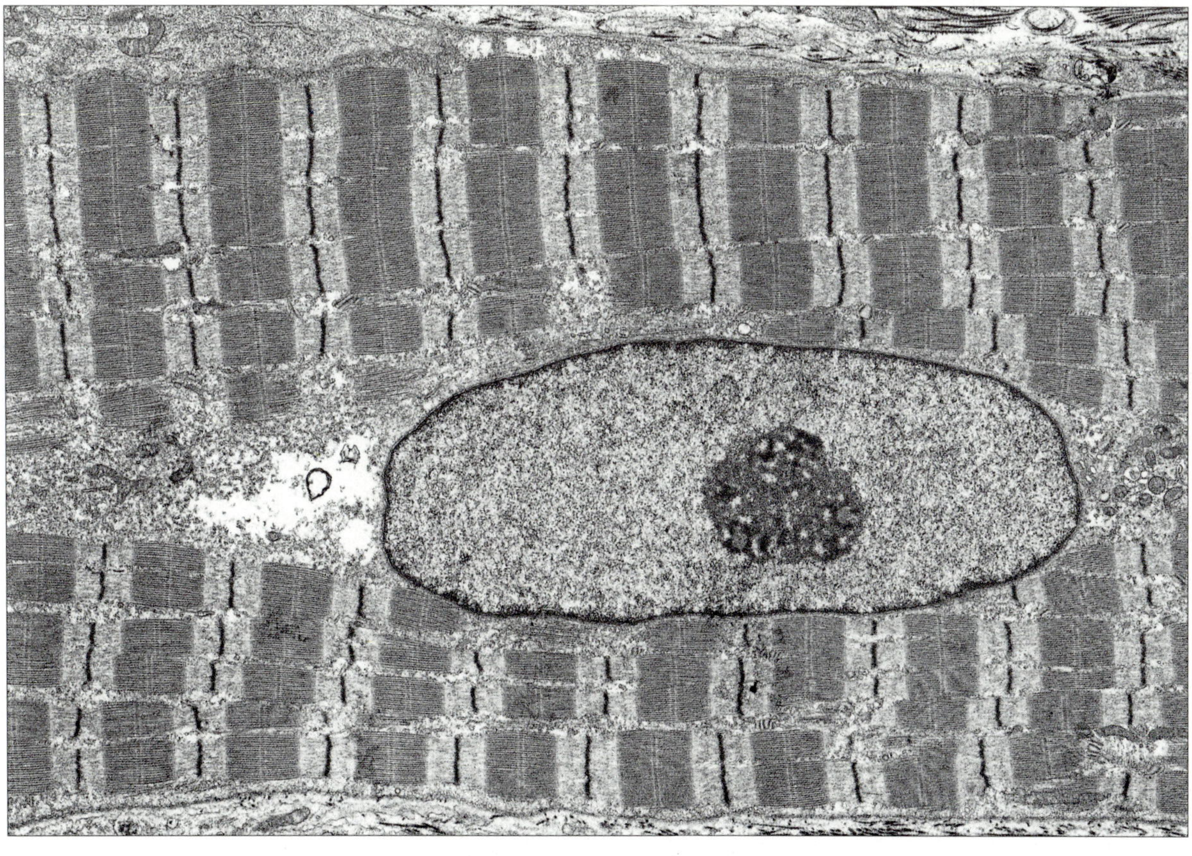

7.36 The nuclei in regenerating fibres lie within the centre of the fibre, and may occur in rows. They are rounded in outline, with prominent nucleoli. At the poles, Golgi apparatus, mitochondria and ribosomes may be seen. (× 8750)

Immunocytochemistry and inflammatory myopathy

The application of immunocytochemical techniques to inflammatory myopathy has greatly expanded knowledge of the aetio-pathogenesis of these conditions. Using the appropriate markers, it is now possible to identify, the constituent cells of the infiltrates (*Table 7.1*).

In PM, the predominant cell type has been shown to be CD8+ (suppressor/cytotoxic T lymphocytes (**7.37**)). There is expression of MHC class I on the surface of muscle cells in the vicinity of the infiltrate.

In DM, both CD4+ (helper/inducer T lymphocytes (**7.38**)) and CD8+ cells are seen, with CD4+ predominating. In addition, B cells (CD22) are seen, and there is expression of C5b-9 (membrane attack complex) on the walls of blood vessels.

In PM/DM, the pathological features described above (inflammatory cell infiltrates, perifascicular atrophy, necrosis, phagocytosis and regeneration) typically occur together, although in varying degrees and stages. The overall appearance is almost pathognomic of these disorders. However, it must be emphasised that each of these features may be seen in isolation, reflecting the patchy nature of the condition.

Treatment of the patient prior to biopsy may reduce the frequency and distribution of the lesions to the point that, in some instances, the muscle biopsy appears normal.

There are no specific features in the myopathology of carcinoma-related PM/DM.

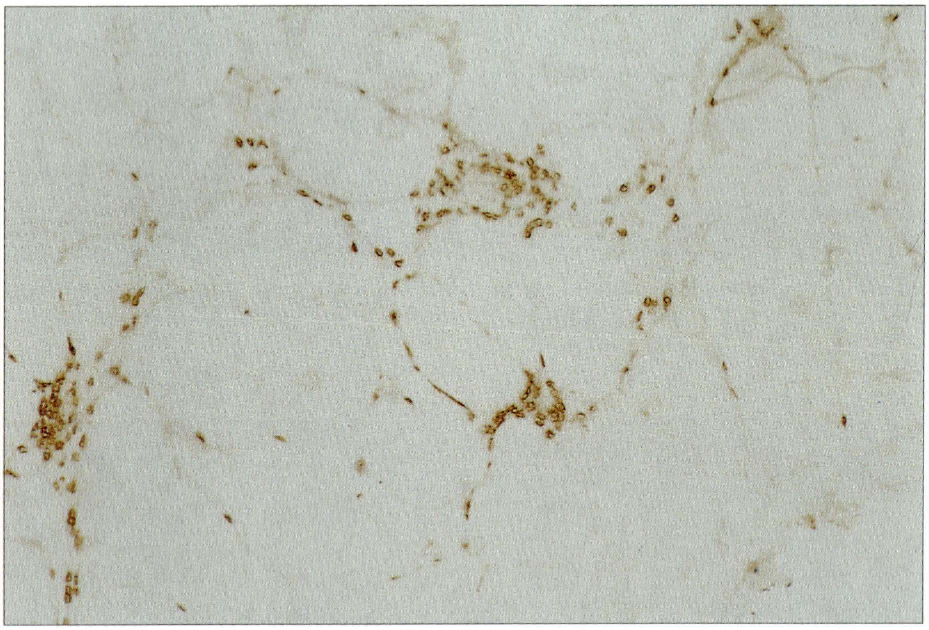

7.37 Inflammatory infiltrate showing CD8+ lymphocytes. (Immunoperoxidase anti-CD8+, ×115)

7.38 Inflammatory infiltrate showing CD4+ lymphocytes. (Immunoperoxidase anti-CD4+, ×115)

IBM

There have been a number of descriptions of a morphological variant of PM/DM in the last decade: so-called inclusion body myositis (IBM). This is now recognised as a separate nosological entity.

The cardinal pathological features are the presence of small rimmed vacuoles (**7.39–7.40**) with ultrastructural containment of fibrillary material (**7.41–7.44**). In addition, some muscle nuclei contain bizarre inclusions (**7.42–7.44**). The pathological features of PM/DM may be seen, varying from absent to prominent.

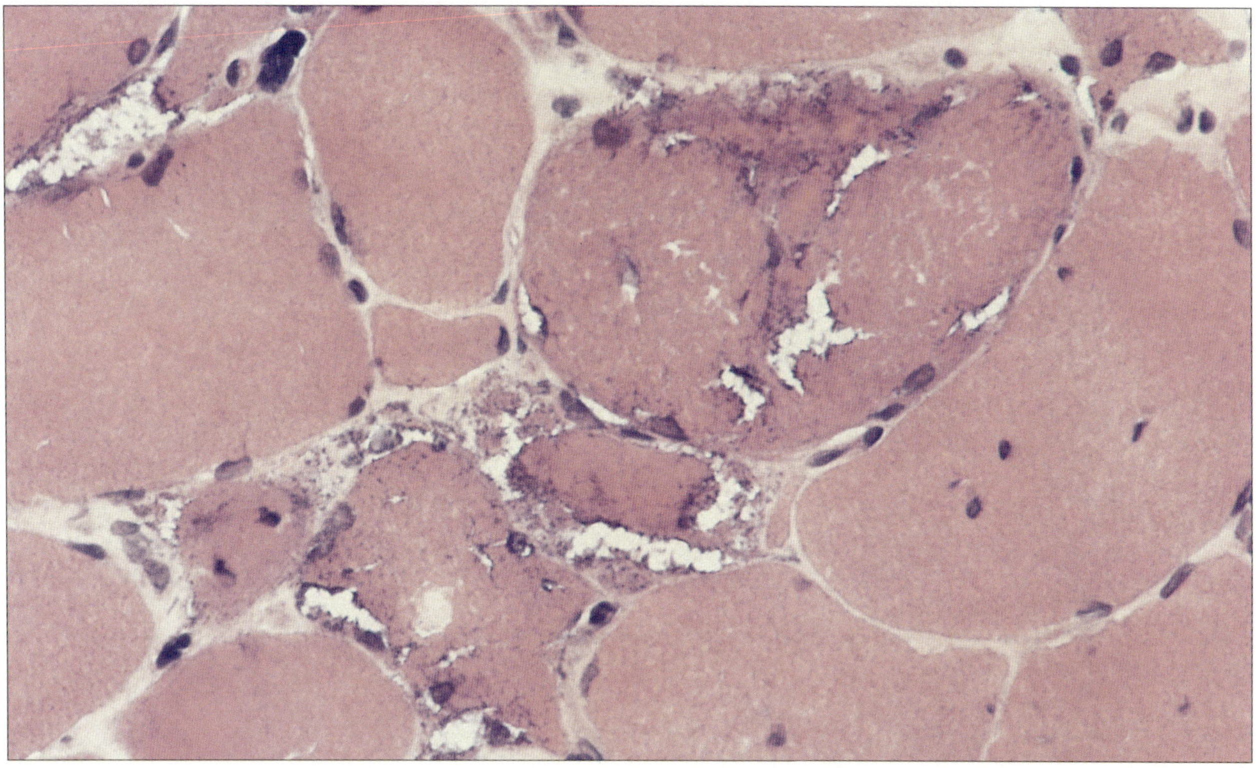

7.39 Rimmed vacuoles are present in some muscle fibres in this case of IBM. (H&E, ×590)

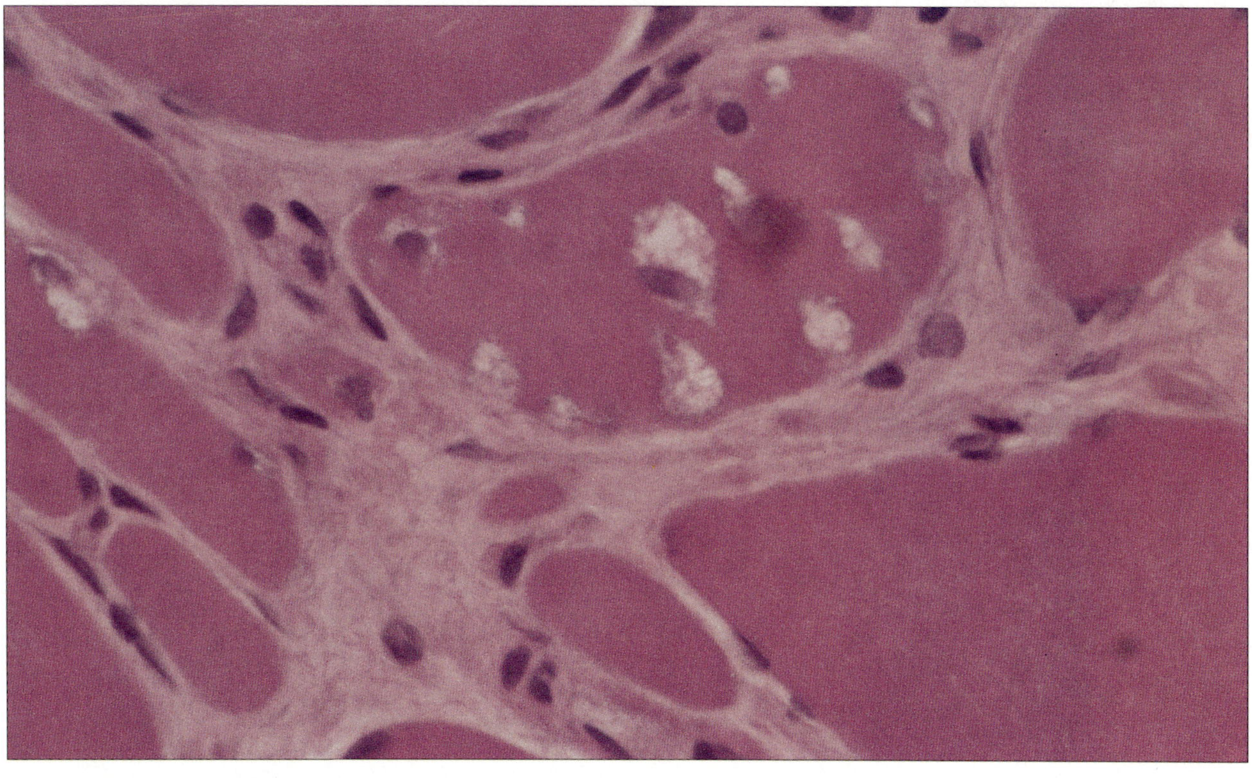

7.40 Vacuolar lesions in IBM that do not show the typical rimmed appearance. (H&E, ×614)

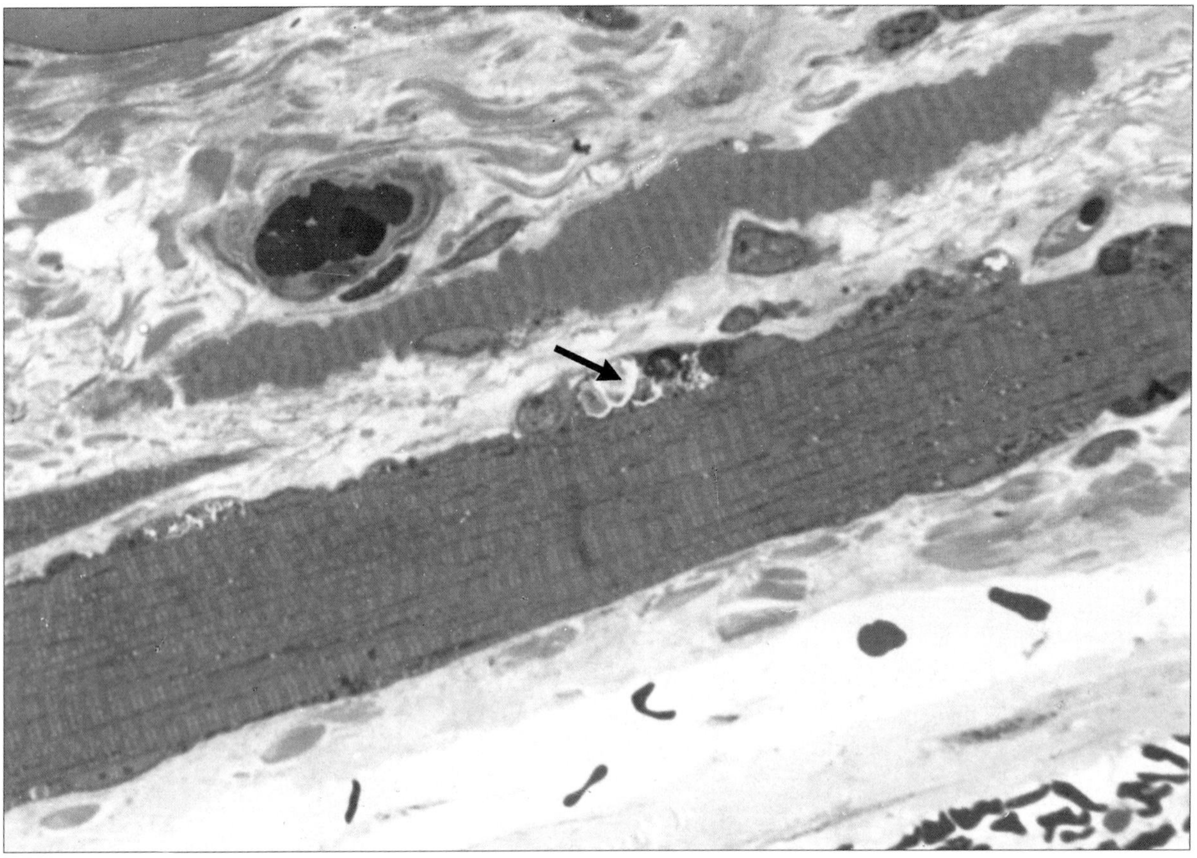

7.41a Intracellular inclusions (arrowed) may be identified in toluidine blue sections. Those illustrated are situated beneath the sarcolemma. (×1000)

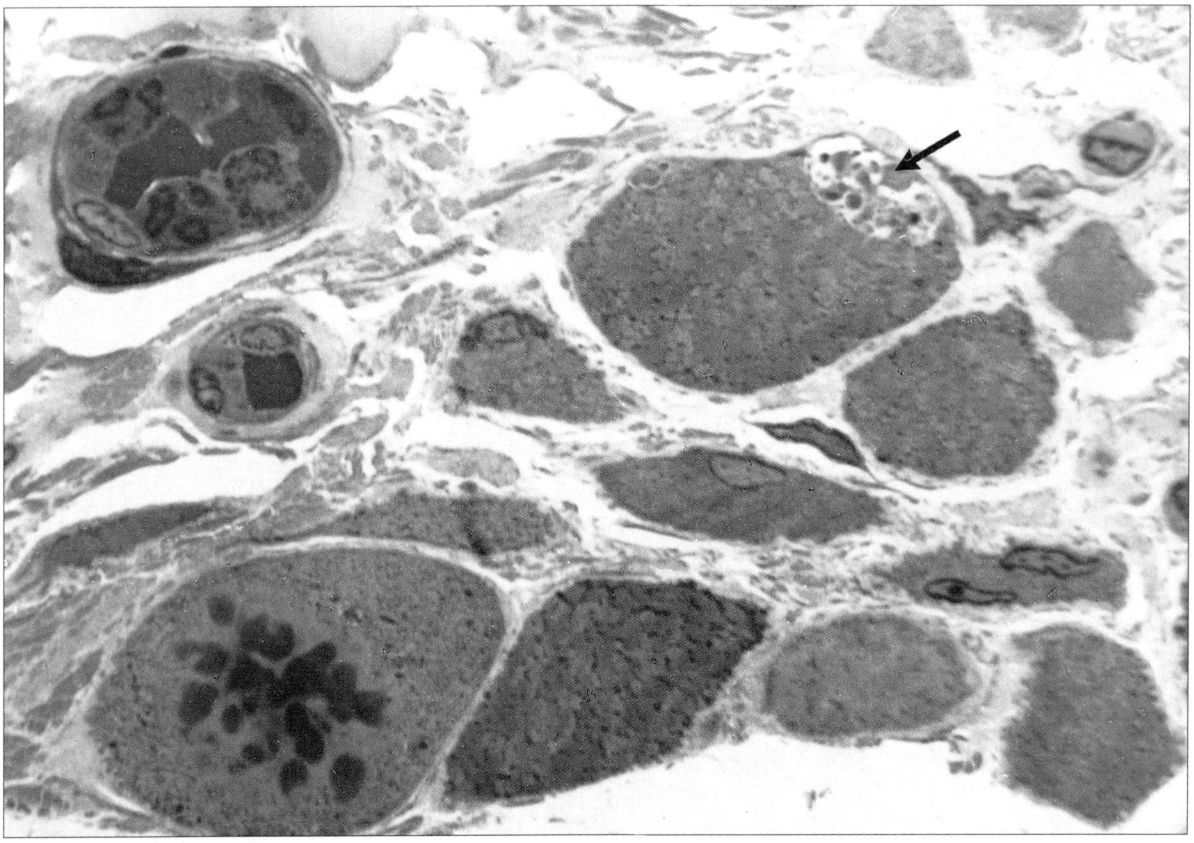

7.41b In transverse section, large sarcoplasmic inclusions may also be seen. Collections of subsarcolemmal vesicles (arrowed) and large darkly staining bodies within the centre of a fibre are visible. (×1200)

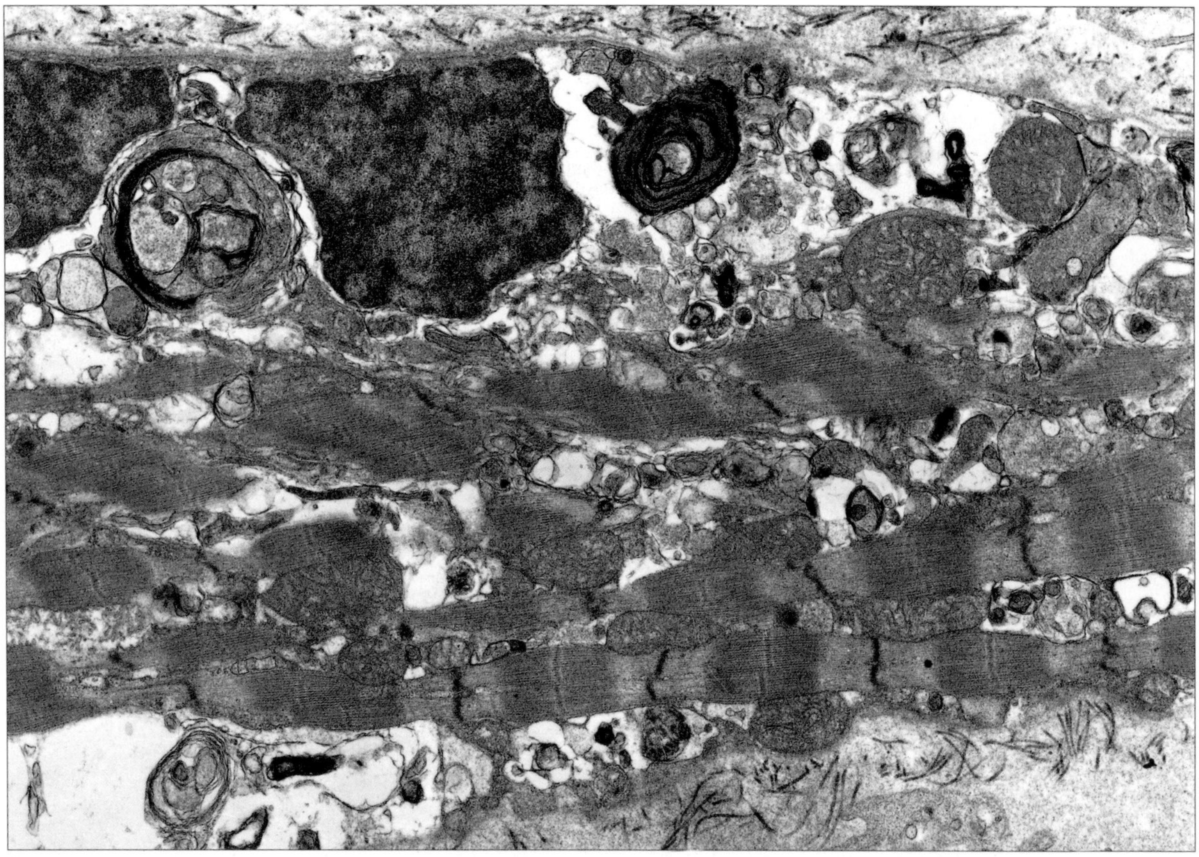

7.42 The subsarcolemmal inclusions are seen to consist of abnormal nuclei, mitochondria, and a wide variety of vesicular bodies and myelin figures. These, together with myofibrillar disruption, are changes that may be found in myositis. (×13750)

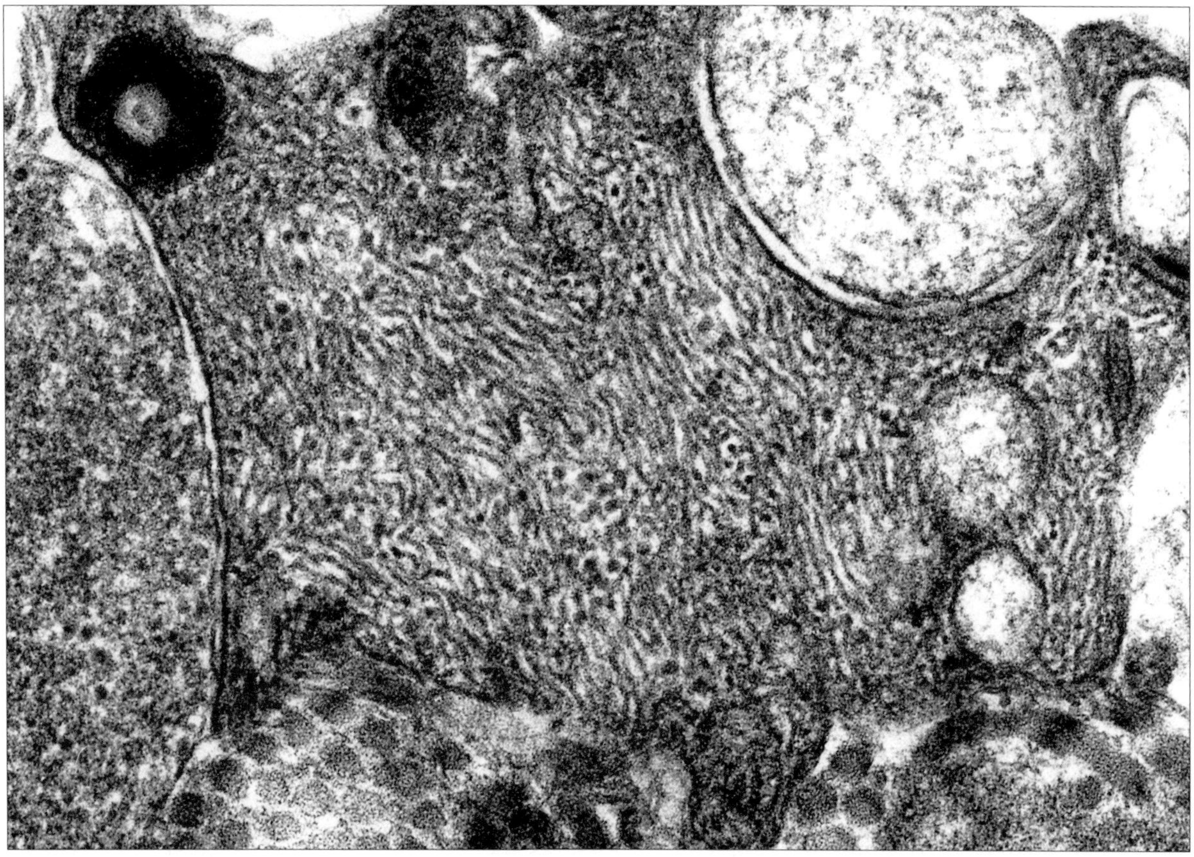

7.43 Filamentous inclusions, either within the cytoplasm or the nucleus, are considered pathognomic of IBM. (×120000)

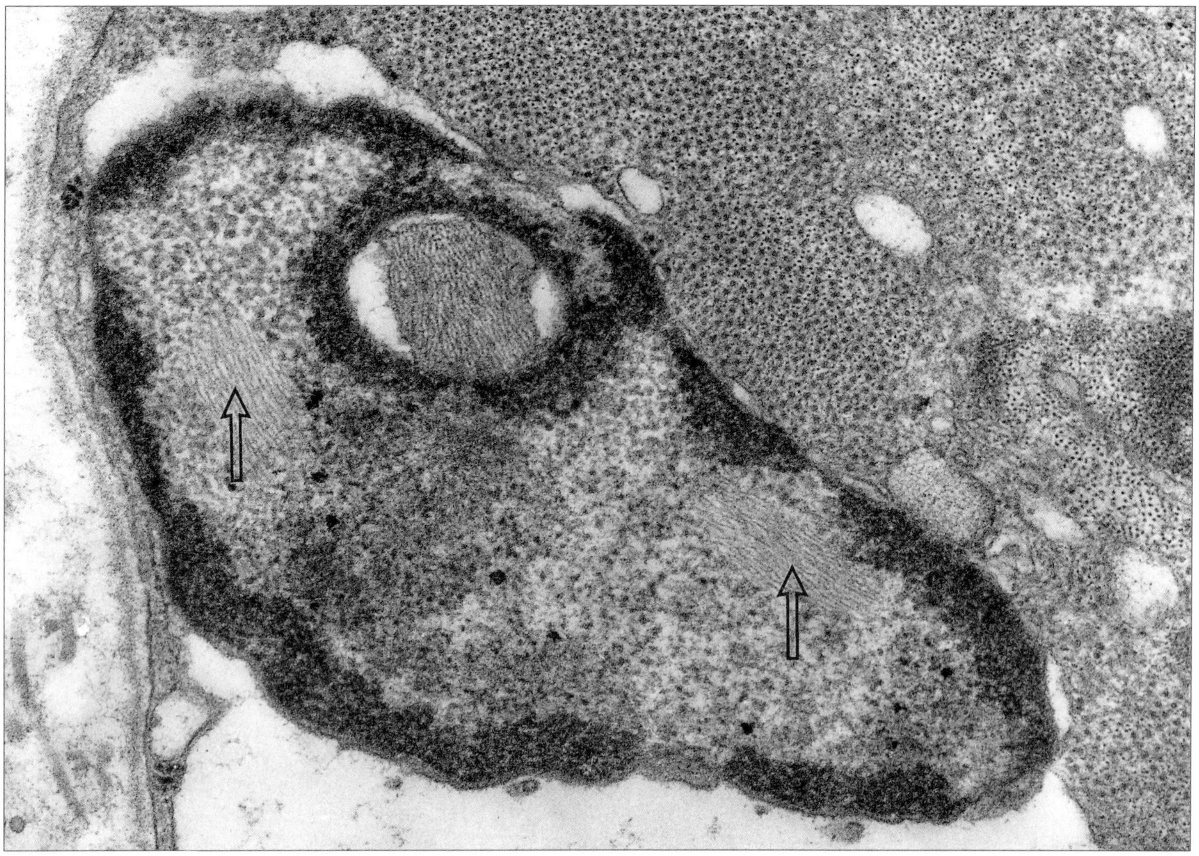

7.44 Filamentous inclusions (arrowed) may be found within the nucleus, which is normally apposed to the plasma membrane and the myofibrils. (×21000)

8

Congenital myopathies

The 'floppy infant' syndrome is a perennial problem for paediatricians generally, and particularly those interested in the investigation and management of children with neuromuscular disorders. In the majority of cases, the pathophysiological basis for their hypotonia, the cardinal clinical feature, remains obscure.

Nonetheless, numerous apparently specific myopathological entities have been described in association with infantile hypotonia, and the list of 'idiopathic' cases continues to shrink. Certainly, exclusion of neurological (cerebral and cerebellar) causes of hypotonia leaves only a small residue of cases with no demonstrable cause in terms of dysfunction of the neuromuscular apparatus.

Some of the more important causes of infantile hypotonia have been described in detail elsewhere (Chapter 6, anterior horn cell disorders and Chapter 9, the metabolic disorders). In this section, we describe the morphological basis for those conditions, which have been the subject of intensive investigation with both the light and electron microscope.

It is difficult to establish clinicopathological correlations with the same degree of precision as in other neuromuscular disorders. One should be aware that a few cases of congenital myopathy are not obviously 'floppy' at first presentation, although they are almost certainly hypotonic physiologically. These are the cases of so-called 'arthrogryposis multiplex congenita' where joint contractures, developing *in utero* fix their joints in the 'Buddha' posture. Some of these children have congenital muscular dystrophy (which is likely to be a non-specific category) and others have a variety of denervating processes, which may be either peripheral or central in origin. The various disease processes described are detailed in *Table 8.1*.

Pathology of the congenital myopathies

The severity of the clinical expression of a congenital myopathy appears to be related to the stage of development and maturation at which the defect occurs (**8.1**) (Fardeau, 1982).

Separation of the various, supposedly specific nosological entities described below has depended in its entirety on the detailed analysis of biopsy material from children with the 'floppy infant' syndrome. In only a few instances has this been accomplished using orthodox histopathological techniques. The application of both enzyme histochemistry and electron microscopy has been essential in most cases.

Given this approach, and the fact that some of these disorders have useful clinical correlates outwith the neuromuscular apparatus (see *Table 8.1*), it should be possible to make a precise diagnosis in most floppy infants in whom central nervous disease has been excluded.

It has been estimated that there will be some 20% of affected children undiagnosed after the application of these techniques. In about half of these, the use of the appropriate antibody staining techniques demonstrate the presence of abnormal light and heavy myosin chains, and other structural proteins (e.g. desmin), to account for their hypotonia.

It is probable that developments in the fields of molecular embryology and genetics will further the elucidation of the aetio-pathogenesis of these disorders (e.g. X-linked centronuclear myopathy which is located at Xq28, central core disease located at 19q13.1 and nemaline myopathy located at 1q21.23).

Table 8.1 Clinicopathological correlation in congenital myopathies

Disease entity	Clinical features (other than hypotonia)
Congenital muscular dystrophy	Joint contractures
Congenital fibre type disproportion (CFTD)	Respiratory insufficiency
Centronuclear (myotubular) myopathy	Eye movement disorders
Nemaline myopathy	Gross hypotonia in early infantile cases with cardiorespiratory insufficiency. 'Marfanoid' body habitus in adolescent and adult cases
Core myopathies	Nil
Type I fibre hypotrophy	Nil
Intra-uterine maturational delay	Nil

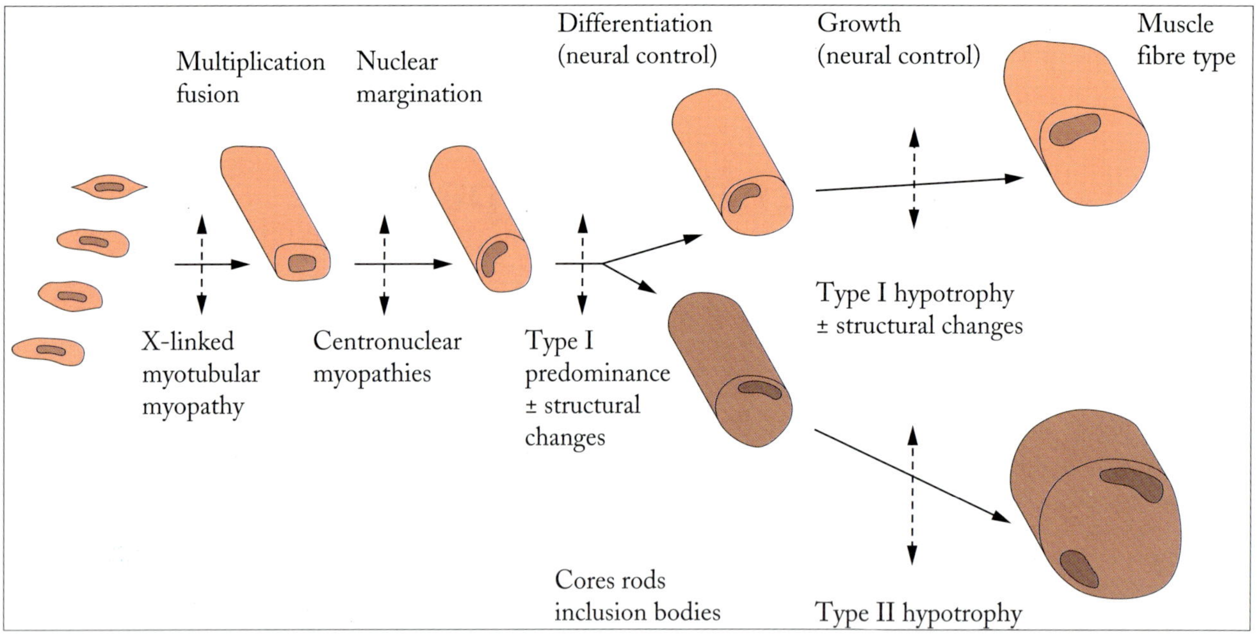

8.1 Operational diagram of the pathophysiology of the congenital myopathies. The different developmental phases of the myofibres are schematically separated. Structural and histochemical changes characteristic of the congenital myopathies may be related to impairment of these phases. However, the mechanism responsible is not fully understood. X-linked myotubular myopathy may be due to an abnormally small pool of myoblastic cells, but there is also lack of margination of the nuclei and reduced growth of the myofibre. In the centronuclear myopathies, margination of most of the muscle nuclei is impaired, myofibre growth is reduced, but fibre type differentiation is also often abnormal. In core and rod myopathies, impairment, with or without structural changes of the myofibre, may be caused by functional, segmental, and motor neuronal changes. Impairment of Type II fibre growth is mainly due to a supraspinal nervous dyfunction; structural changes of the myofibres are exceptional in this latter case.

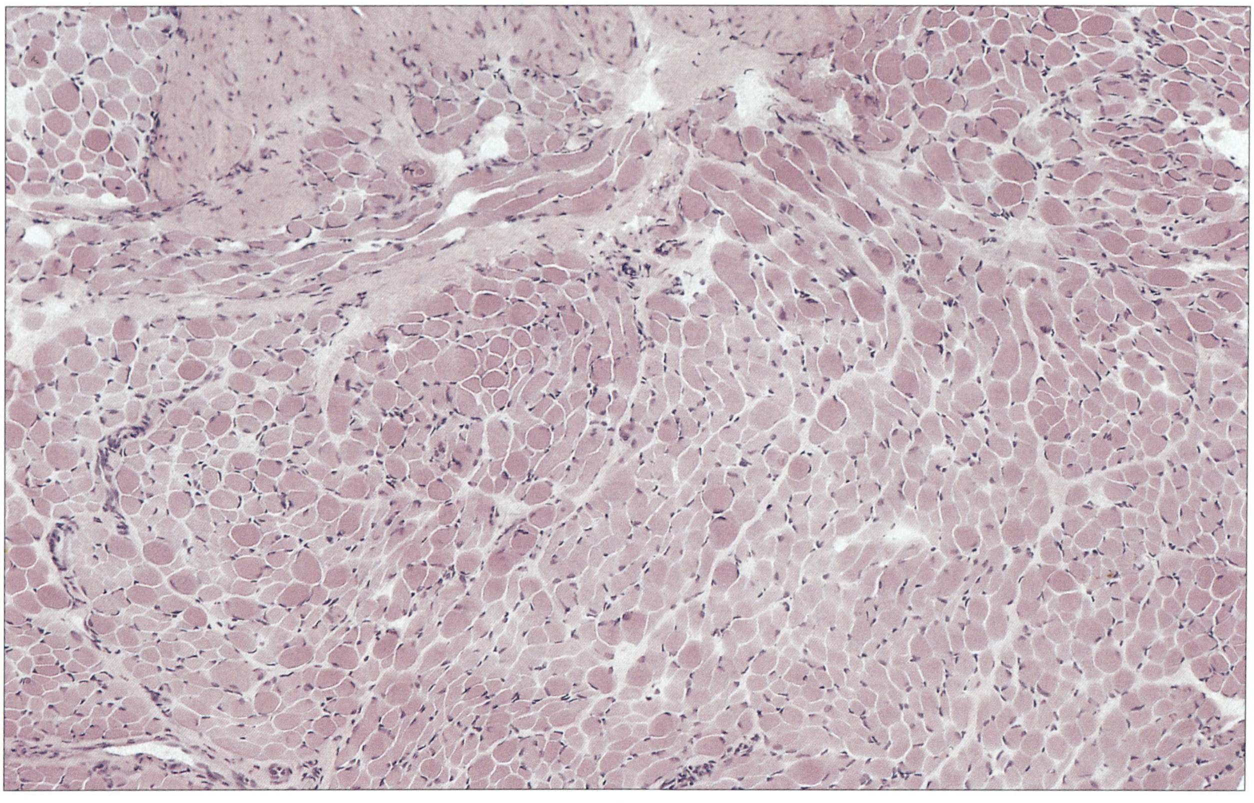

8.2 Congenital dystrophy of unspecified type, showing increased connective tissue, increased variation in fibre size and necrosis. (H&E, ×40)

Congenital muscular dystrophy

This disorder, or group of disorders, are so-named because of the histological appearance of skeletal muscle from affected infants (**8.2**). There is no absolute certainty that we are dealing with a single nosological entity, as the clinical expression of congenital dystrophy seems to vary significantly from case to case, with some families showing evidence of central nervous system involvement (Fukuyama type).

Nevertheless, there is no denying the 'dystrophic' appearance of the biopsy material in these cases (**8.2**), although the chronic features readily visible are not accompanied by the active necrobiotic changes seen in DMD. Indeed, the histopathology of CMD is reminiscent of late stage DMD (age 9 years onwards, see Chapter 5), except that immunolabelling for dystrophin shows that the fibres have a normal content and distribution. Paradoxically, the prognosis in these children is quite good in the long term, provided they survive early infancy with its risk of ventilatory dysfunction associated with hypotonia.

Congenital fibre type disproportion

Congenital fibre type disproportion was described initially as an entity in its own right, manifesting clinically as infantile hypotonia associated with ventilatory dysfunction, which may be severe in some cases. The highly characteristic morphological abnormality is best seen in myosin ATPase preparations (**8.3**). The gross discrepancy between the size of the groups of tiny Type II fibres and the much larger Type I fibres is clear.

The functional significance of this anomaly is unclear, but there is some evidence that it is a nonspecific reaction to abnormal motor unit innervation in very young children (**8.1**). Repeat biopsies, taken months or years later, may show quite clearly that these children are suffering from a form of SMA.

The biopsy in **4.2**, shows characteristic 'grouped' atrophy, and was taken from the mother of the case illustrated in **8.3**.

Centronuclear myopathy

The light microscopic appearance of biopsy material from this group of disorders is highly characteristic, with 80% or more of the fibres in any one low power field contain internal nuclei that are surrounded by a 'halo' (**8.4a**, **8.5a**). The gross interstitial fibrosis, which is another characteristic of the condition, is also illustrated.

The internal location of the myonuclei is obvious in the ATPase preparation (**8.4b**, **8.5b**), as is the loss of myofibrillar material, which is confirmed by ultrastructural studies (**8.6a–b**).

The NADH-Tr preparations reveal dense central deposits of reaction products (**8.5c**). These fibres often show radial spokes in NADH-Tr (**8.5c**) and desmin (**8.5d**) stained sections.

Electron microscopy shows that the space between the centrally-placed nuclei is occupied almost entirely by mitochondria.

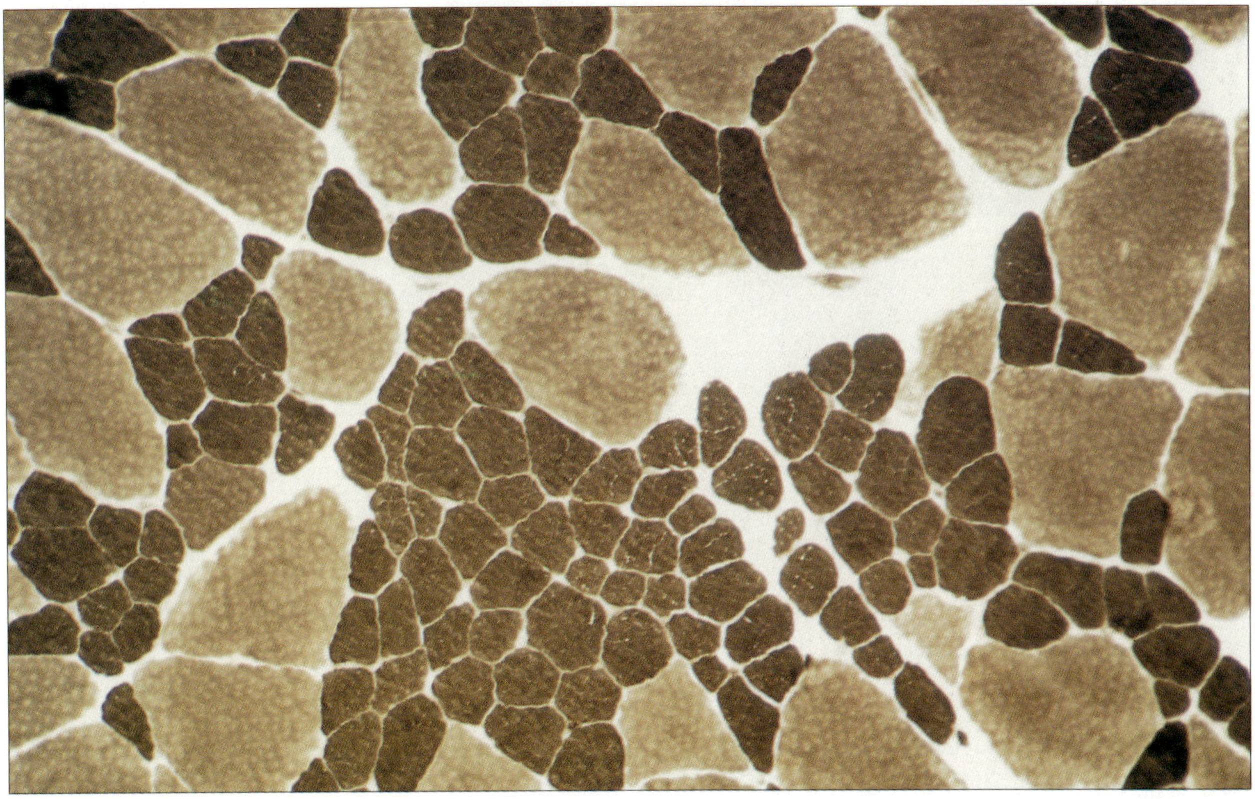

8.3 Congenital fibre type disproportion, showing marked hypotrophy and excess of Type I fibres. (Myofibrillar ATPase, pH 4.35, ×236)

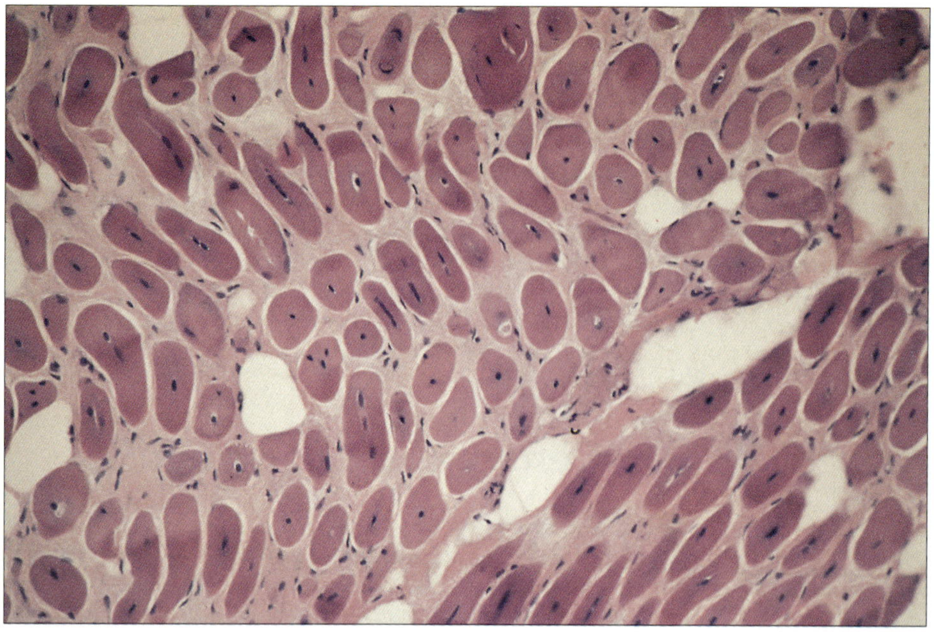

8.4a Centronuclear myopathy showing excess connective tissue and small rounded muscle fibres with centrally located chains of nuclei.
(H&E, ×175)

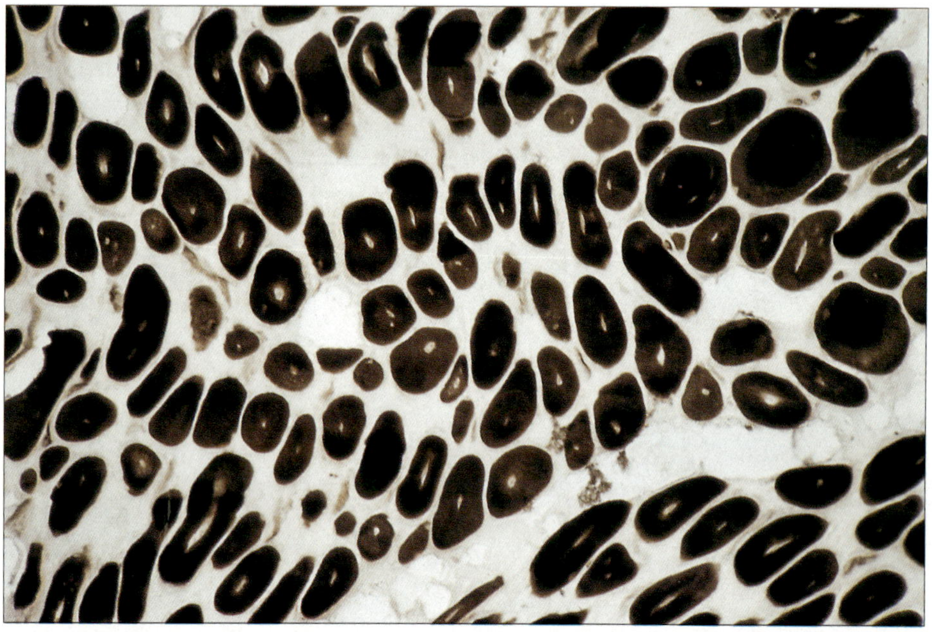

b) Sequential section to that of **8.4a**, showing the lack of fibrils in the central cores occupied by nuclei, and a lack of fibre type differentiation.
(Myofibrillar ATPase, pH 4.35, ×175)

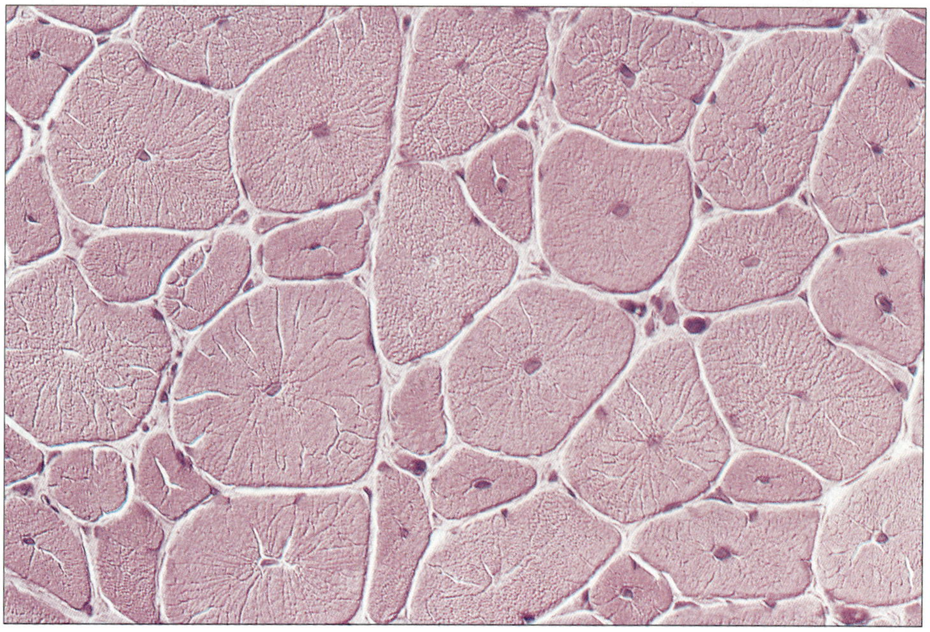

8.5 Adult centronuclear myopathy.
a) H&E, ×263;

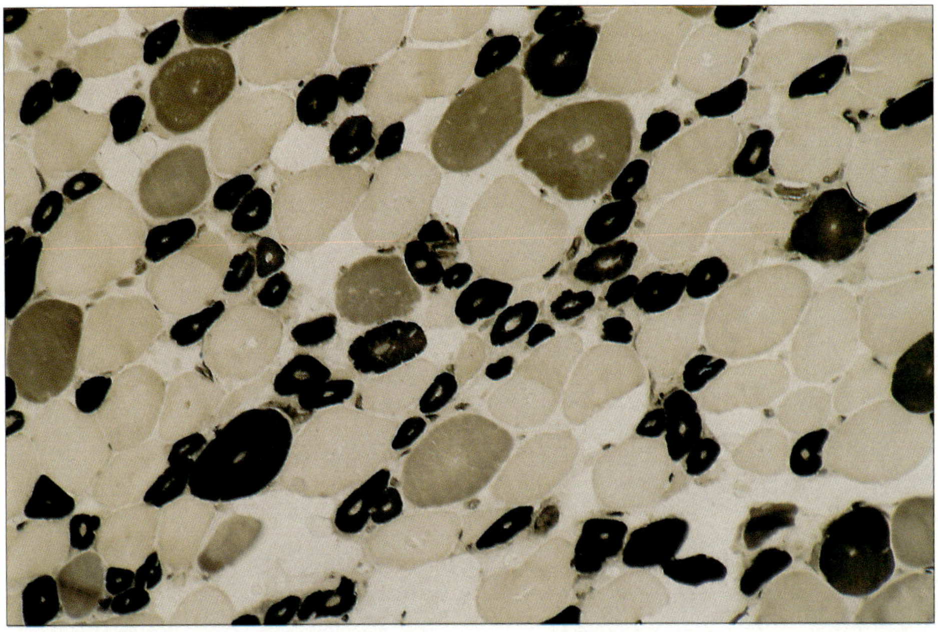

b) ATPase, pH 4.35, ×105;

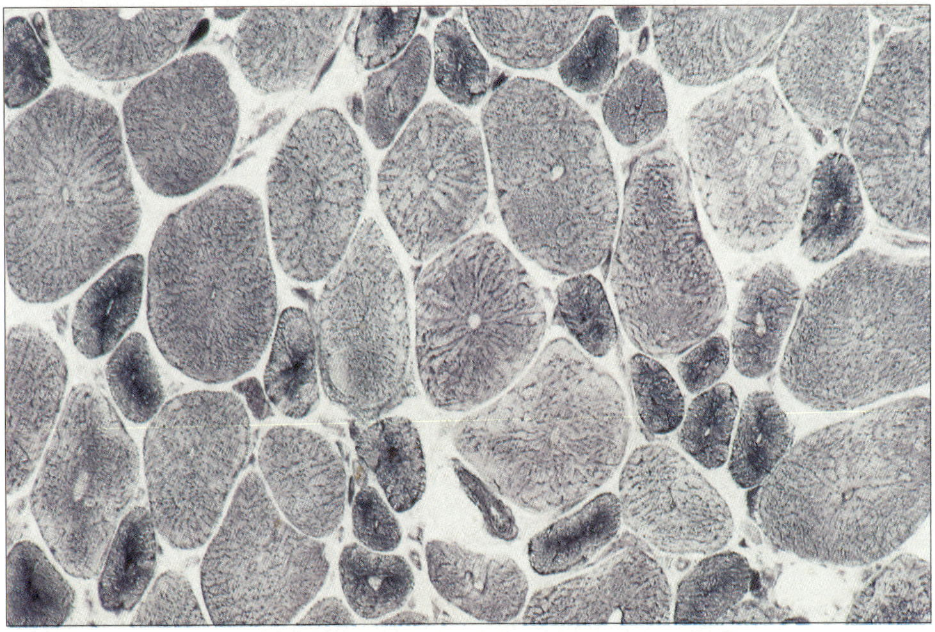

c) NADH-Tr, ×210;

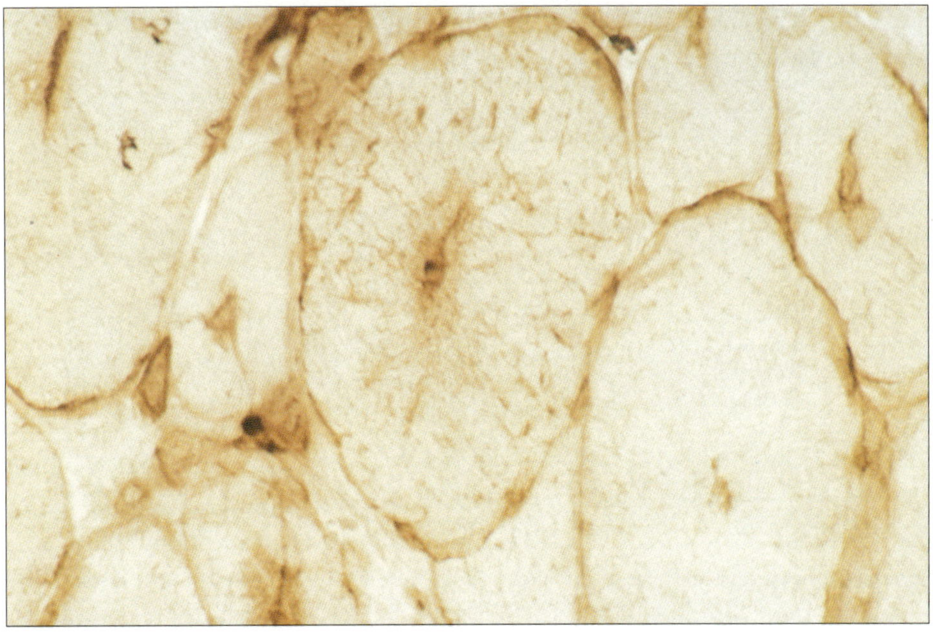

d) desmin, ×210.

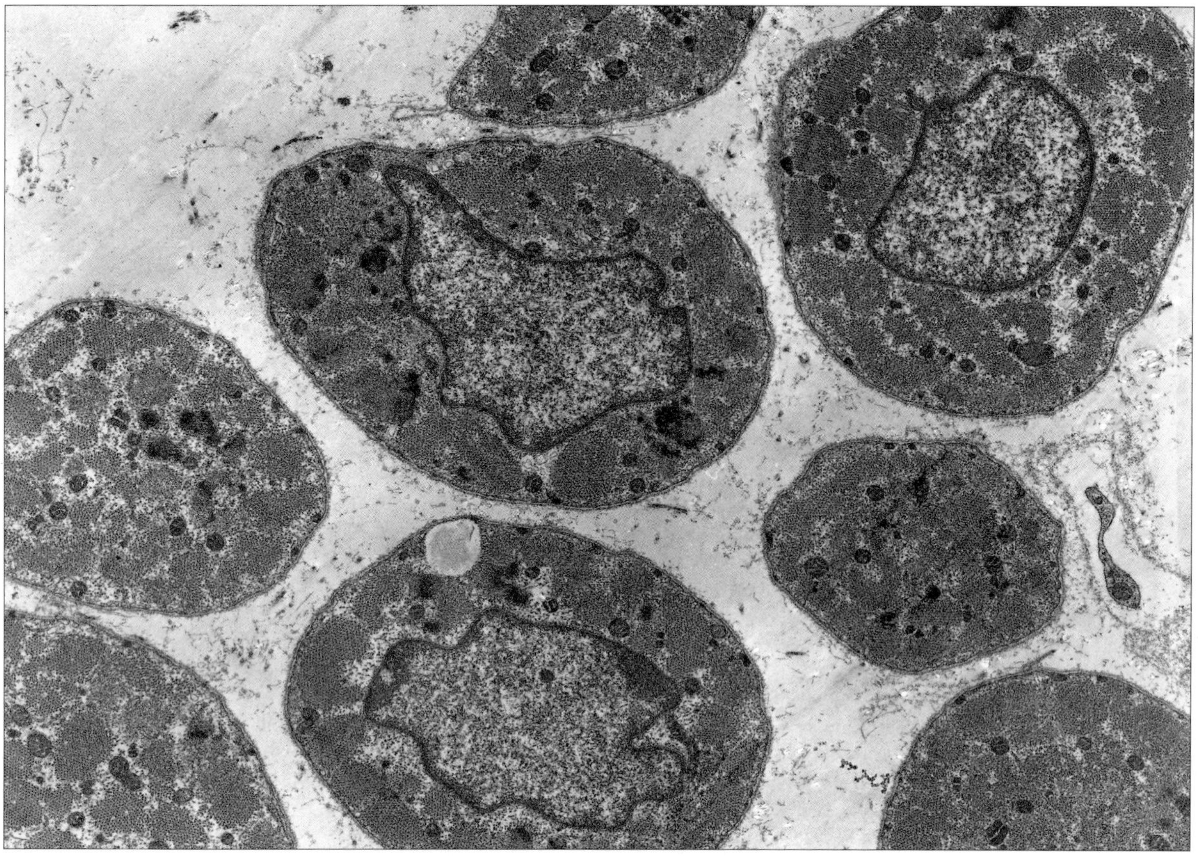

8.6a In centronuclear myopathy, the nuclei are situated in the centre of small rounded muscle fibres. The chromatin appears dispersed, and stains more homogeneously than normal resting nuclei. (×12500)

8.6b In longitudinal section, myofibrils are present on both sides of the nuclei. Small nucleoli may be seen, but little dense chromatin is visible. (×12500)

Nemaline myopathy

The rod-like bodies from which this myopathy derives its name may be difficult to detect in H&E stained sections, but are best shown by phase or interference contrast microscopy, where they appear as refractile bodies. The best screening stain for detecting rod bodies is the modified Gomori trichrome, in which they can be seen as clumps of dark red granules, usually located in the corners of small, angulated fibres (**8.7–8.8**) which are, without exception, Type I in profile.

The fine structure, and their derivation from the Z-disc in this condition are best seen with the electron microscope (**8.9–8.11, 4.70–4.72**).

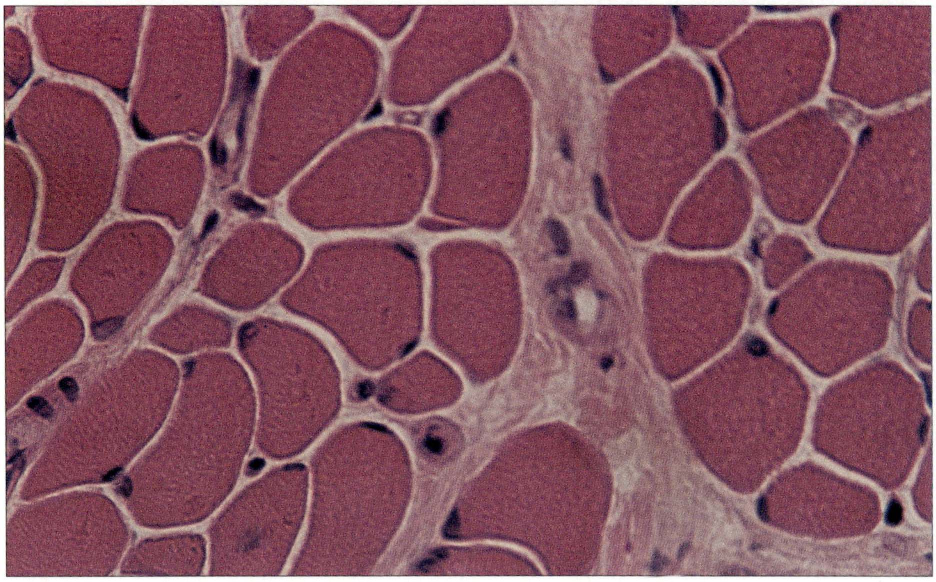

8.7 Abnormal filamentous material, in this case of nemaline myopathy, is only marginally visible with H&E staining, but can be easily recognised in the serial section shown in **8.8**. (H&E, ×298)

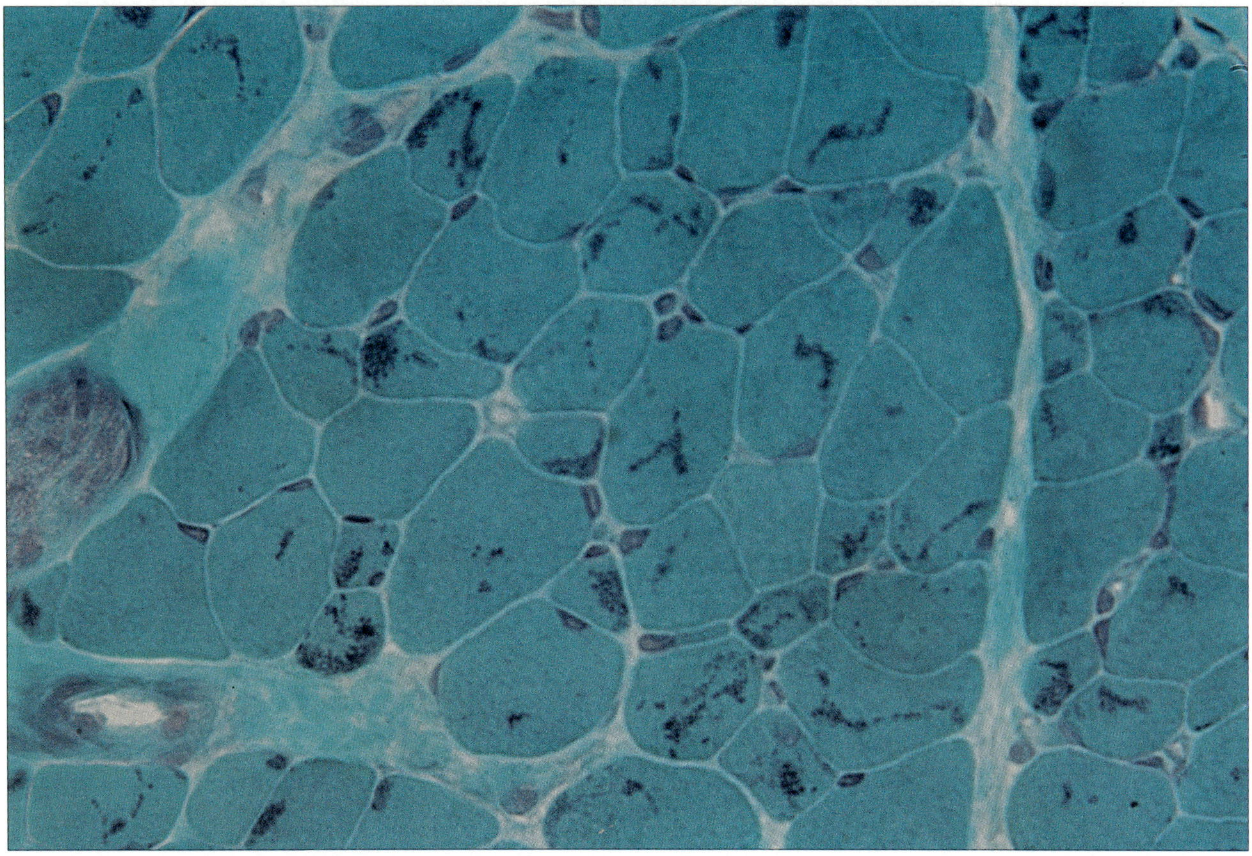

8.8 Nemaline myopathy illustrating nemaline rods. The nemaline rods are present in most fibres of this biopsy, and are easily recognised by their dark staining appearance. (Modified Gomori trichrome, ×378)

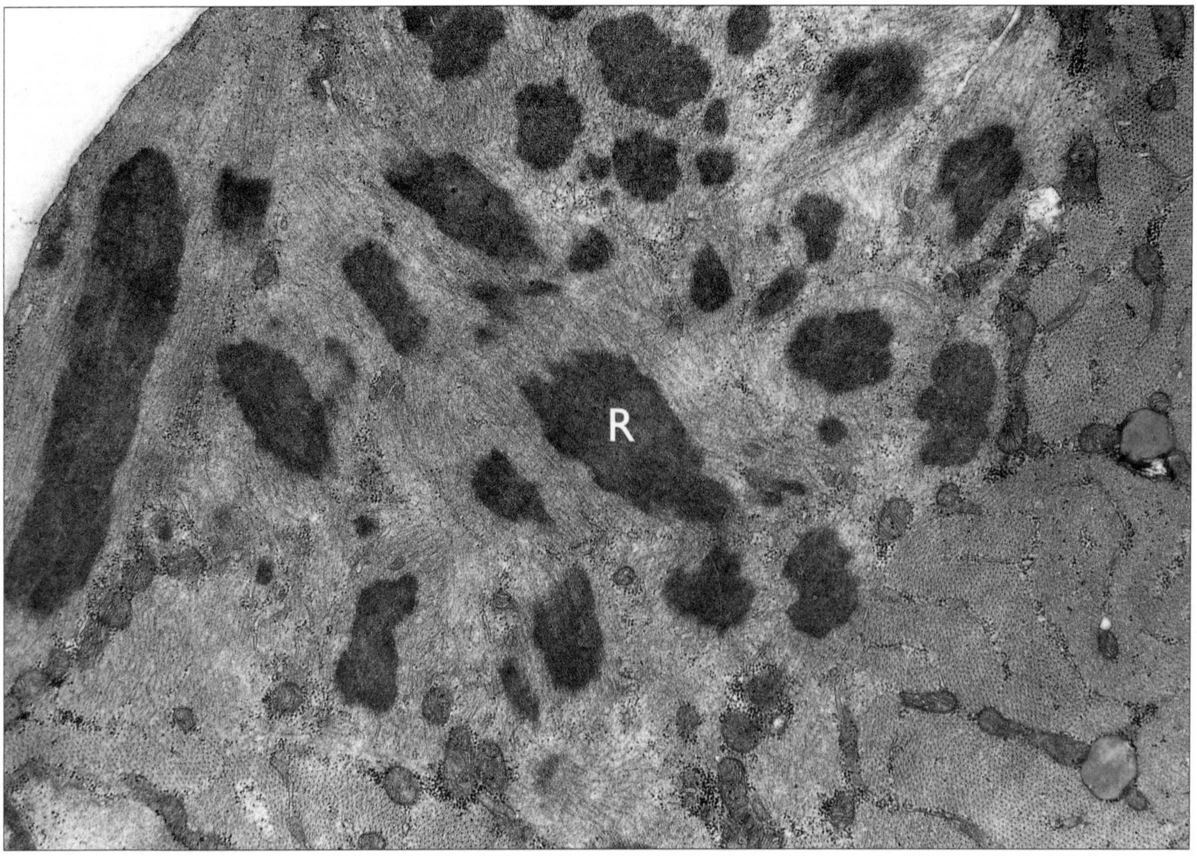

8.9 Nemaline rods (R) may be found as randomly orientated subsarcolemmal aggregates of dense irregular bodies. (×9600)

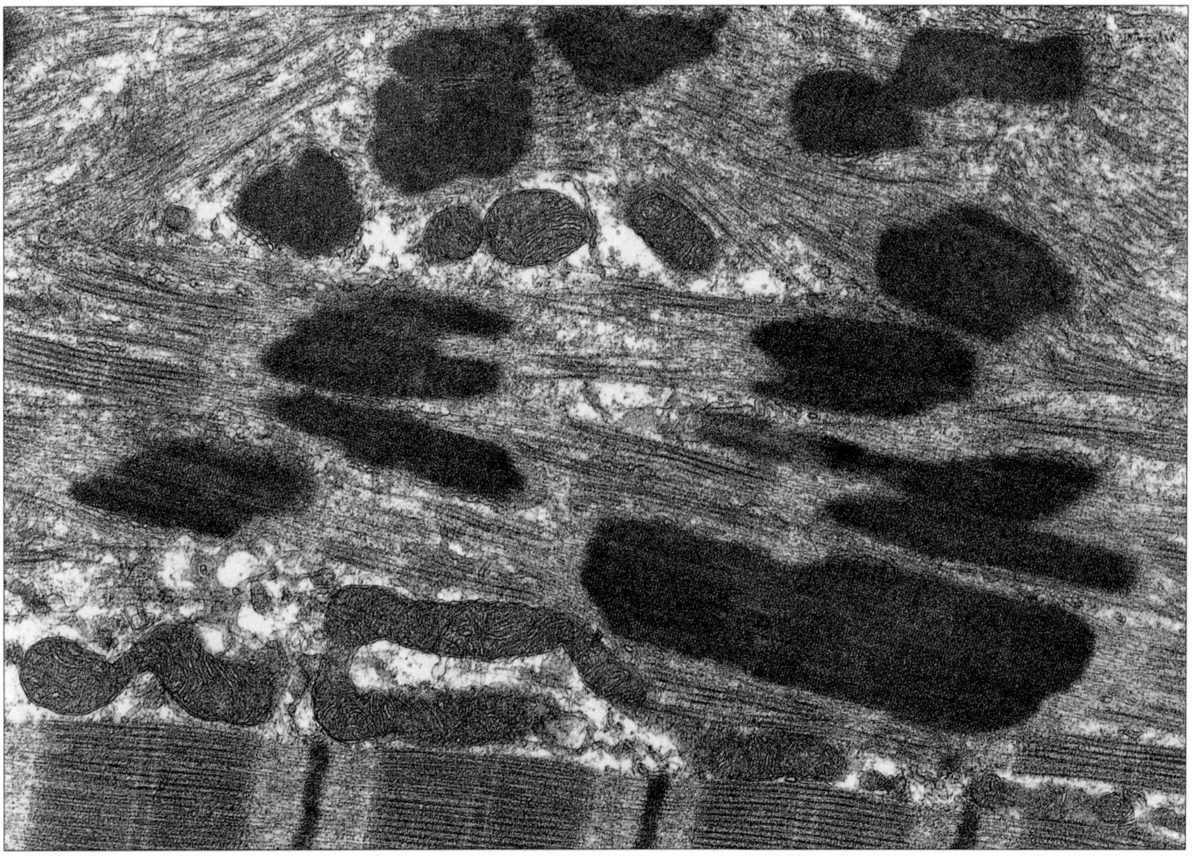

8.10 Longitudinal sections show variability in rod length.They exhibit a regular periodicity. Thin filaments extend from each end, sometimes interdigitating with thicker filaments. See also **4.71–4.72**. (×26500)

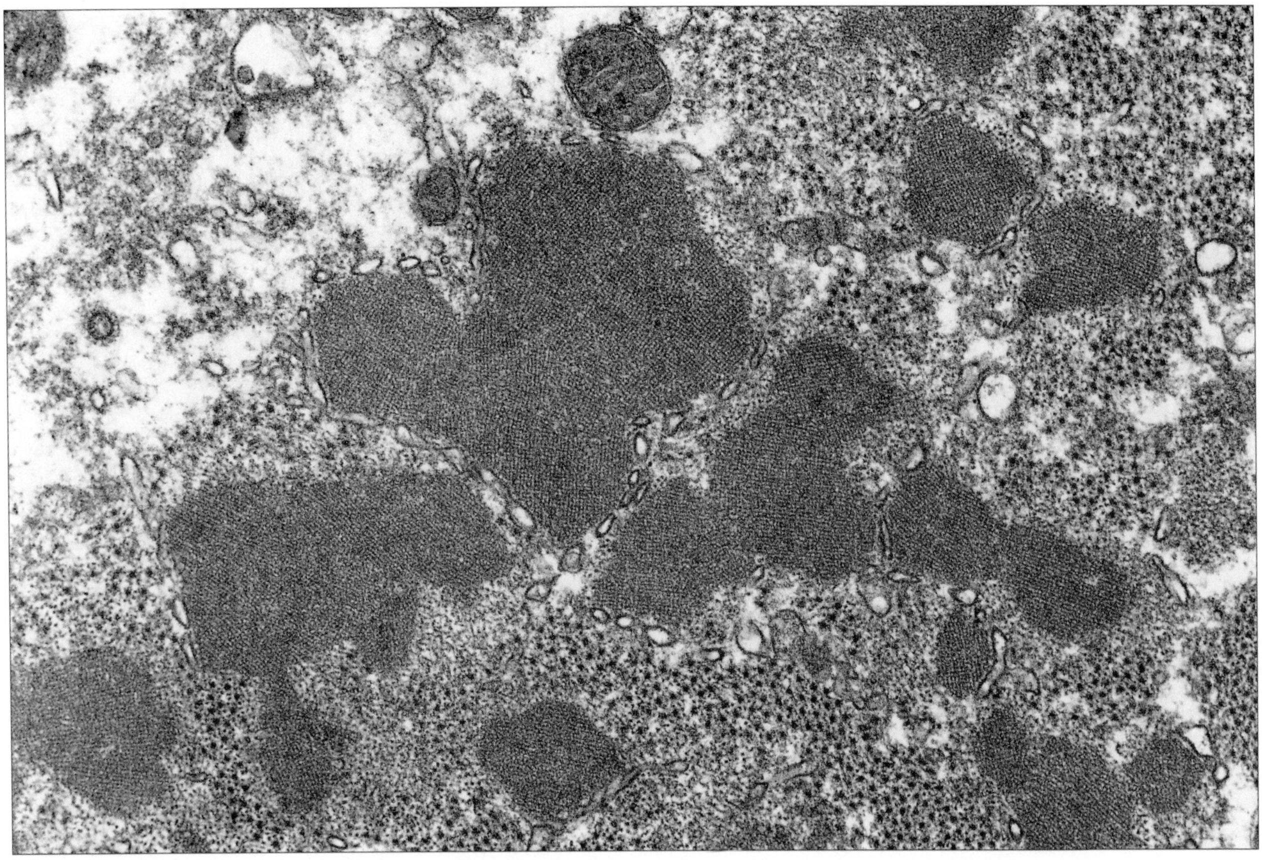

8.11 Cut transversely, the rods are difficult to distinguish from normal Z-discs. (×50000)

Core myopathies

Central core disease was the first 'specific' pathological entity to be associated with infantile hypotonia. The abnormality is illustrated in **4.57–4.58**, with the illustrated core being 'unstructured' (that is to say there is no identifiable ATPase activity and the contractile elements are abnormal) (**4.56**). In contrast, no abnormality can be identified in 'structured' cores, which can only be identified by the lack of oxidative activity in NADH-Tr preparations.

A number of other disorders have been variously labelled minicore, multicore or multi/minicore myopathies (**8.12–8.13**, **4.59**). There is considerable uncertainty about the nosological status of these conditions, and it may well be that they represent no more than variations in the morphological expression of a single disorder.

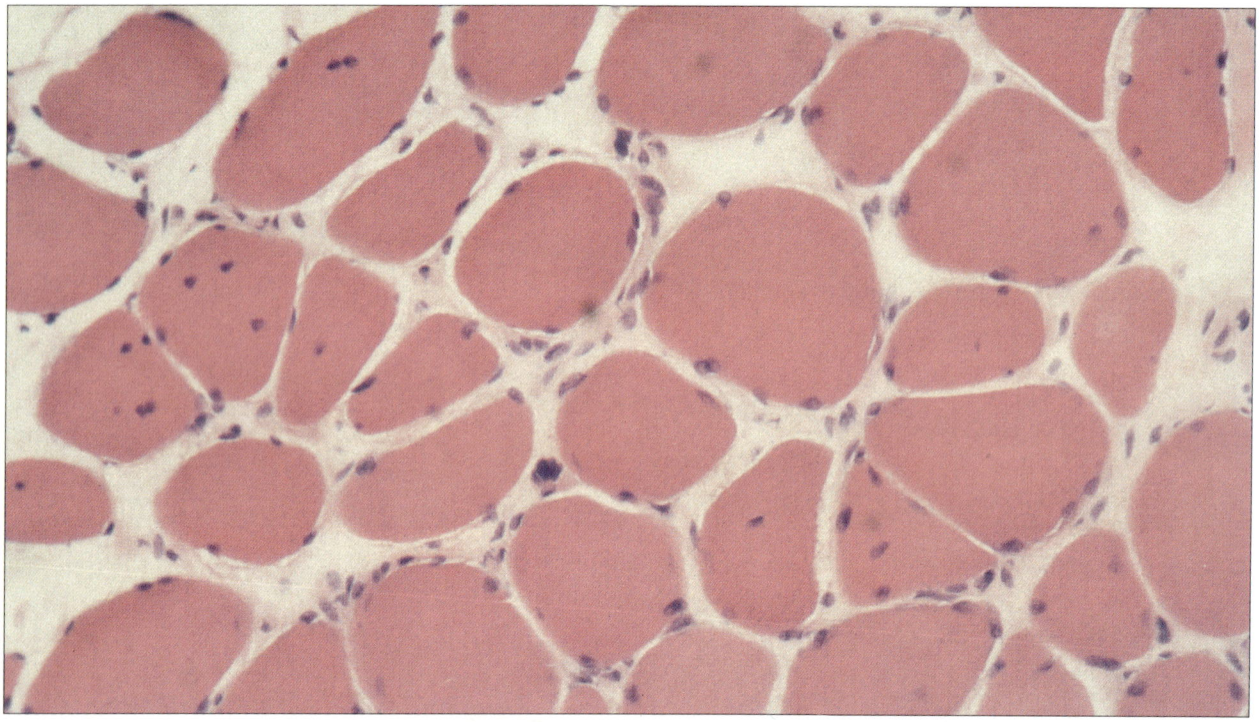

8.12 Although abnormal fibre size and shape are readily observed in this biopsy, the characteristic central cores of this disorder are only recognised with histochemical staining. (H&E, ×293)

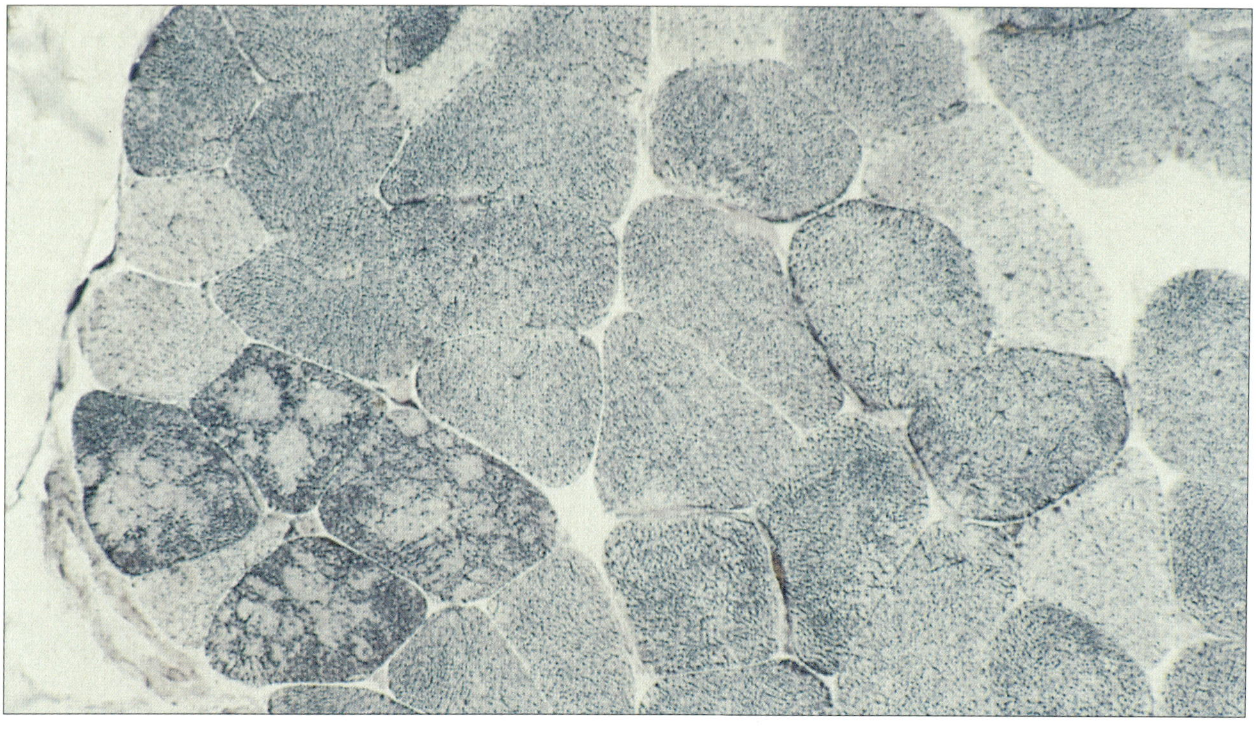

8.13 Central cores (multicores or minicores), in this case of central core disease, are shown as areas devoid of staining in this section from the same specimen as **8.12**. (NADH-TR, ×293)

Intra-uterine maturational delay

The normal intra-uterine development of skeletal muscle, in terms of morphology and fibre type differentiation (**8.1**), should be nearly complete by the 28th week of gestation. However, occasional large eosinophilic fibres of indeterminate histochemical status (they appear to have intense NADH-Tr activity and are difficult to type with myofibrillar ATPase) often persist in extra-uterine life, usually disappearing by the end of the first year (Wohlfart type B fibres; **8.14–8.15**).

Persistence of an excessive number of these fibres (greater than 2%) is often associated with mild infantile hypotonia, and is believed to imply delay in the normal intra-uterine development of these fibres.

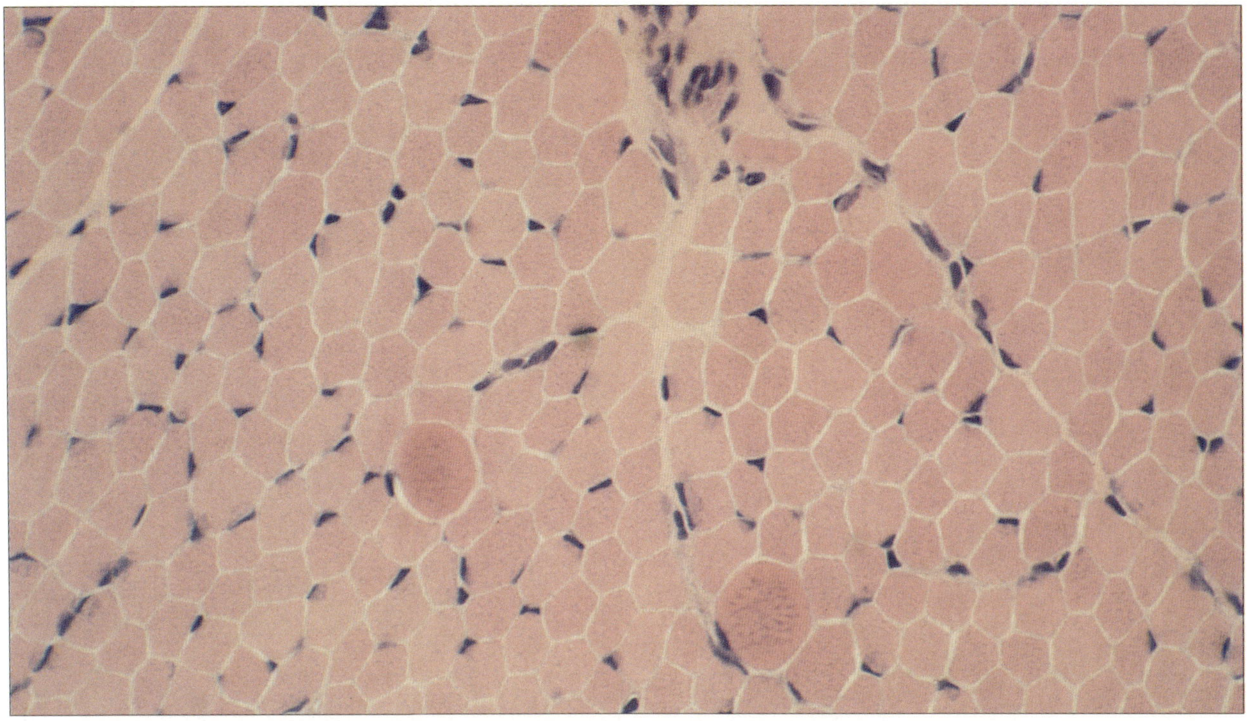

8.14 Amongst the closely-packed hypotrophic muscle fibres, two larger and more densely stained (Wohlfart B) fibres are visible in this case of retarded muscle development. (H&E, ×473)

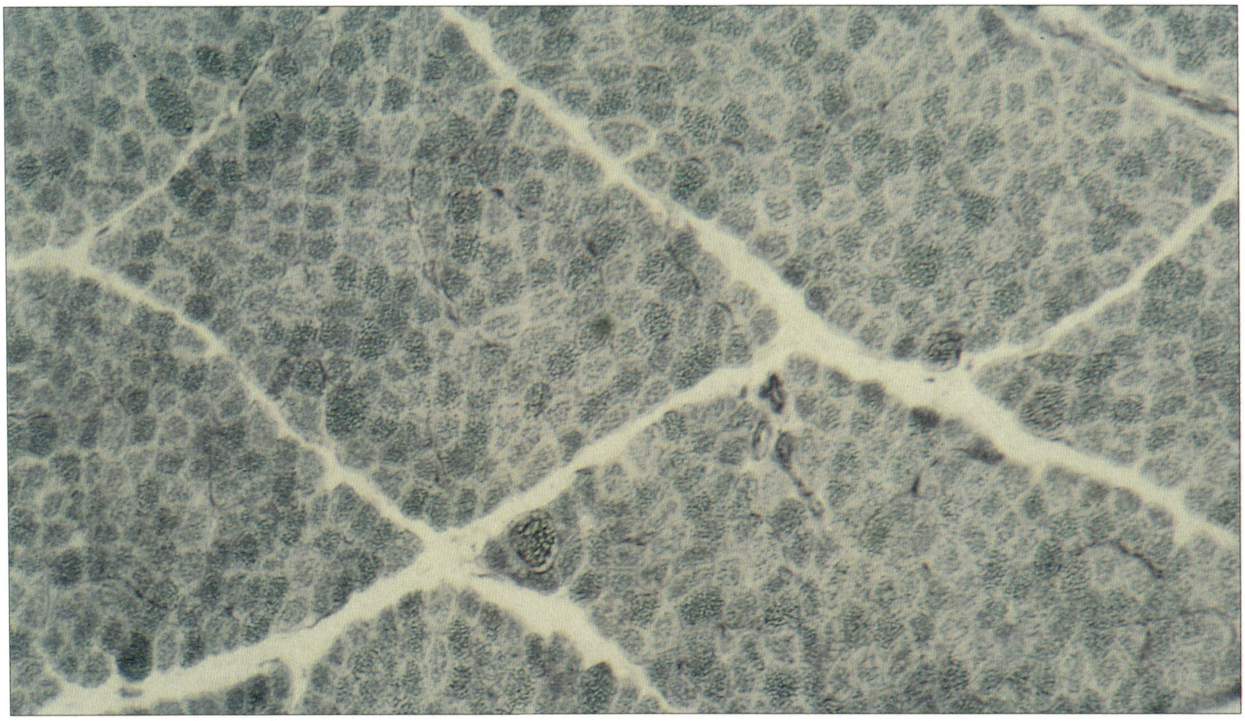

8.15 Retarded muscular development, showing lack of fibre type differentiation and a larger, rounded, and more granular Wohlfart B fibre near the centre of the field. From the same specimen as **8.14**. (NADH-Tr, ×293)

Type I fibre hypotrophy

Occasional cases of infantile hypotonia are associated with the apparently unique morphological/ histochemical abnormality (hypotrophy) illustrated in **8.16–8.17**.

In this condition, the distribution of the fibre types appears to be normal, and the Type II fibres are either normal or slightly increased in size. The Type I fibres are, without exception, tiny, and usually contain large, vesicular central nuclei (**8.16**). The discrepancy between the size of the Type I and II fibres is best seen in the myofibrillar ATPase preparation (**8.17**), and is assumed to be due to a functional rather than structural abnormality in motor unit innervation.

A variety of other, sometimes quite striking morphological abnormalities have been reported in single cases of infantile hypotonia. These include trilaminar myopathy, finger-print body myopathy, concentric laminar inclusion body myopathy, and zebra body myopathy.

The relevance of these findings to the clinical syndrome in individual cases and their relationship to the disorders described above (studied in more detail, often in families) remains unclear to say the least. Some authorities dismiss such findings as wholly non-specific and of no pathogenetic significance.

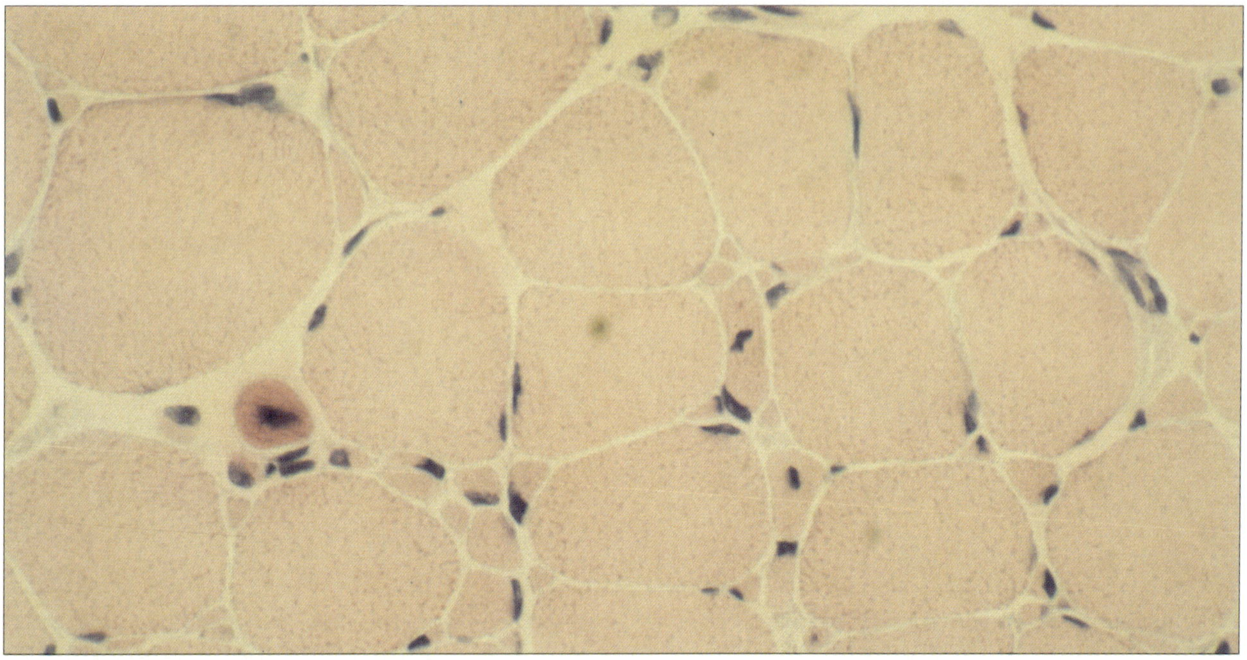

8.16 Congenital myopathy, showing collections of extremely small fibres (some with central nuclei) amongst normal muscle fibres. See also **8.17**. (H&E, ×591)

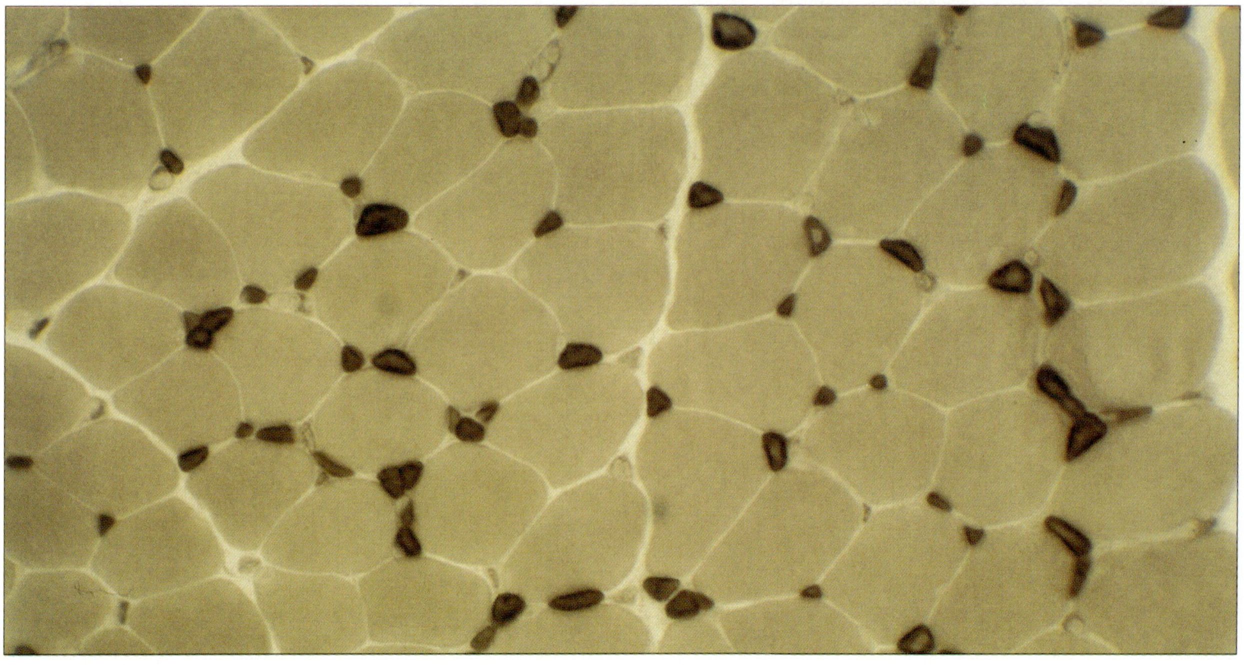

8.17 Type I fibre hypotrophy is the predominant feature of this congenital myopathy. From the same specimen as **8.16**. (Myofibrillar ATPase, pH 4.6, ×473)

9

Metabolic disorders

The metabolic myopathies may be divided into three broad categories: disorders of glycogen storage, disorders of lipid metabolism, and disorders of mitochondrial structure and function.

There appear to be two basic clinical patterns in the metabolic myopathies: an otherwise non-specific, gradually progressive, limb-girdle myopathy on the one hand, and cramps in muscle induced by exercise on the other, with the exception of children who present as infantile hypotonia.

Confirmation of the 'metabolic' nature of a myopathy is provided by the histochemical techniques applied to the muscle biopsy. For disorders where muscle biopsy is informative, a list of the metabolic myopathies with their underlying defects is given in *Table 9.1*. For further consideration of the metabolic pathways involved, the reader is referred to the detailed review by DiMauro *et al.* (1992).

Pathological reactions in metabolic myopathies

The pathological reactions are manifest as the presence of vacuoles within the fibres, the loss of specific enzyme activity, the presence of excessive amounts of glycogen and lipid, changes in oxidative enzyme activity, and the presence of abnormal (e.g. ragged-red) fibres.

At the ultrastructural level, the presence of abnormal amounts of glycogen and lipid can be confirmed. Mitochondrial abnormalities may be seen, as well as structurally abnormal fibres.

In most instances, the further elucidation of the underlying defect is biochemical. In some cases, analysis of mitochondrial DNA will aid in diagnosis.

Table 9.1 Metabolic muscle disorders

Disorder	Clinical	Deficiency	Storage
Glycogenoses			
Type II			
Pompe's disease	Infantile - profound hypotonia	α-1-4 glucosidase	Glycogen
Acid maltase deficiency	Adolescent/adult - LG syndrome		
Type IV			
Andersen's disease	LG syndrome	Brancher enzyme	Acid mucopolysaccharide
Brancher	Haemolytic anaemia in some cases		Glycogen
Type V			
McArdle's disease	Exercise induced muscle pain and cramp	Myophosphorylase B	Glycogen
Lipid Disorders			
Carnitine palmityl - transferase deficiency	Muscle pain and cramp, particularly in the fasted state May produce myoglobinuria	CPT Type I	Lipid
Carnitine deficiency	LG syndrome	Carnitine	Lipid

Disorders of glycogen storage

Type II – Pompe's disease

The biopsy in the infantile form of the disease is characterised by a gross vacuolar myopathy, affecting both skeletal (**9.1**) and cardiac muscle fibres. PAS staining will show that the fibres contain large amounts of glycogen (**9.2**). At the ultrastructural level (**9.3–9.6**), the glycogen appears to be membrane bound, although there may be evidence of rupture of the membranes, with free glycogen within the sarcoplasm (**9.6**). The amount of stored material may be sufficiently great to occupy virtually all of the muscle fibre (**9.5–9.6**).

The adolescent and adult forms of the disease are usually less aggressive in both morphological and clinical terms (**9.7–9.9**). Histochemical evidence of excessive acid phosphatase activity, associated with functionally abnormal lysosomes, can be seen (**9.10**). The ultrastructural changes differ only in amount from the infantile form (**9.3, 9.5–9.6**)

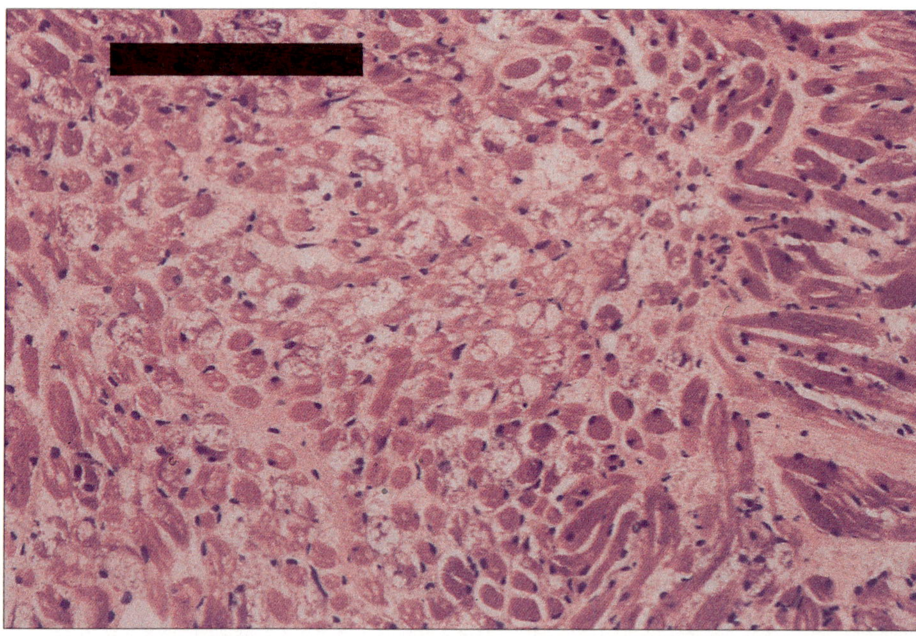

9.1 Infantile acid maltase deficiency (glycogenosis type II), showing heavily vacuolated fibres which also stain with PAS. (H&E, ×280)

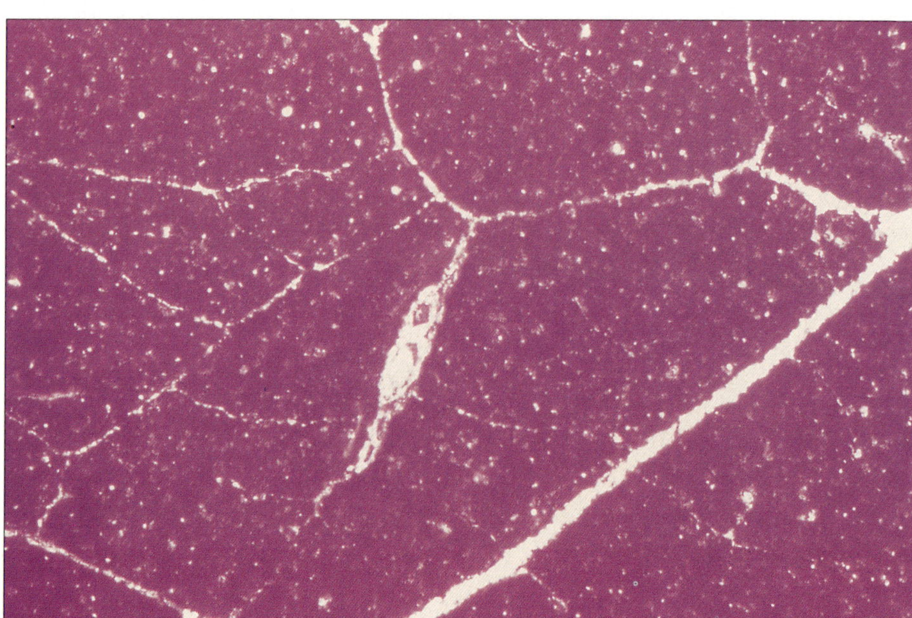

9.2 Excess glycogen storage in a case of infantile acid maltase deficiency (glycogenosis type II). (PAS celloidin, ×280)

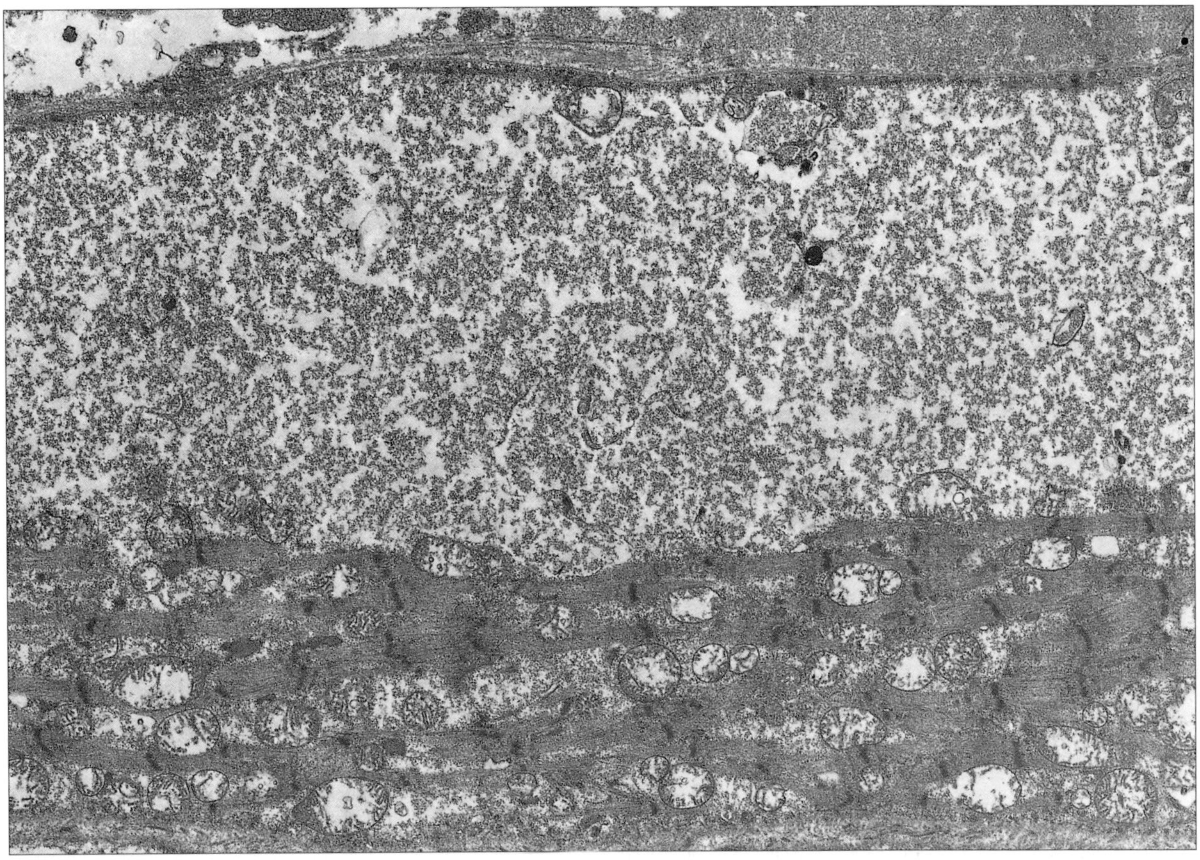

9.3 Abnormal amounts of stored glycogen may occupy a large proportion of the fibre volume. In such fibres only a few abnormal myofibrils remain. (×10000)

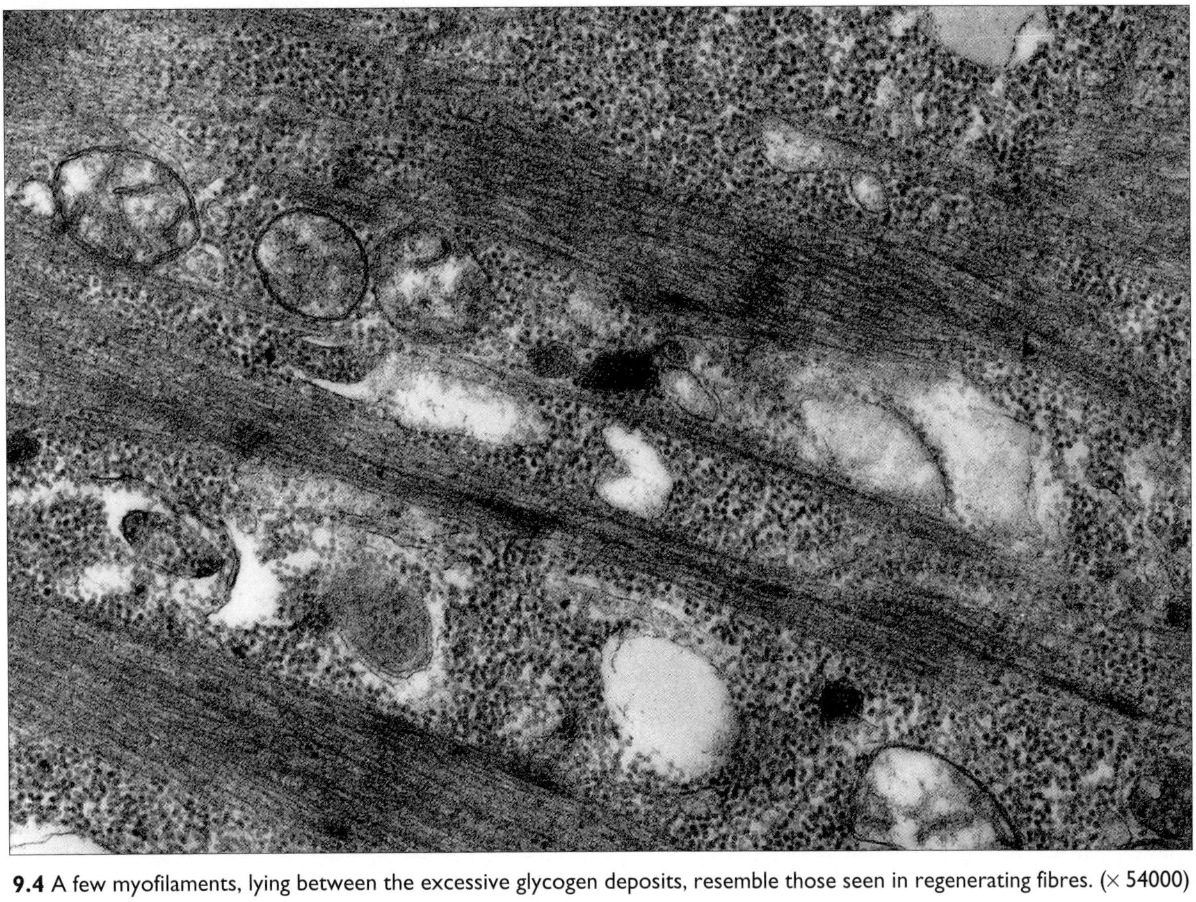

9.4 A few myofilaments, lying between the excessive glycogen deposits, resemble those seen in regenerating fibres. (× 54000)

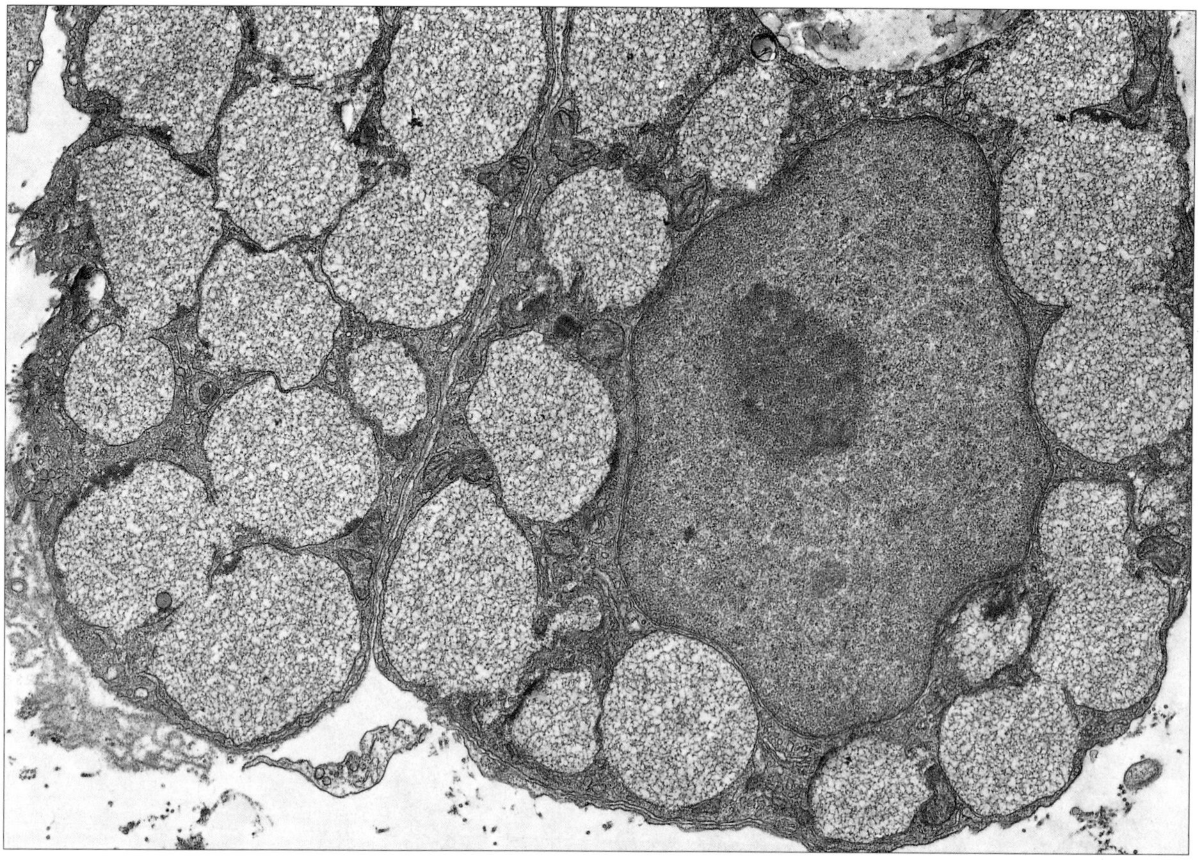

9.5 The stored glycogen may be membrane-bound, in discrete compartments within the fibre. (×12800)

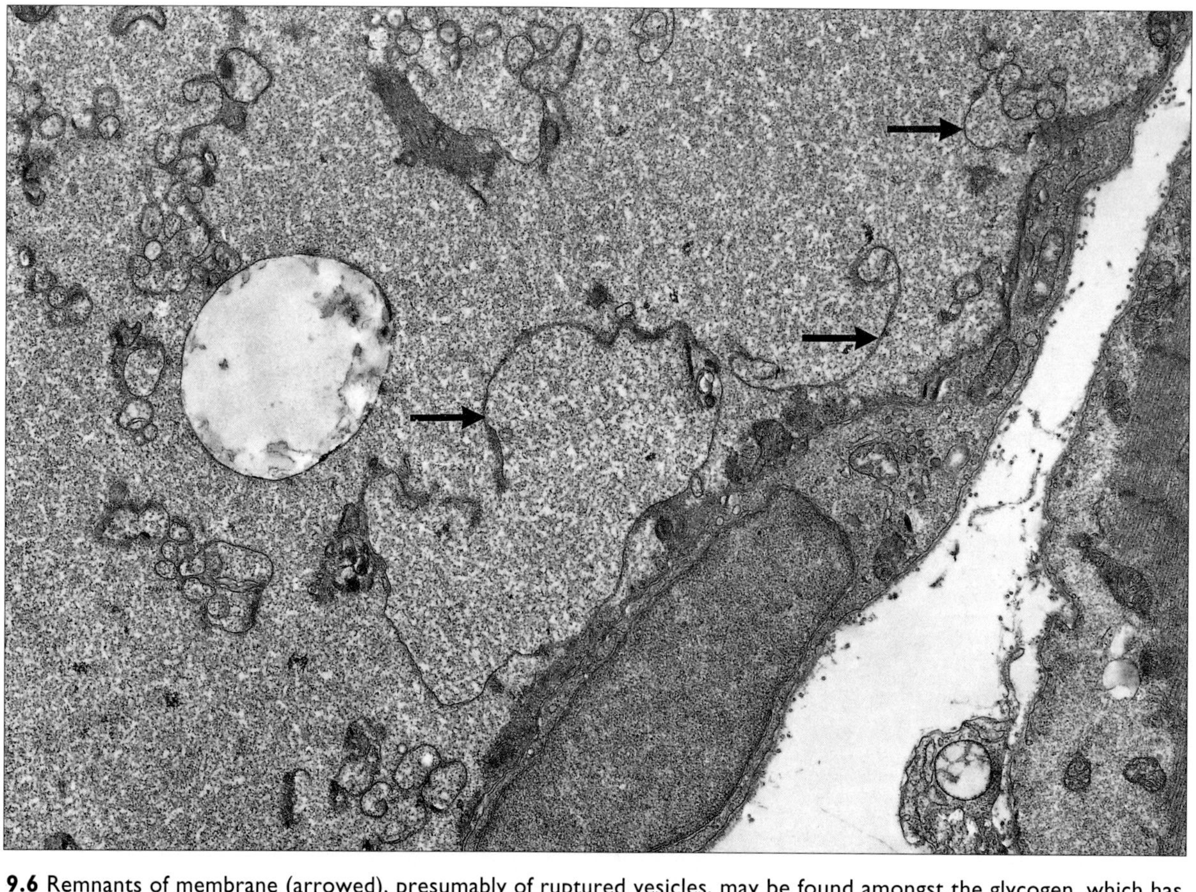

9.6 Remnants of membrane (arrowed), presumably of ruptured vesicles, may be found amongst the glycogen, which has completely replaced the myofibrils within the fibre. (×12800)

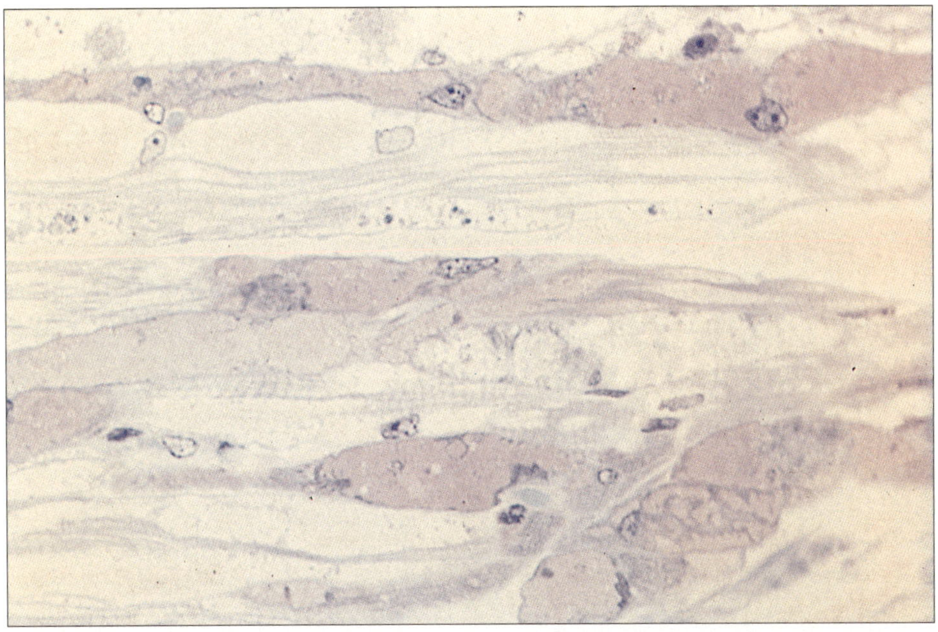

9.7 Juvenile acid maltase deficiency (glycogenosis type II), showing elongated vacuolated regions within the muscle fibres. (1 μm thick plastic section, toluidine blue, ×350)

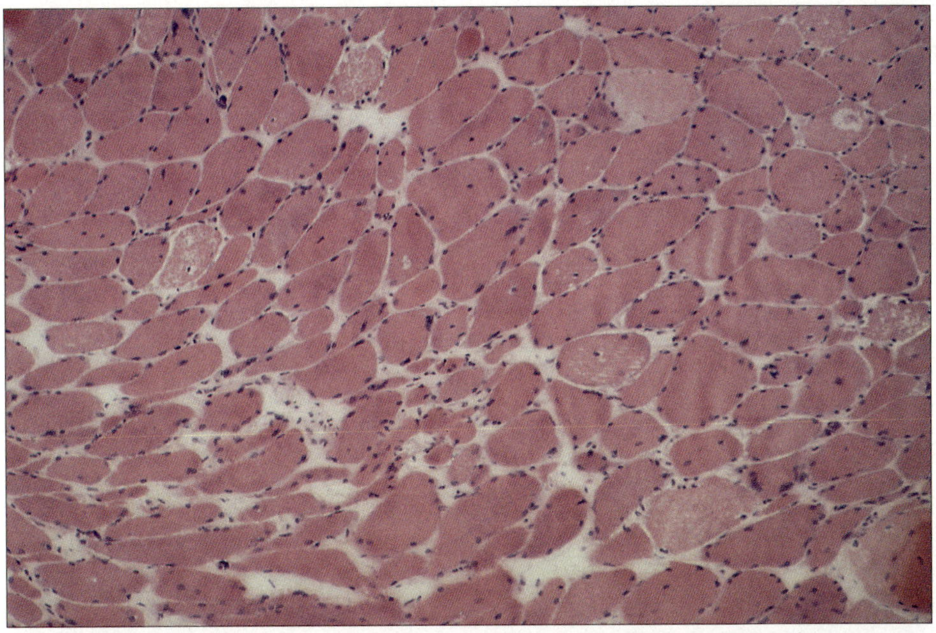

9.8 Adult-onset acid maltase deficiency (glycogenosis type II or Pompe's disease), where a number of the larger fibres have numerous small vacuoles. Other myopathic features, such as increased variation of fibre size and the presence of internal nuclei, are also visible. (H&E, ×88)

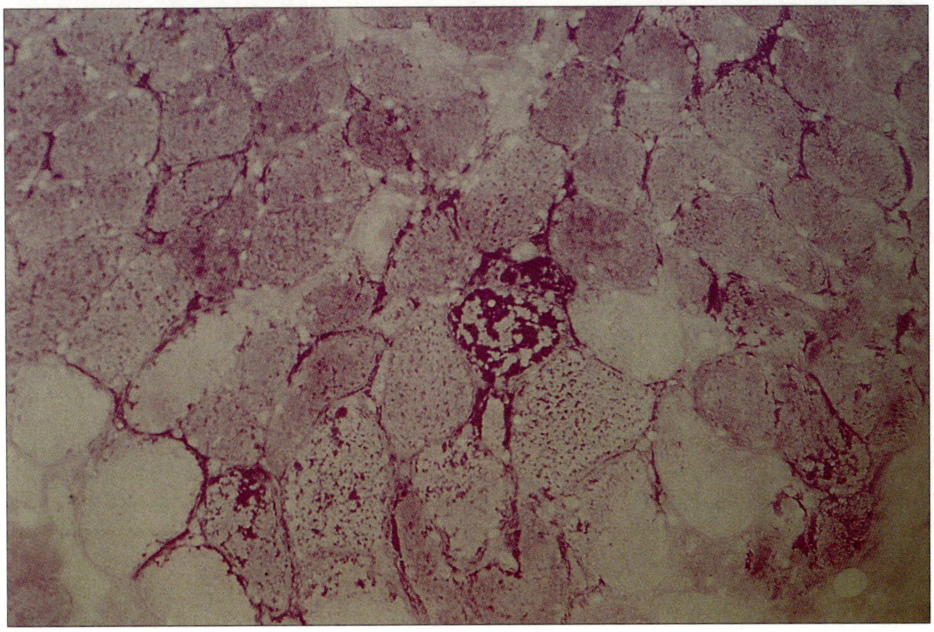

9.9 Section from the same specimen as **9.8**, showing excess glycogen storage in the vacuolated muscle fibres. (PAS, ×175)

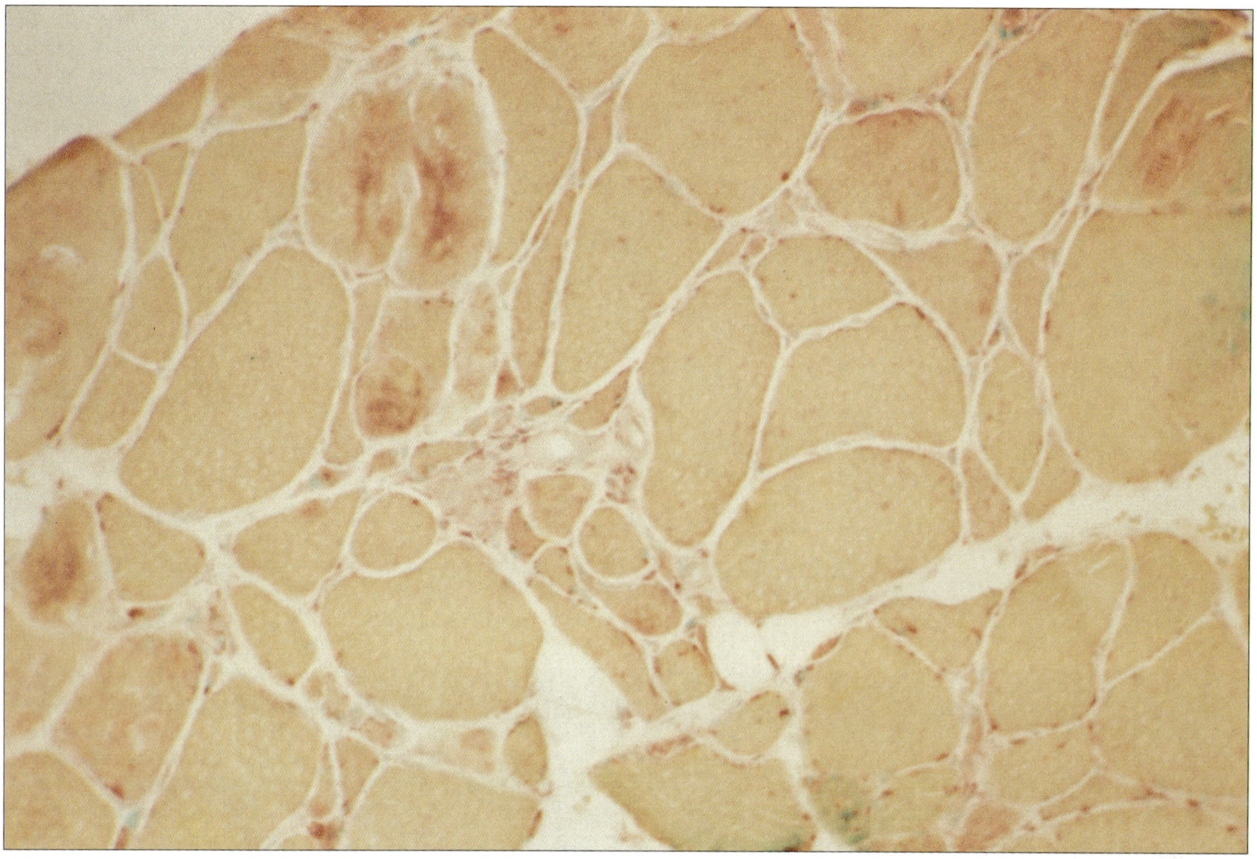

9.10 Excess lysosomal material is indicated by the red staining of certain muscle fibres. Adult-onset acid maltase deficiency (glycogenosis type II). (Acid phosphatase, ×189)

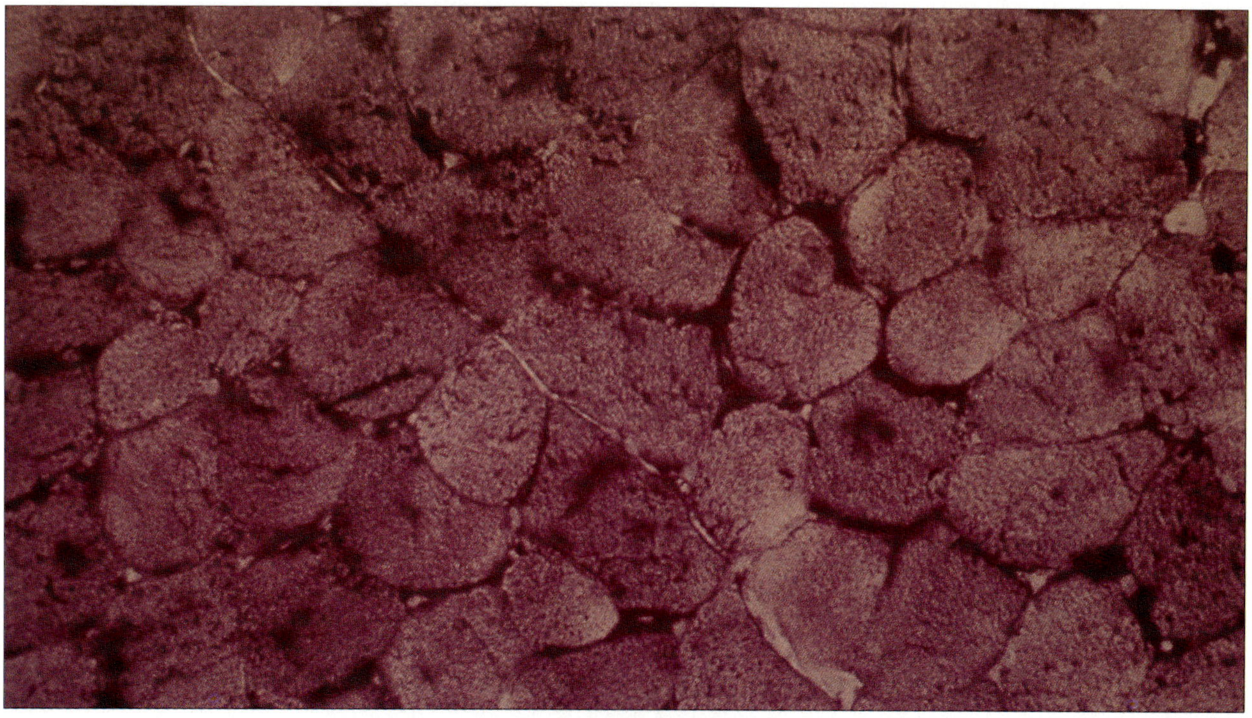

9.11 Phosphorylase deficiency (glycogenosis type V or McArdle's disease), showing excess glycogen storage without the dense masses typical of acid maltase deficiency. (PAS celloidin, ×284)

Type V – McArdle's disease

This is marked by the presence of subsarcolemmal crescents or vacuoles, which appear to be devoid of content in H&E stained preparations. However, they contain abnormal accumulations of glycogen in both PAS stained sections and the corresponding electron micrographs (**9.4**, **9.11–9.12**).

Myophosphorylase preparations will show that the activity of the enzyme is absent (**9.13**).

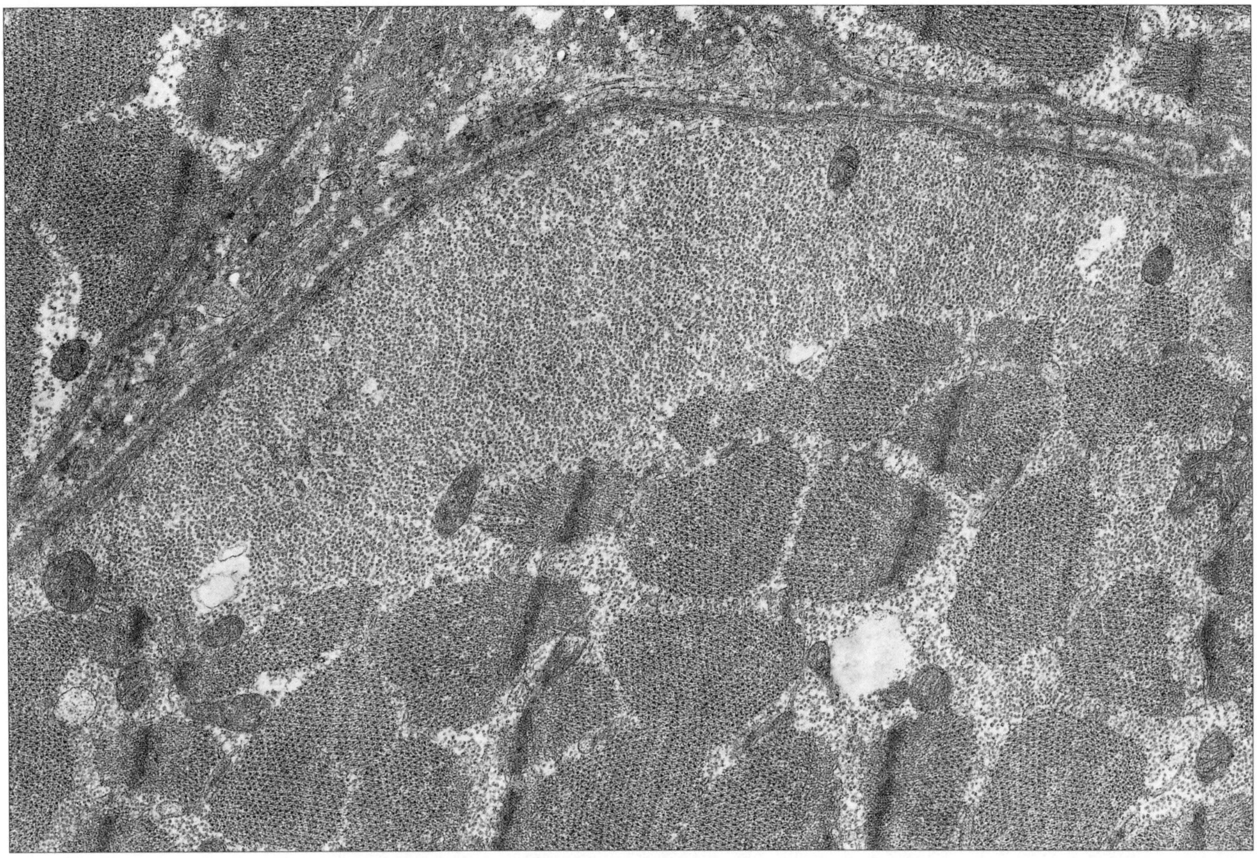

9.12 Glycogen storage may be less severe (as in McArdle's disease), and can occur as accumulations beneath the sarcolemma and disseminated between the myofibrils. (×19688)

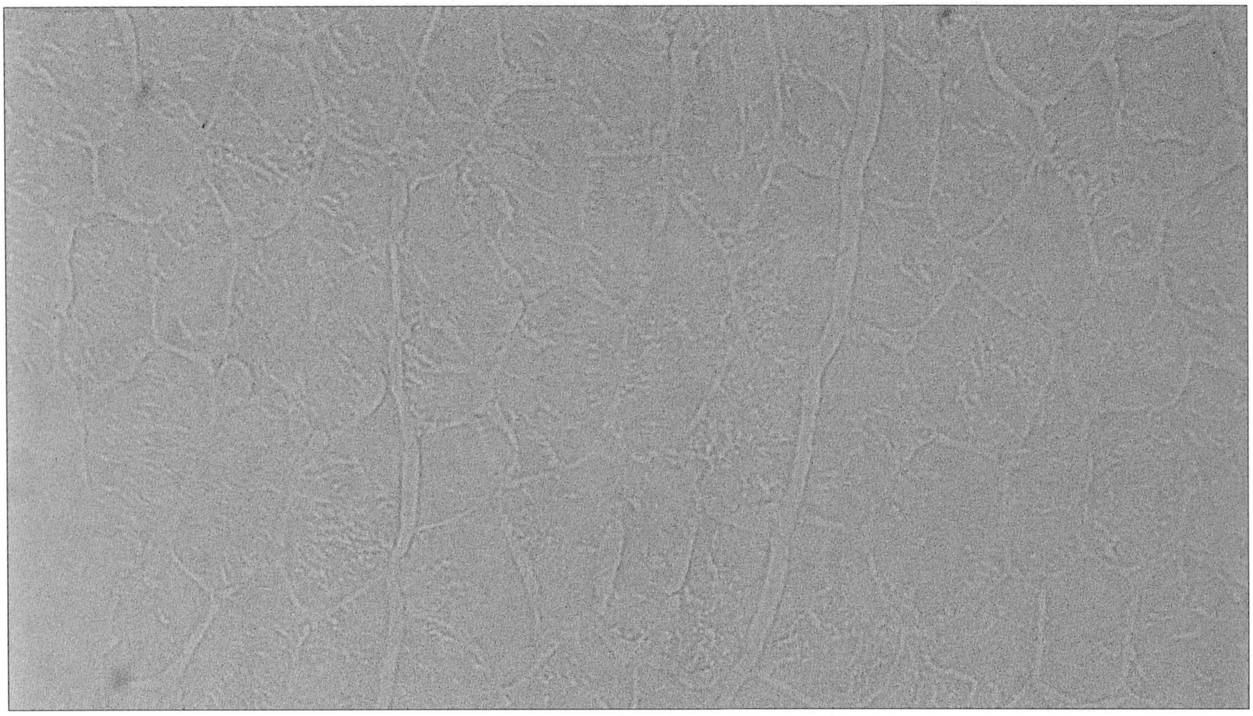

9.13 Complete lack of enzyme staining in a case of phosphorylase deficiency (glycogenosis type V). (Phosphorylase, ×284)

Type IV – Andersen's disease

This is an unusual cause of slowly progressive limb girdle myopathy. There is impaired function of brancher enzyme, leading to the accumulation of unbranched glycogen within the muscle fibre.

The areas of glycogen accumulation are seen as vacuoles within the fibre (**9.14**), lying in either a subsarcolemmal or central location. They are confined to Type I fibres (**9.15**).

Since the glycogen is structurally abnormal, it cannot be completely removed by predigestion with diastase (**9.16**), and the mucopolysaccharide thus produced is visible by staining with Alcian blue (**9.17**). At the ultrastructural level (**9.18**), the absence of a limiting membrane is apparent, as is the fibrillar nature of the glycogen.

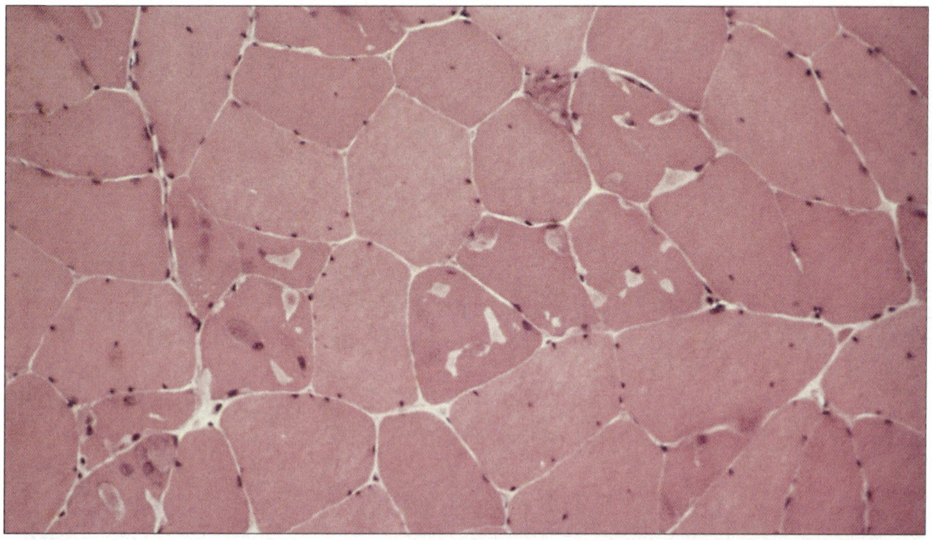

9.14 Figures **9.14–9.17** are from an adult patient with branching enzyme deficiency (glycogenosis type IV or Andersen's disease). The unstained vacuoles were present in 30% of fibre profiles.
(H&E, ×175)

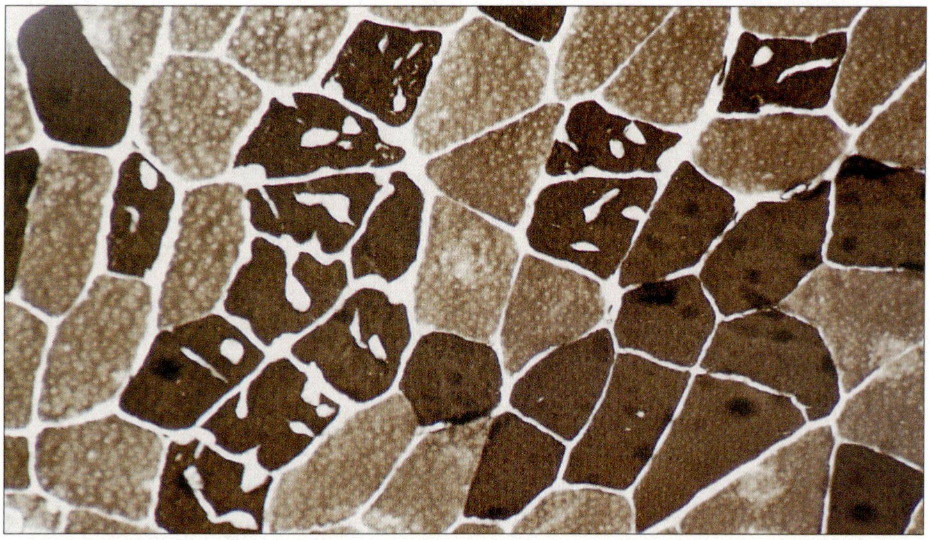

9.15 The vacuoles are present only in Type I fibres. (See caption to **9.14**).
(Myofibrillar ATPase, pH 4.35, × 175)

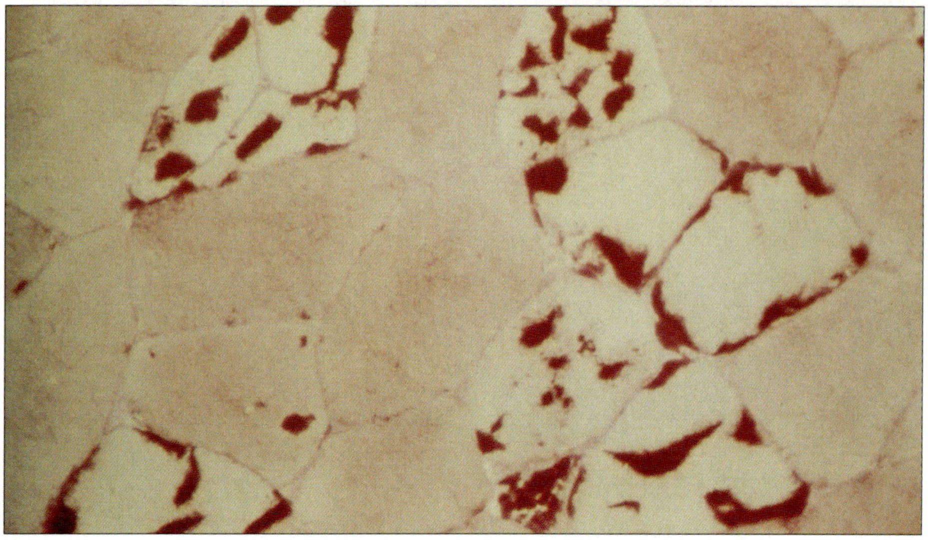

9.16 Densely stained vacuoles are present in fibres otherwise devoid of their normal glycogen content. The vacuoles were resistant to amylase digestion. (See caption to **9.14**).
(PAS, ×238)

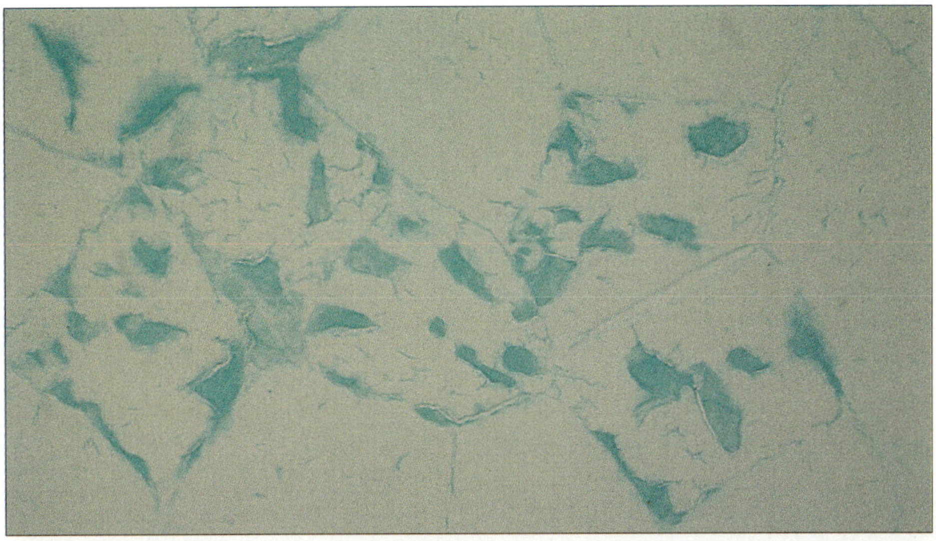

9.17 The vacuoles are stained intensely with alcian blue at acid pH, suggesting the presence of sulphated mucopolysaccharide. Their ultrastructure is shown in **9.18**. (See caption to **9.14**). (Alcian blue, pH 1.0, ×350)

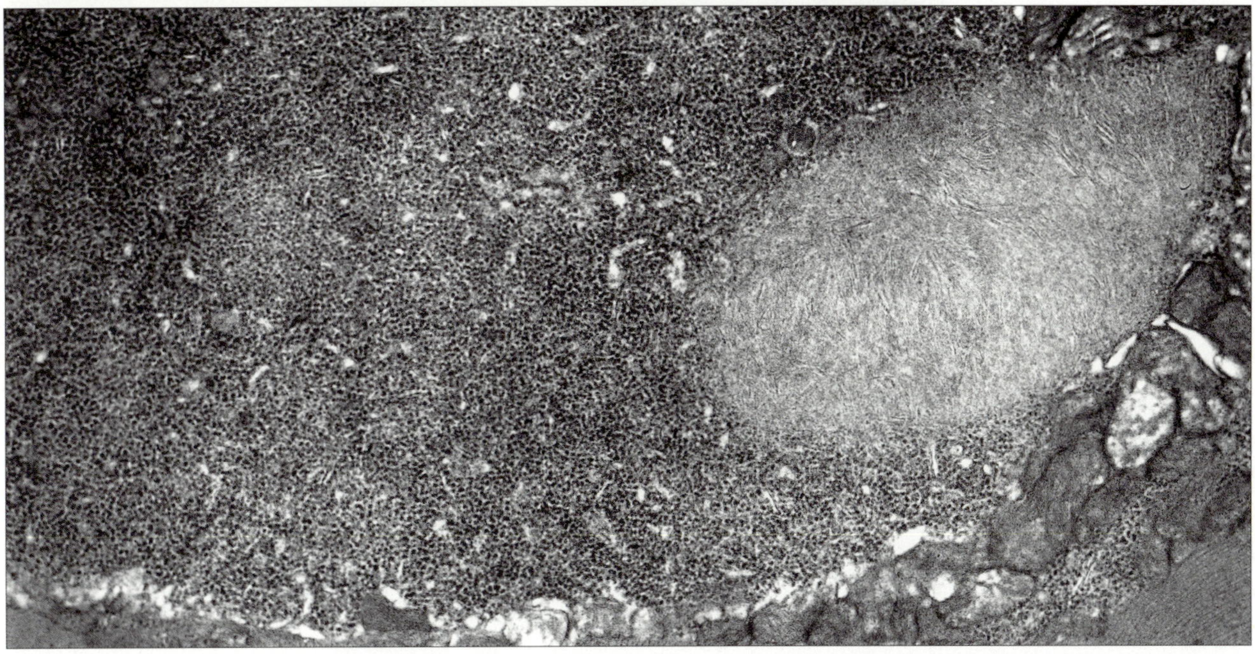

9.18a In this illustration, there is a filamentous inclusion within the mass of accumulated glycogen. (×21000)

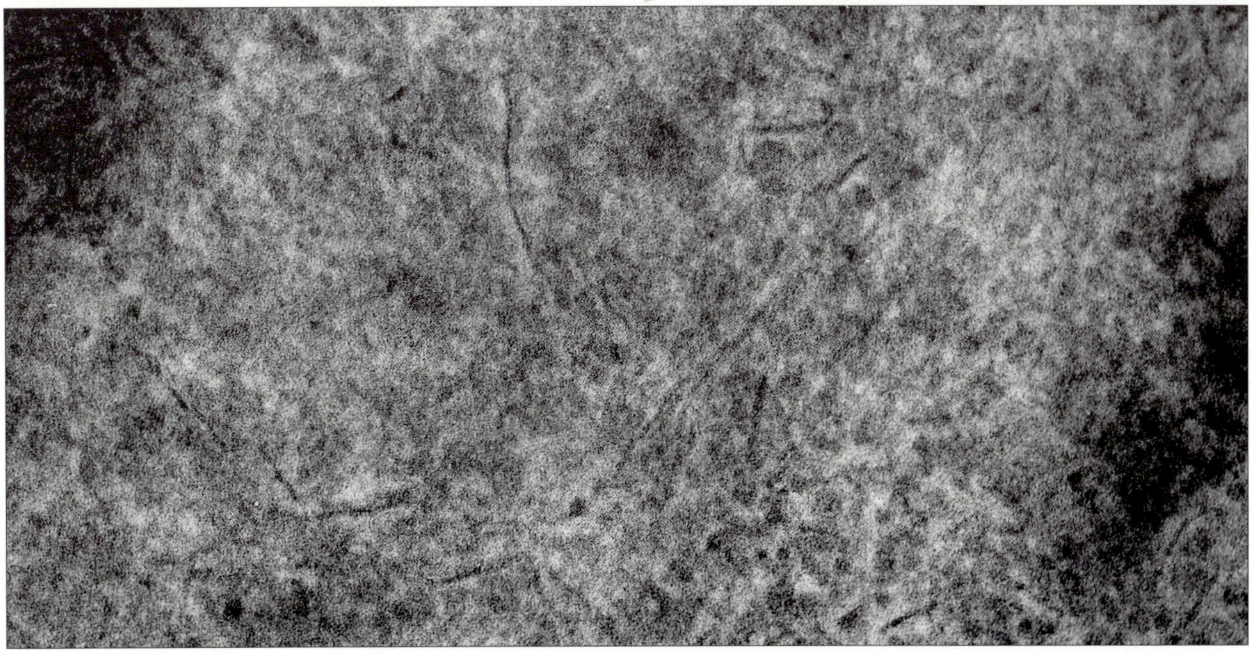

9.18b At higher magnification, the sparse filaments can be seen to be randomly orientated. (×210000)

Lipid myopathies

These present clinically as either a LG syndrome or exercise induced muscle pain and cramp.

Carnitine deficiency

Carnitine deficiency is associated with a slowly progressive LG myopathy, the histopathological feature of which is gross storage of lipid in vacuoles within the myofibre (**9.19**). The vacuoles may be sufficiently large and numerous to obliterate the myofibrillar apparatus. This degree of lipid storage is commonly seen in the face of normal mitochondrial function, and occurs to a greater degree in Type I fibres (**9.20**).

When lipid accumulation occurs secondarily to mitochondrial dysfunction, it is reflected in abnormalities seen on NADH-Tr and usually a much less severe lipid accumulation. The presence of lipid within the vacuoles is identified by staining with Sudan black or Oil red O (**9.21**).

At the ultrastructural level, large lipid droplets are seen (**9.22**– **9.23**), and the usually associated fibre disruption is more obvious than at the light microscope level.

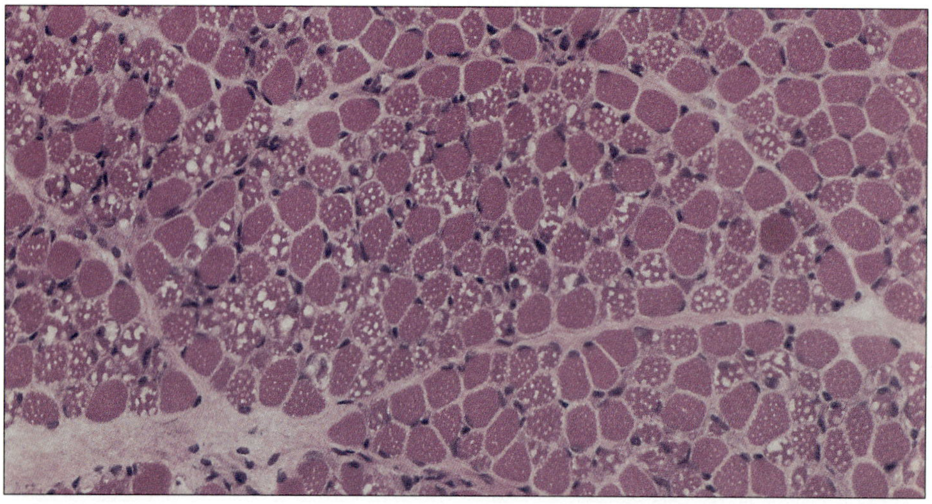

9.19 Carnitine deficiency myopathy (Multiple acyl CoA deficiency with lipid storage) in a child aged 5 years. Most fibres are vacuolated to some degree (See also **9.20**–**9.21** from the same case). (H&E, ×280)

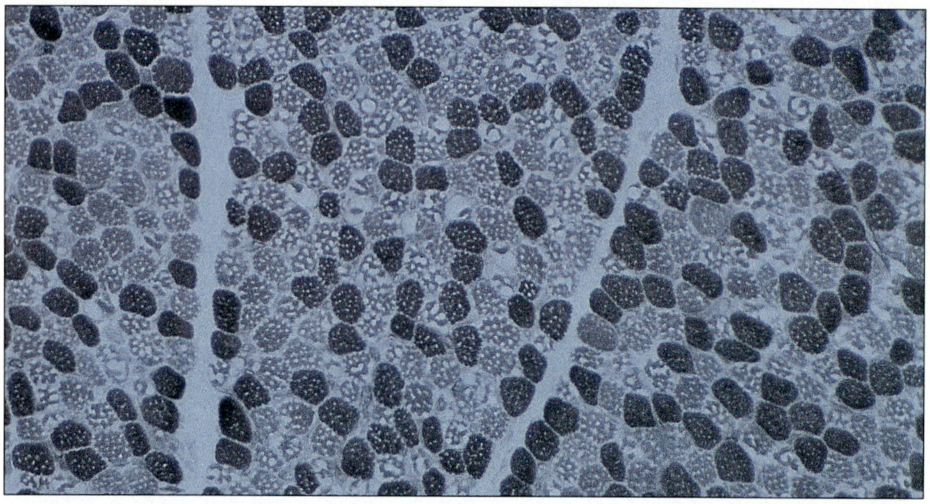

9.20 Type I fibres show the most extensive vacuolation, although vacuoles are present in nearly all fibres in the biopsy. (See caption to **9.19**). (Myofibrillar ATPase, pH 9.5, ×280)

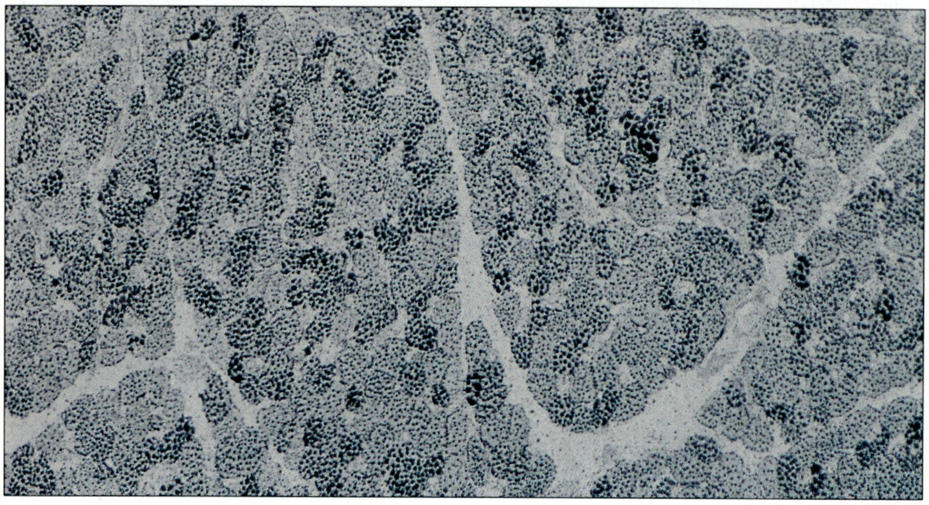

9.21 Excess lipid storage is visible in most fibres, although particularly large masses are present in Type I oxidative fibres. (See caption to **9.19**). (Sudan black B, ×280)

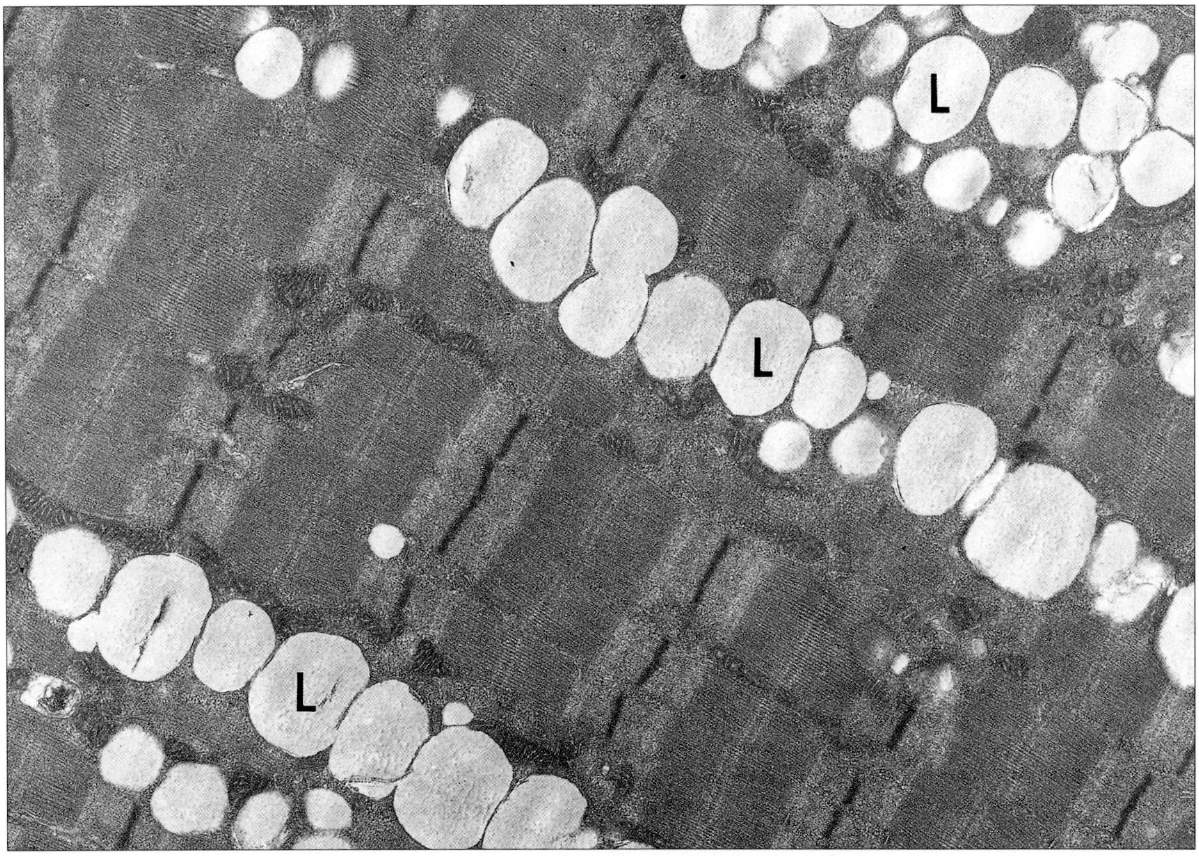

9.22 Excess lipid is seen as rows of droplets (L) lying between normal myofibrils. (×13750)

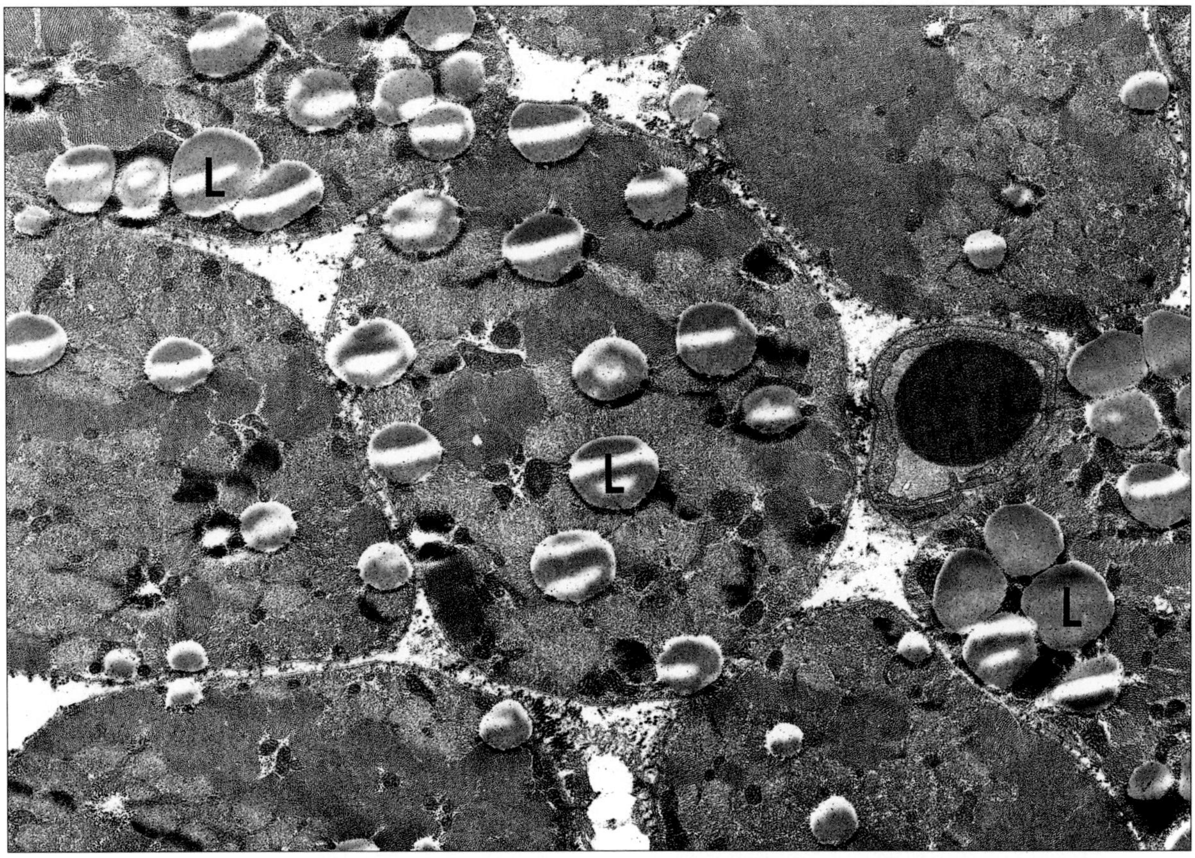

9.23 Cut in transverse section, lipid droplets (L) can be seen to have a random distribution within the fibres. (×8150)

CPT deficiency

This is not associated with a myopathy, but rather with exercise induced muscle pain and cramp. Under conditions of food depletion, this may result in myoglobinuria.

The degree of lipid storage is not as great as in carnitine deficiency (**9.24**, **4.45–4.46**), where the appearance of the H&E may be mistakenly interpreted as ice-crystal artefacts unless routine histochemical studies are undertaken.

The association of CPT with the mitochondrion is illustrated by the appearance of excess mitochondrial activity on NADH-Tr (**9.25**). The lipid filled vacuoles are obvious in Type I fibres (**4.46**). The excess lipid is again demonstrated by appropriate lipid stains (**9.22–9.23**, **9.26**, **4.47**).

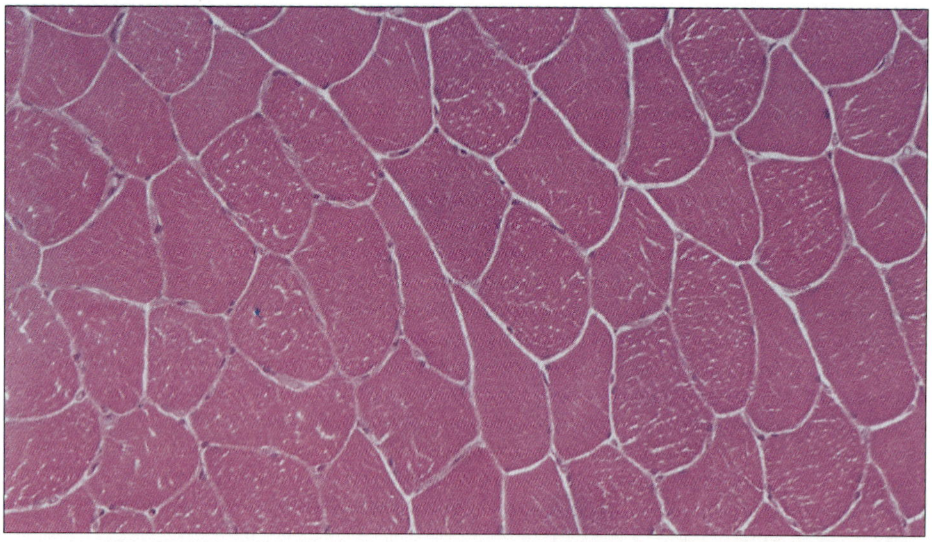

9.24 An adult case of CPT deficiency, showing a finely vacuolar myopathy. (H&E, ×175)

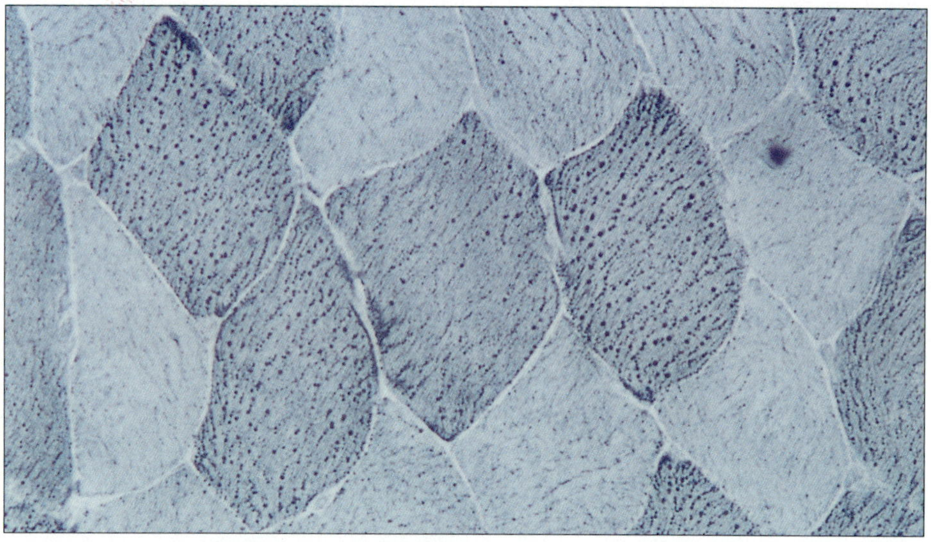

9.25 Dense, particulate reaction product is visible in Type I oxidative fibres (from the same specimen as in **9.24**). (NADH-Tr, ×280)

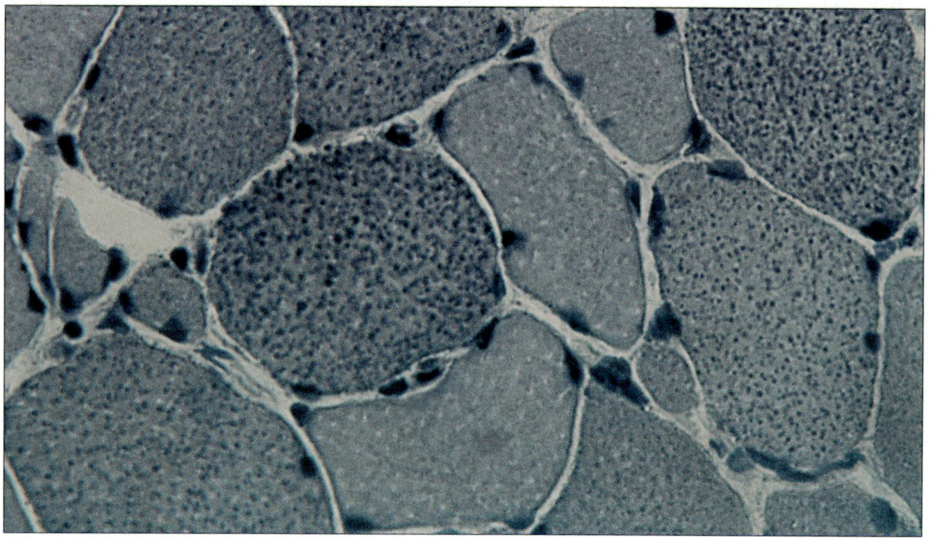

9.26 Lipid storage, indicated by the dense black staining deposits, is restricted to the larger oxidative muscle fibres in this case of CPT deficiency. (Sudan black, ×280)

Mitochondrial myopathies

For many of these disorders, defects in mitochondrial DNA (either as deletions or point mutations) have been described, and developments in this field will eventually clarify the underlying pathogenesis of these conditions.

The clinical presentation of mitochondrial myopathies is varied, and includes chronic progressive external ophthalmoplegia (CPEO) and its systemic variants (e.g. Kearn-Sayre's syndrome), LG and FSH myopathies, cardiomyopathy, exercise induced muscle pain and cramp (occasionally with paralytic attacks), and severe infantile hypotonia.

The initial suspicion that there may be an underlying mitochondrial disorder is evident in the H&E preparation, where there is a punctate basophilic cytoplasm (**9.27–9.28**, **4.59–4.50**) and/or marked subsarcolemmal abnormalities (**9.27–9.32**, **4.50**). That these subsarcolemmal deposits are mitochondria, is shown by staining with SDH (**4.51**), NADH-Tr (**9.31**, **4.52**), or cytochrome oxidase (**9.32**).

The ragged-red fibres in mitochondrial disorders are demonstrated by staining with modified Gomori (**9.30**, **4.53**).

The range of mitochondrial abnormalities that are seen at the ultrastructural level are illustrated in **9.33–9.44**.

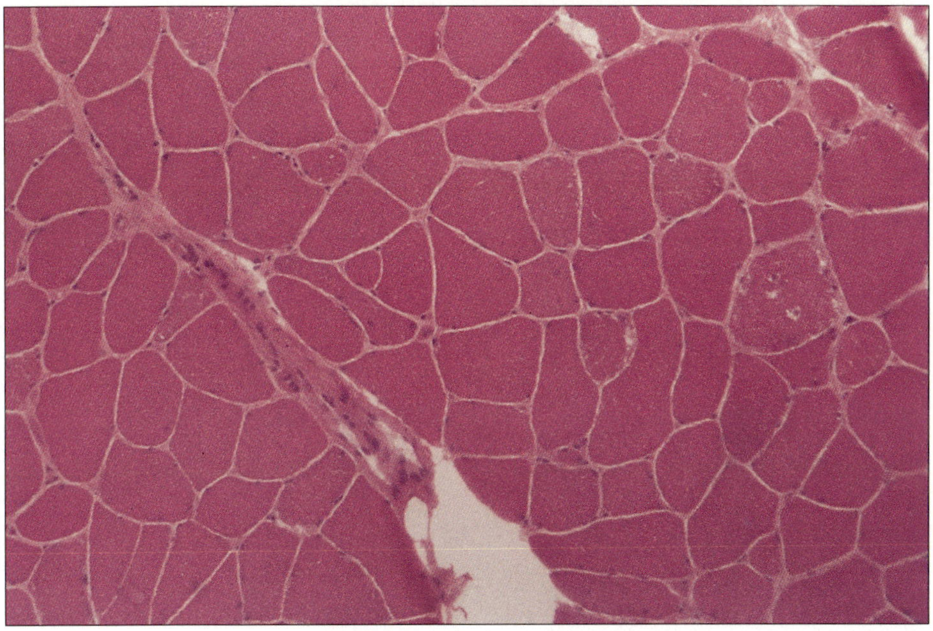

9.27 Mitochondrial myopathy, showing patchy staining of some muscle fibres, especially in the subsarcolemmal regions. Increased variation of fibre size is also a feature of this myopathy. (H&E, ×140)

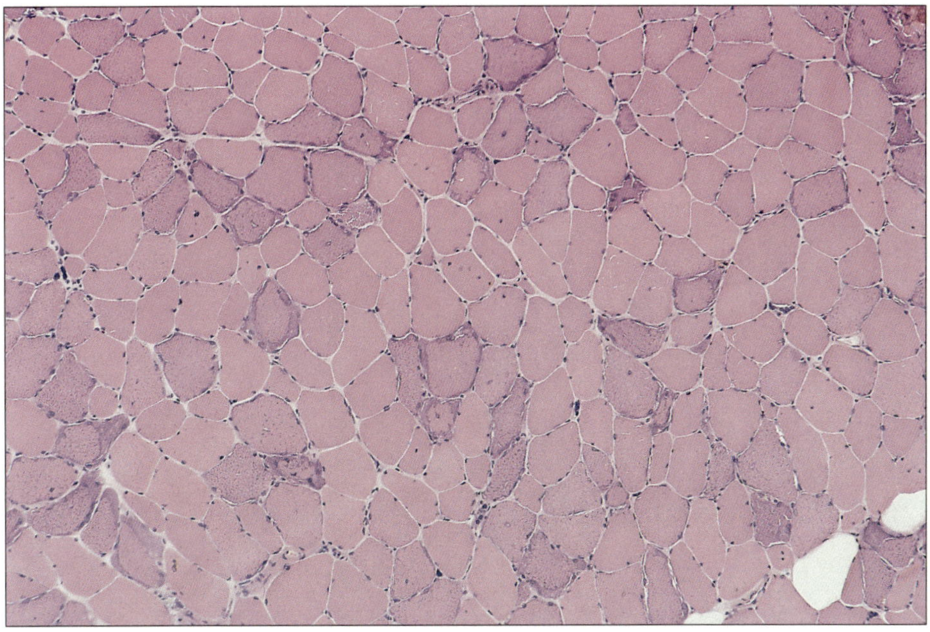

9.28 The subsarcolemmal basophilic areas seen in a few muscle fibres are typical of a mitochondrial myopathy. These regions stain darkly with oxidative enzymes. This biopsy was taken from a patient with Kearn-Sayre's syndrome. (H&E, ×88)

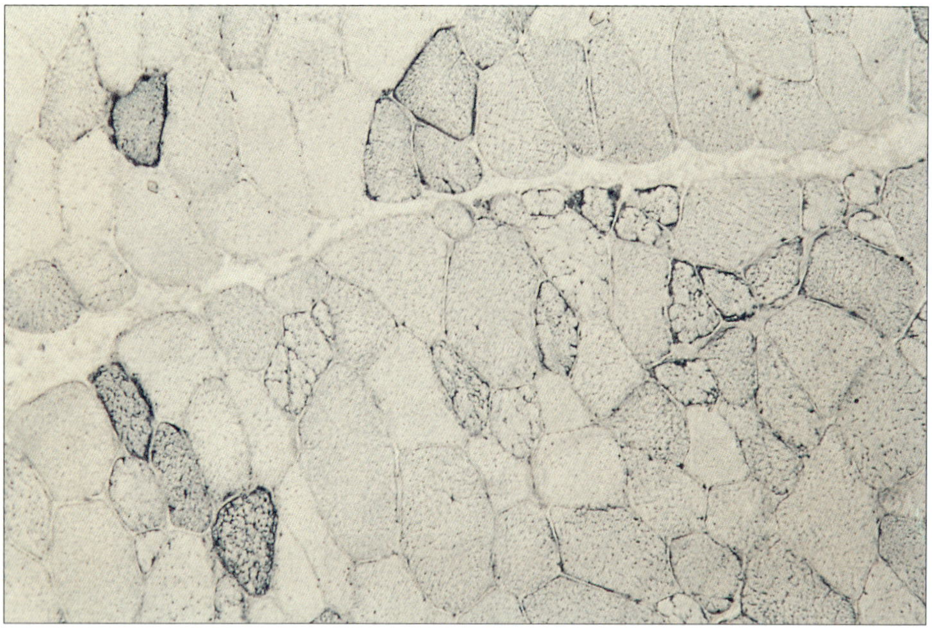

9.29 Dense accumulations of oxidative enzyme reaction product within muscle fibres, particularly in the sub-sarcolemmal regions, are a major feature of a mitochondrial myopathy.
(NADH-Tr, ×140)

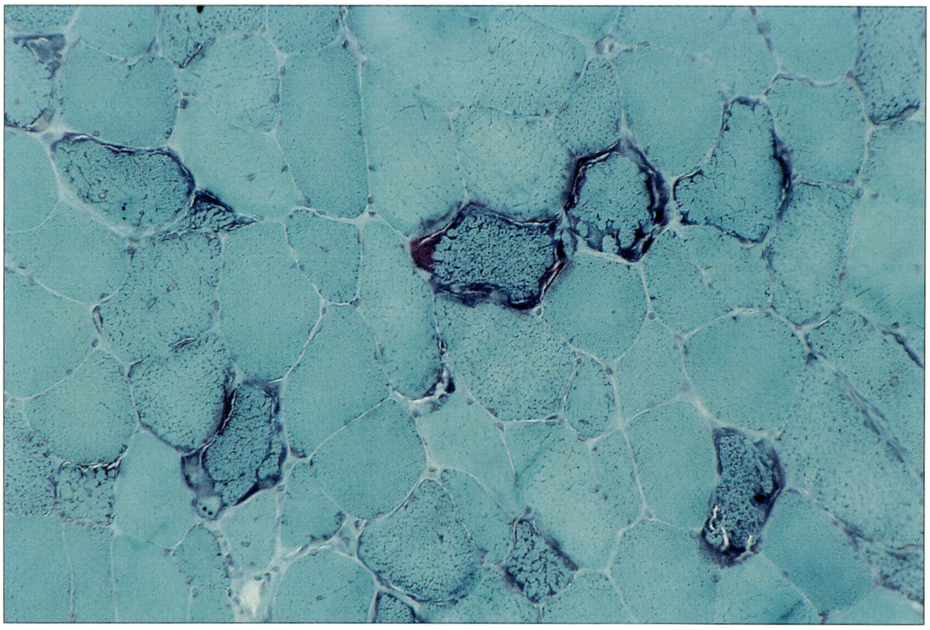

9.30 Ragged-red fibres are prominent.
a) Gomori trichrome, ×175;

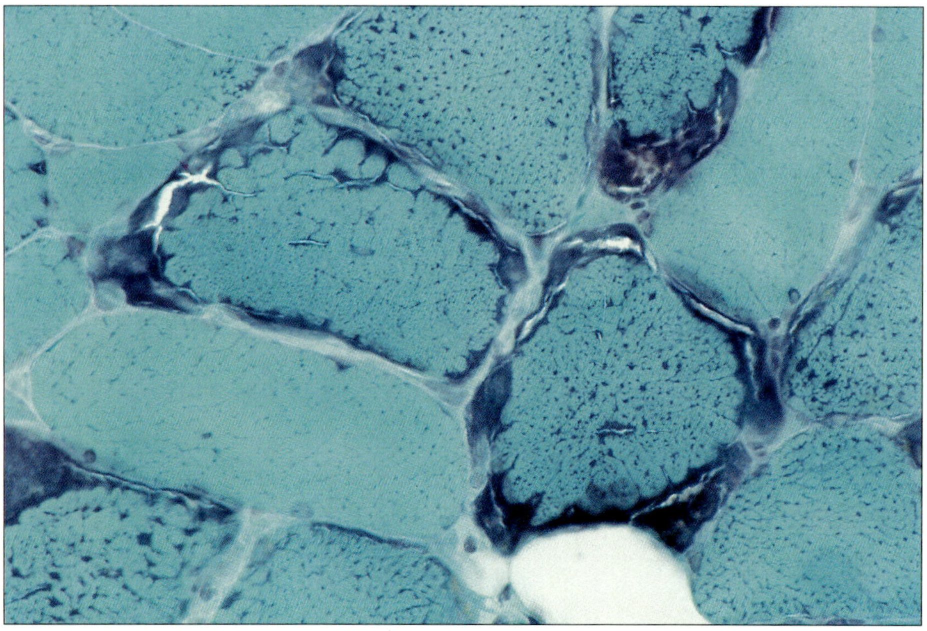

b) Gomori trichrome, ×350.

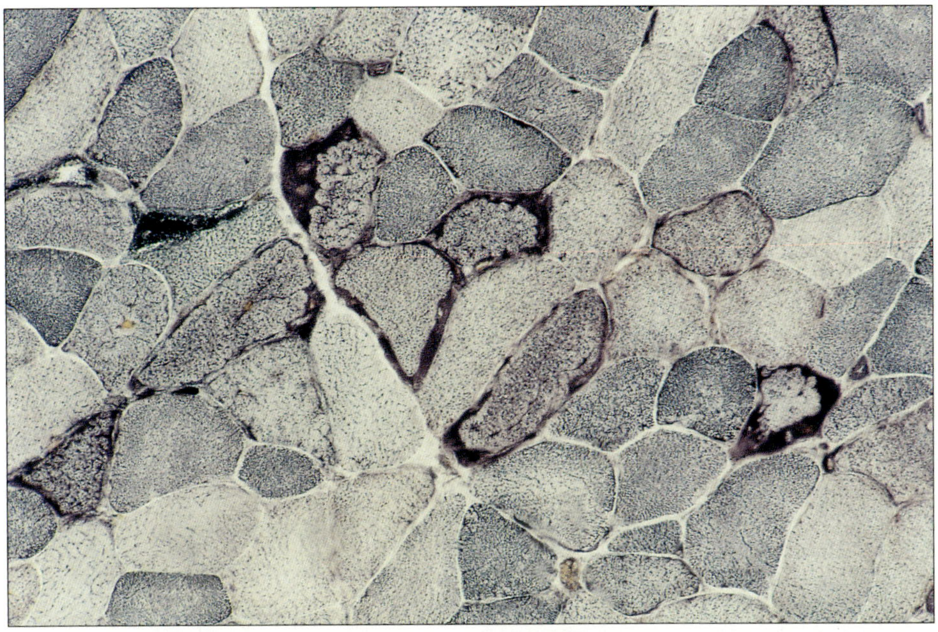

9.31 Dense subsarcolemmal accumulations of enzyme. (NADH-Tr, ×175)

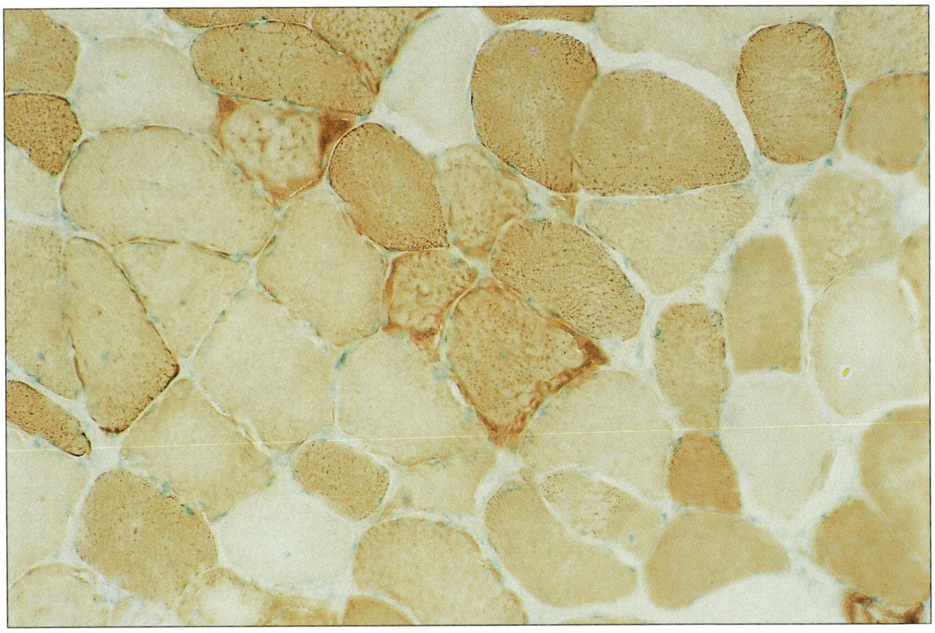

9.32 Mitochondrial myopathy in which some fibres show increased enzyme activity, particularly in the sub-sarcolemmal regions, and others show relatively poor staining..
a) Cytochrome oxidase, ×175;

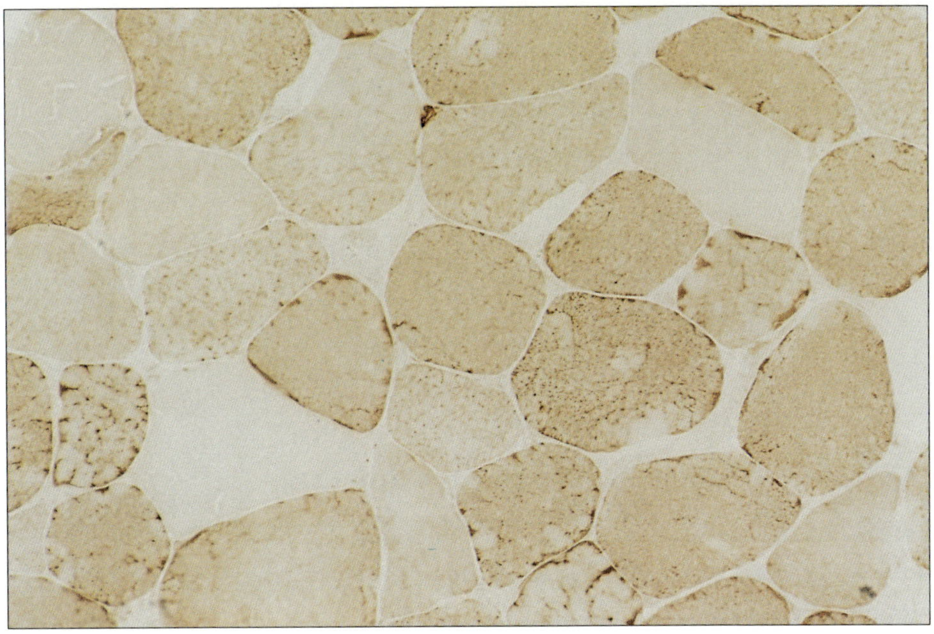

b) Cytochrome oxidase, ×263)

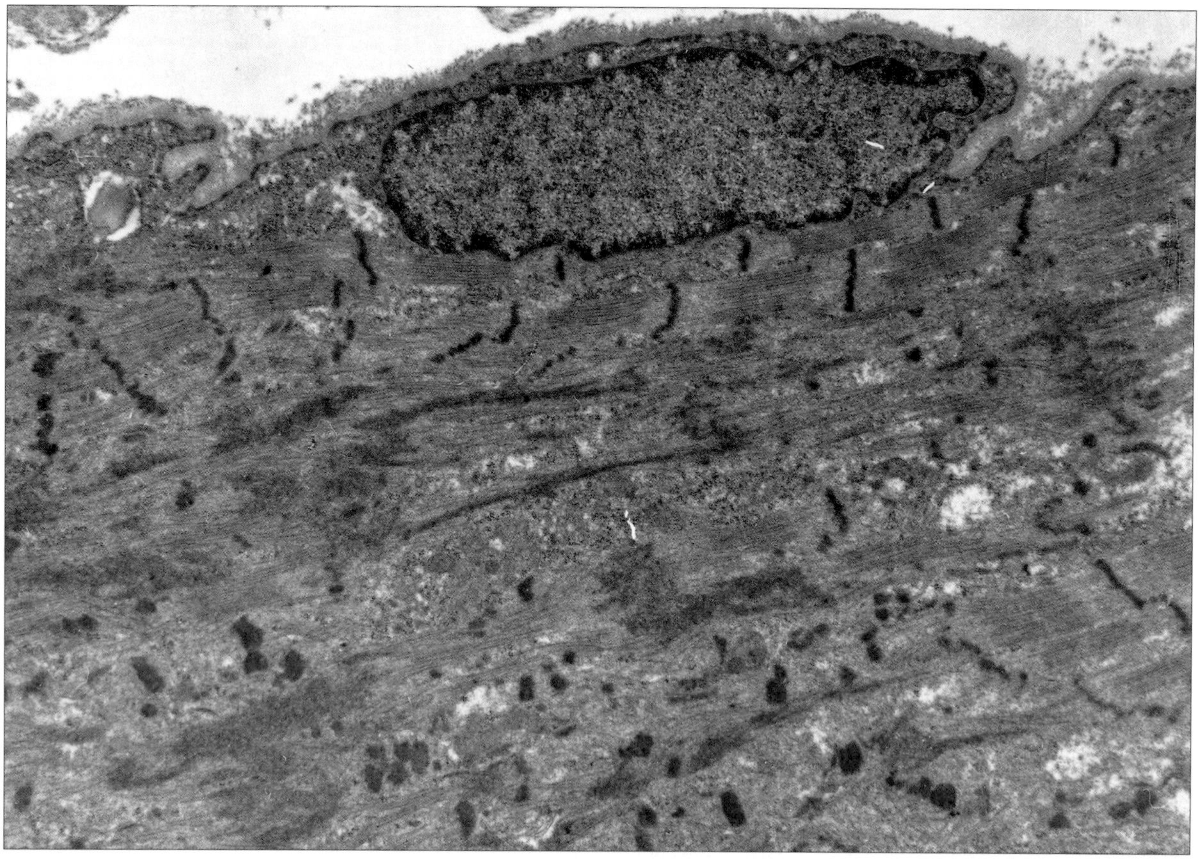

9.33 Myofibrillar disruption, severe Z-band streaming, and loss of mitochondria are seen as signs of fibre degeneration. (×11100)

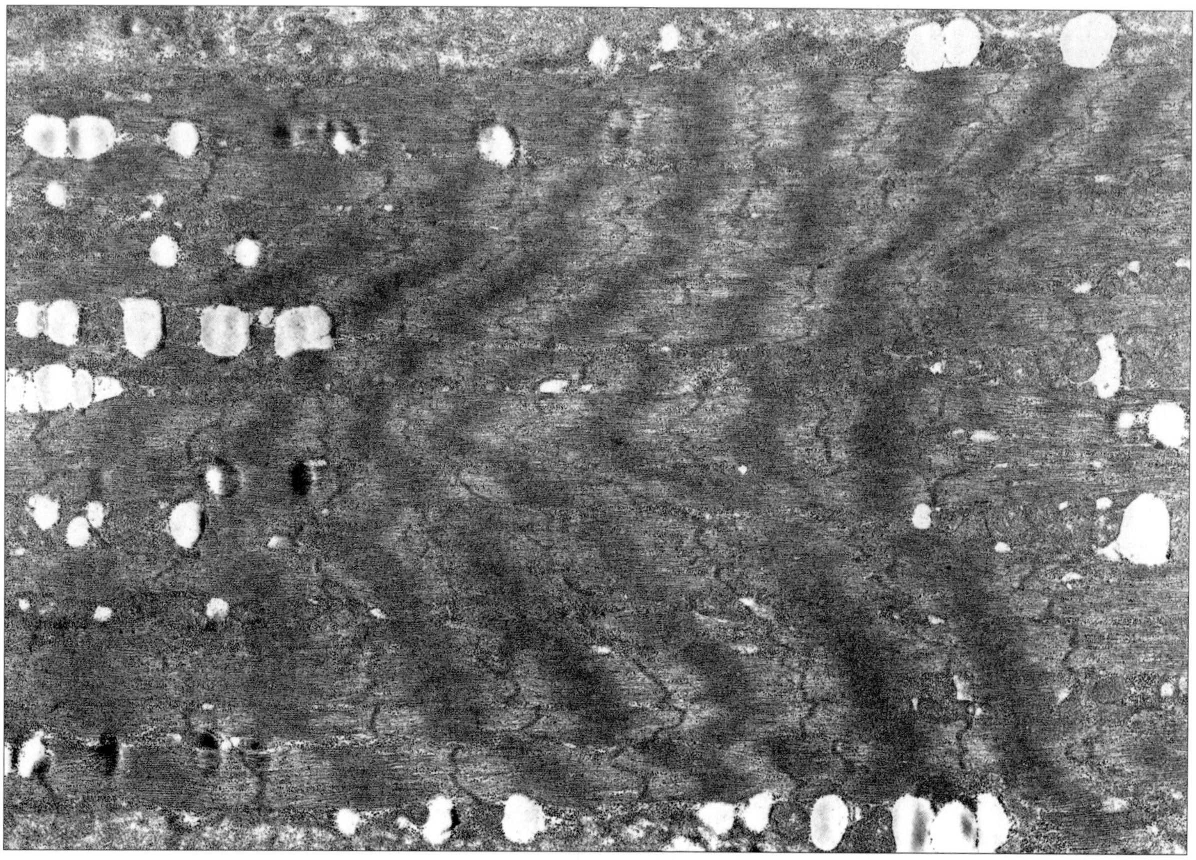

9.34 Z-band misalignment, together with excessive amounts of glycogen and lipid droplets are also signs of fibre degeneration. (×10000)

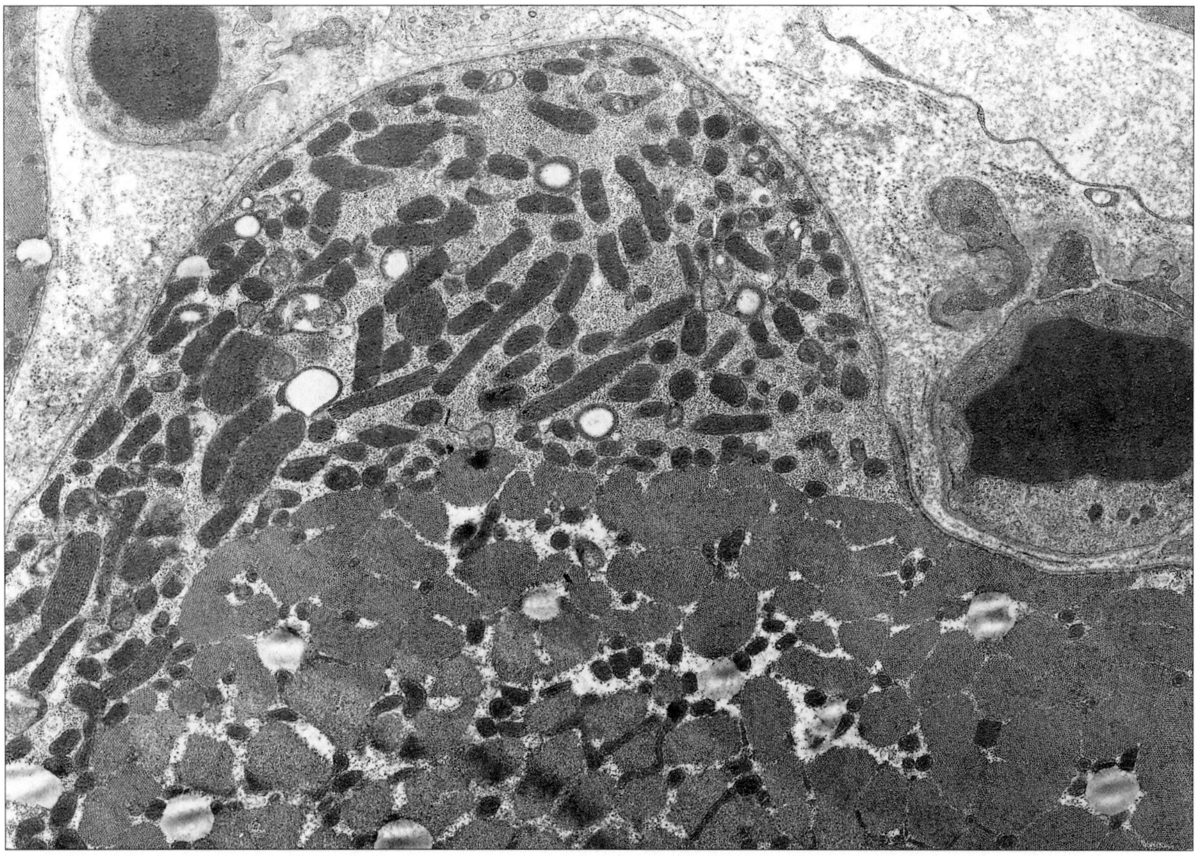

9.35 Increased numbers of mitochondria may be found within the fibres. They are found most commonly as large subsarcolemmal aggregates. In this illustration, the mitochondria are large and abnormal. (×9600)

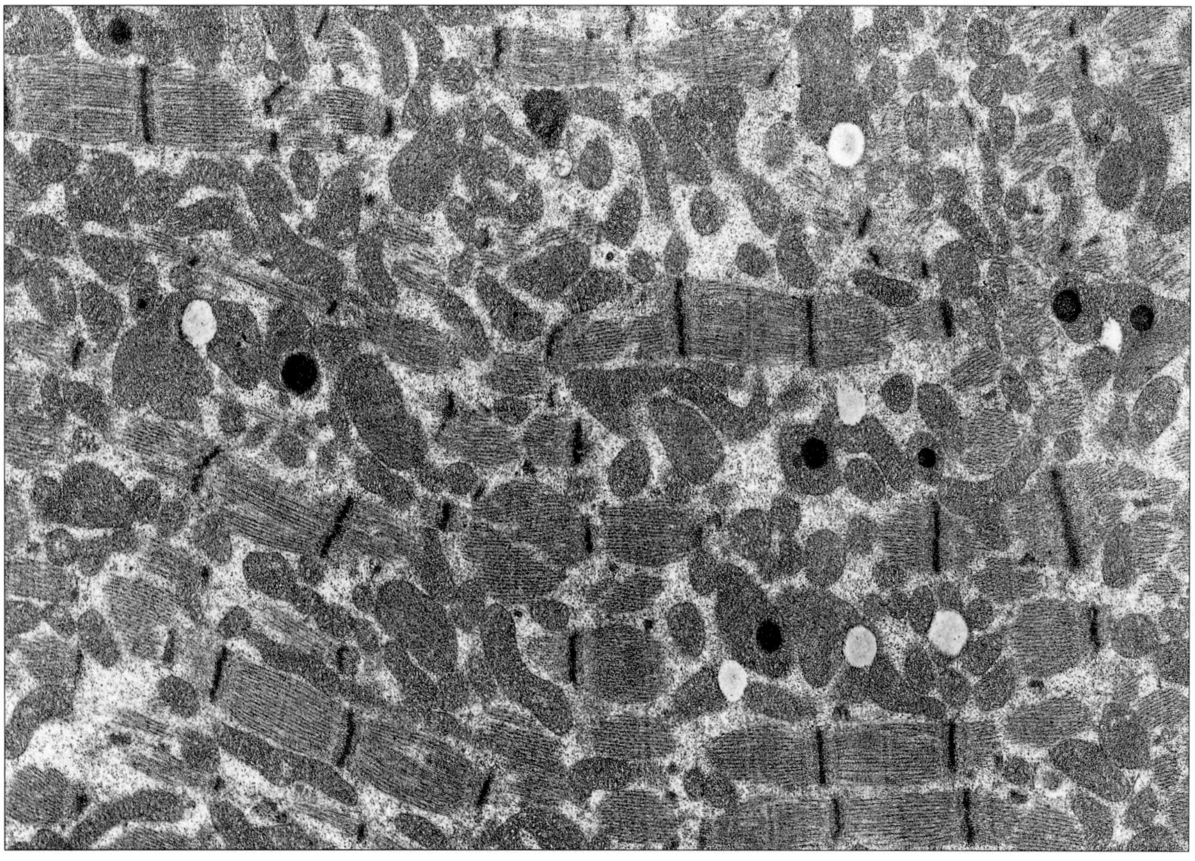

9.36 Randomly distributed mitochondria may also be observed. Those illustrated lie between misaligned myofibrils, and have electron-dense inclusions. (×14000)

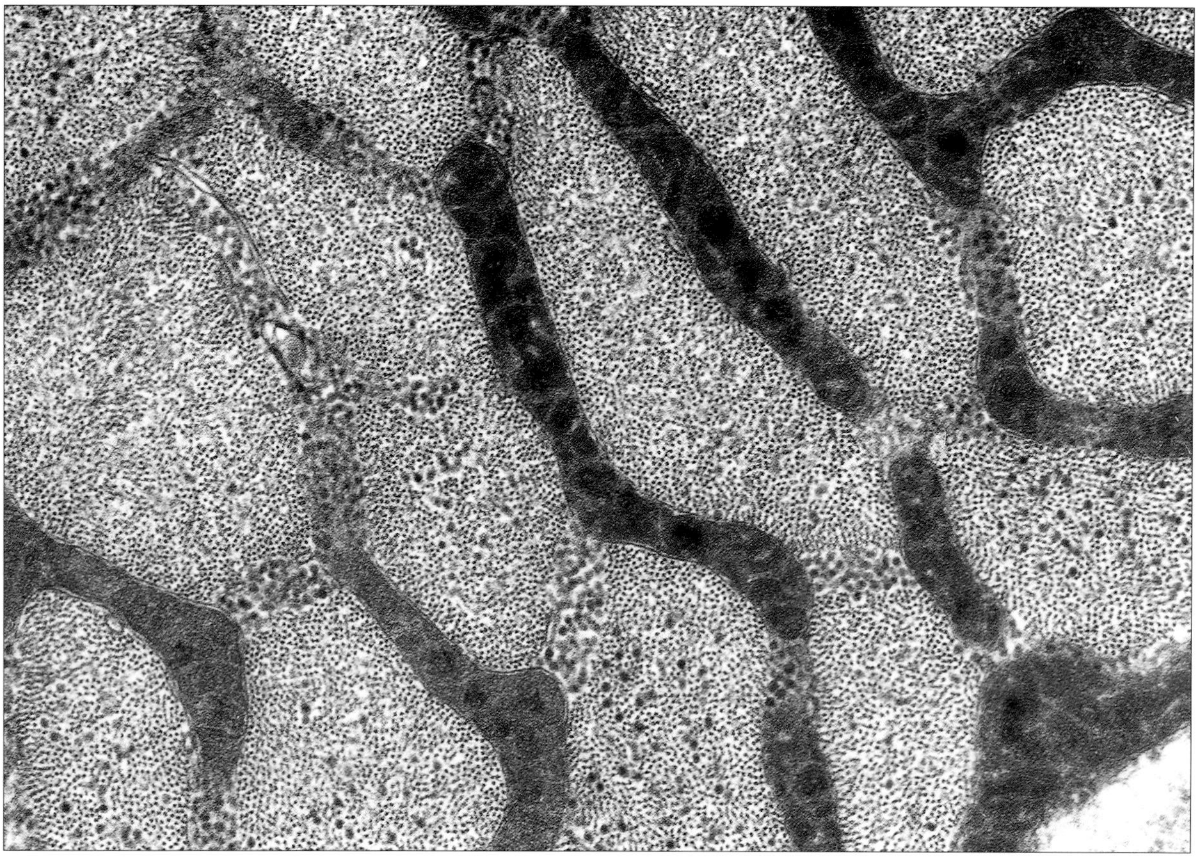

9.37 Mitochondria in which the cristae have been replaced by electron-dense structures and glycogen-like granular material. (×60000)

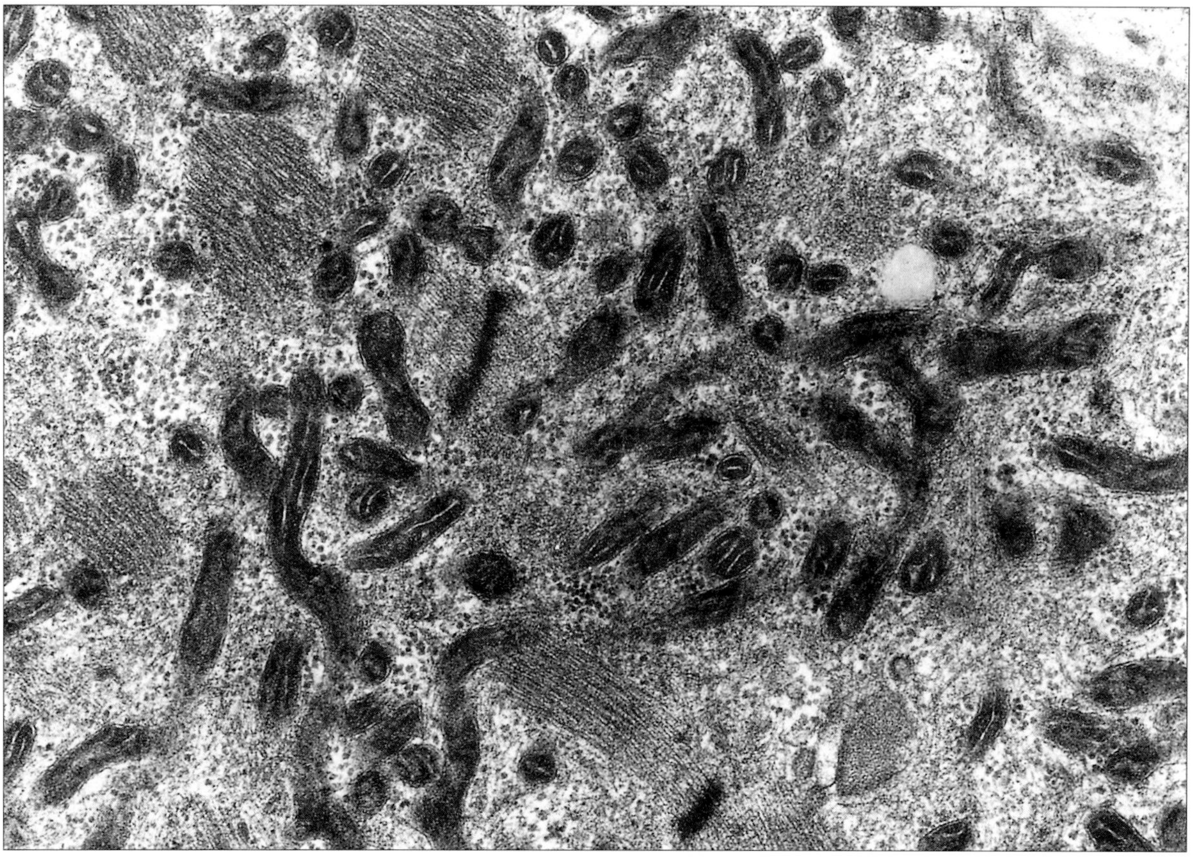

9.38 Electron-dense, thickened and disorientated cristae may be observed. (×37500)

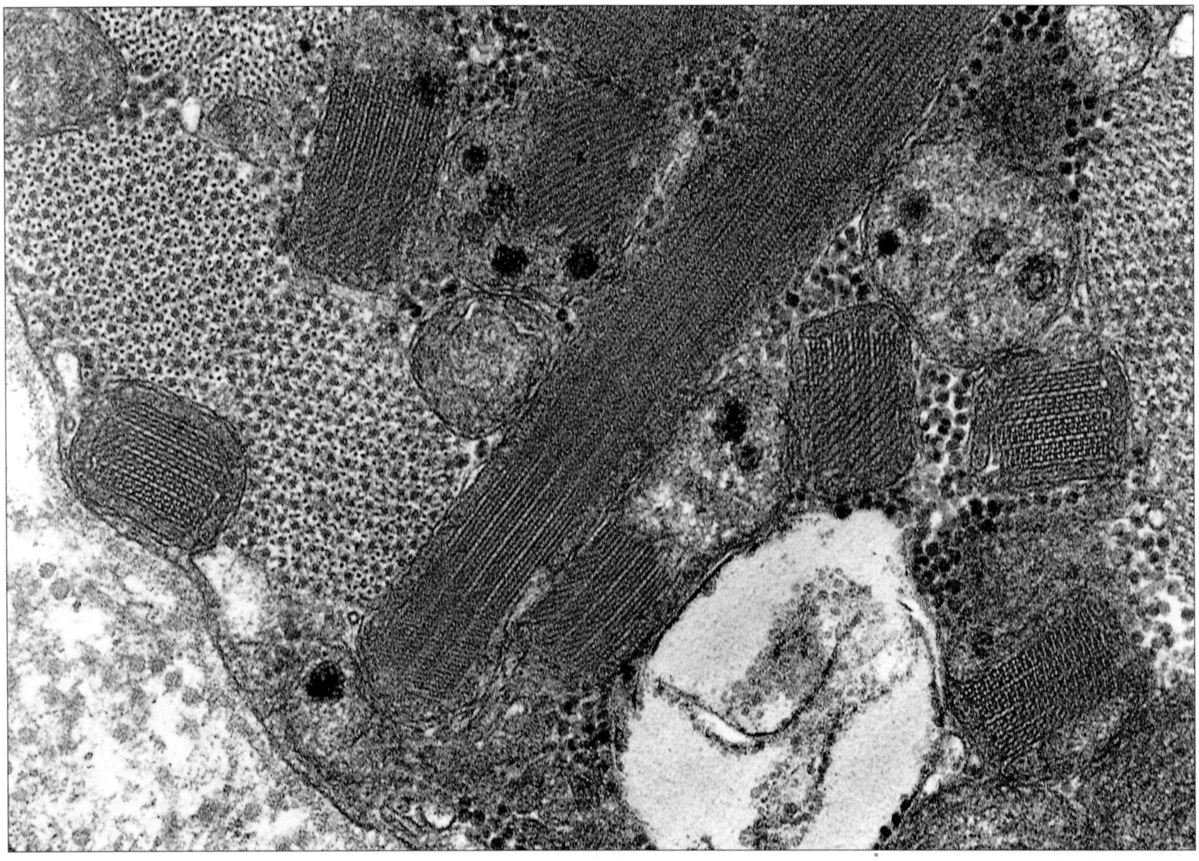

9.39 Paracrystalline inclusions within the mitochondria are considered to be due to accumulated creatine kinase. The inclusions vary in length and orientation. (×82500)

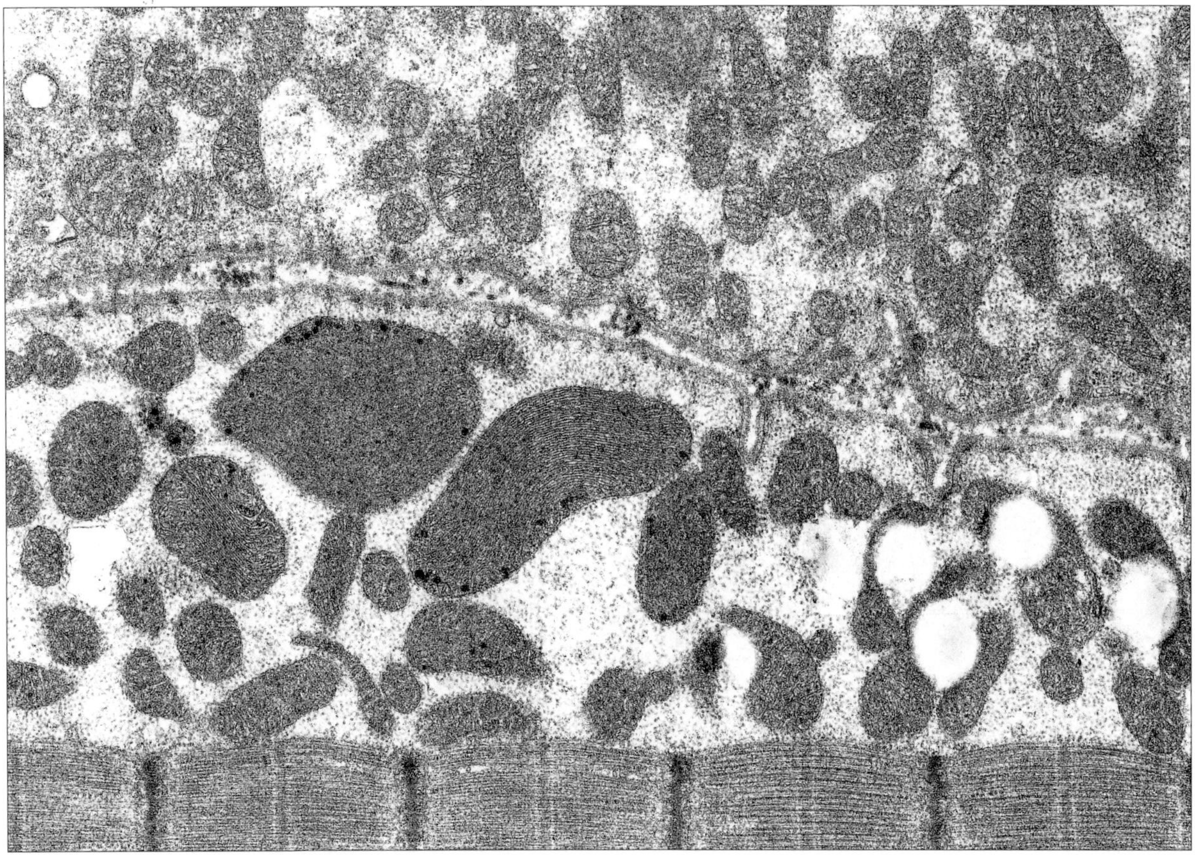

9.40 Abnormal mitochondria may be observed to be present in one fibre, whilst those in neighbouring fibres appear normal. (×24000)

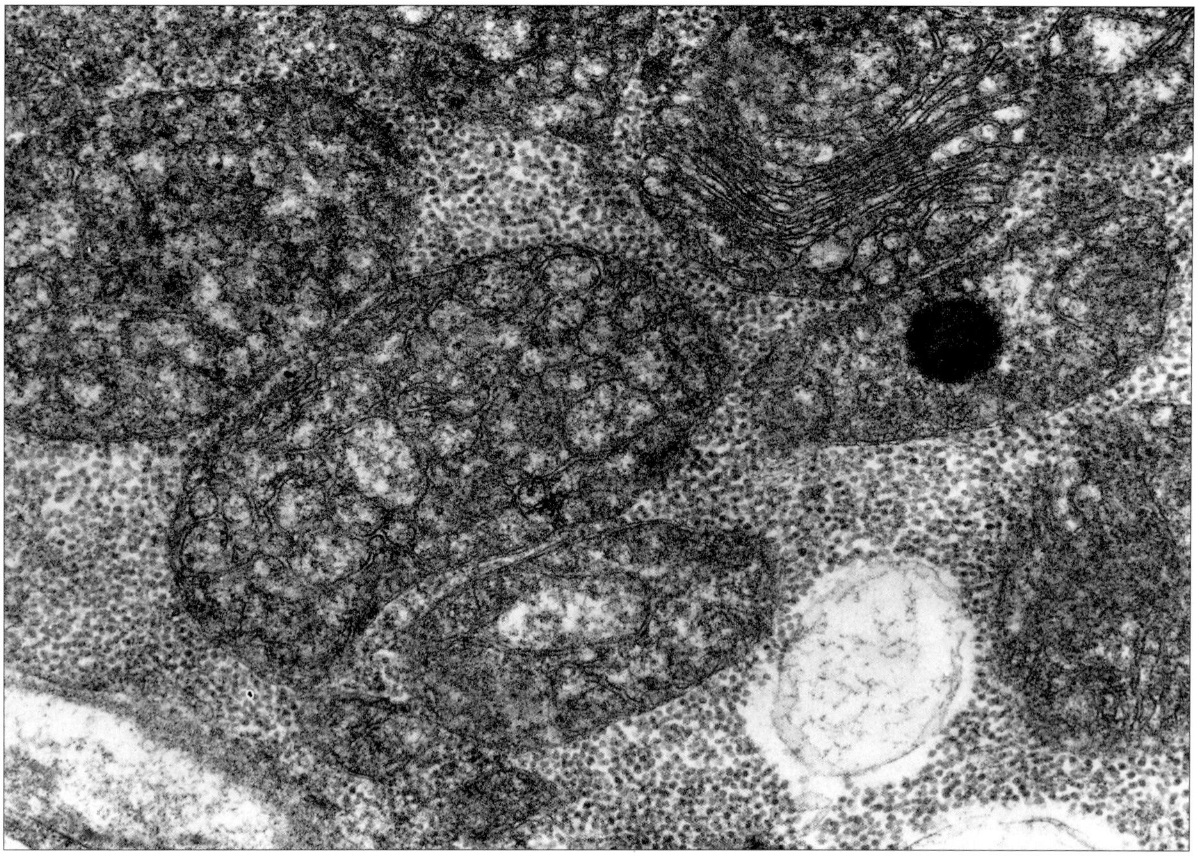

9.39 Paracrystalline inclusions within the mitochondria are considered to be due to accumulated creatine kinase. The inclusions vary in length and orientation. (×82500)

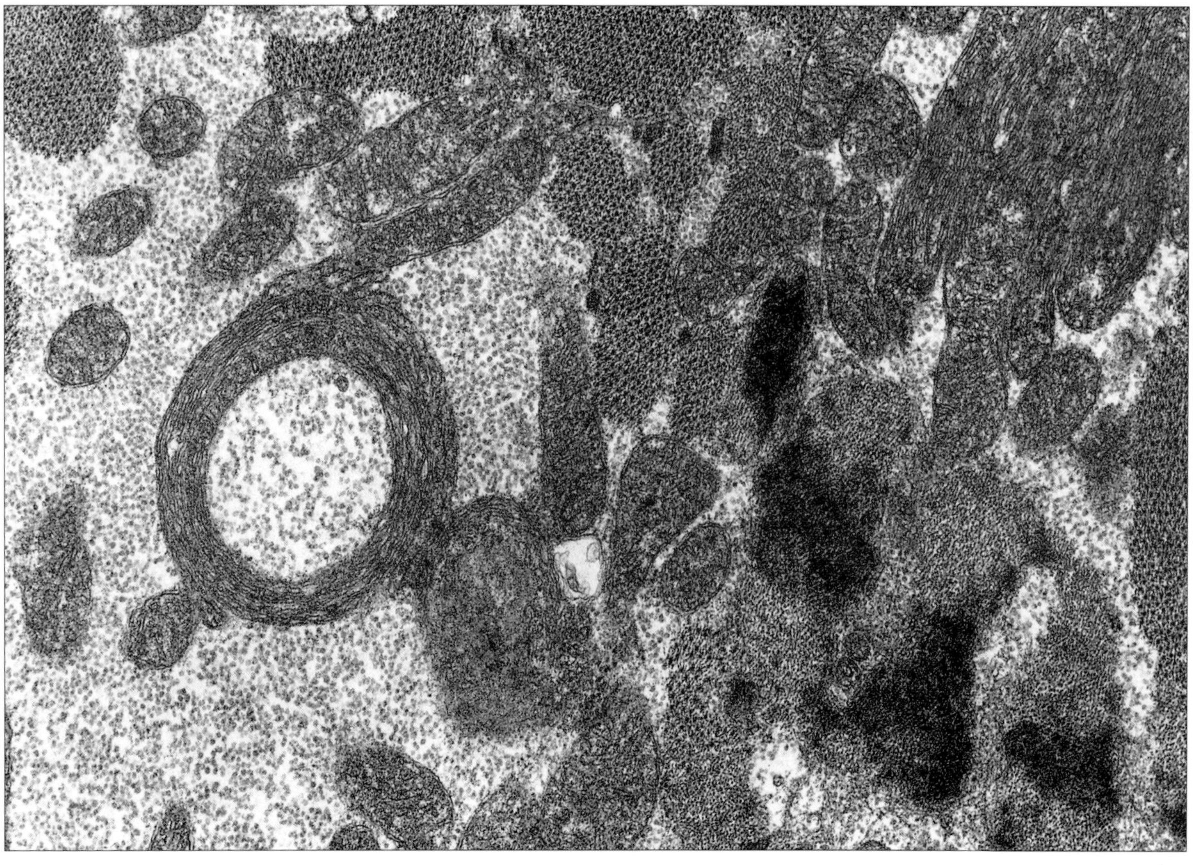

9.40 Abnormal mitochondria may be observed to be present in one fibre, whilst those in neighbouring fibres appear normal. (×24000)

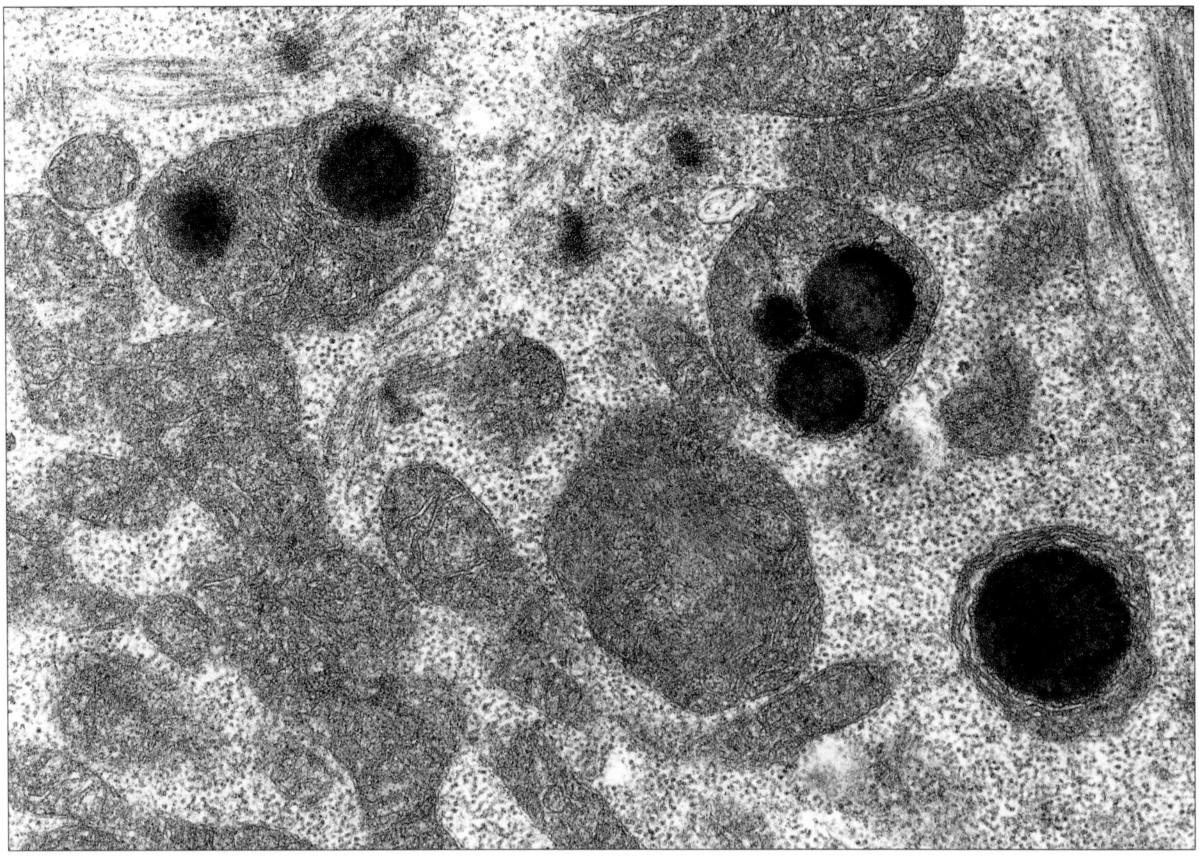

9.43 Multiple electron-dense inclusions may be present, some surrounded by concentric lamelli. In other mitochondria, the cristae are very indistinct. (×45000)

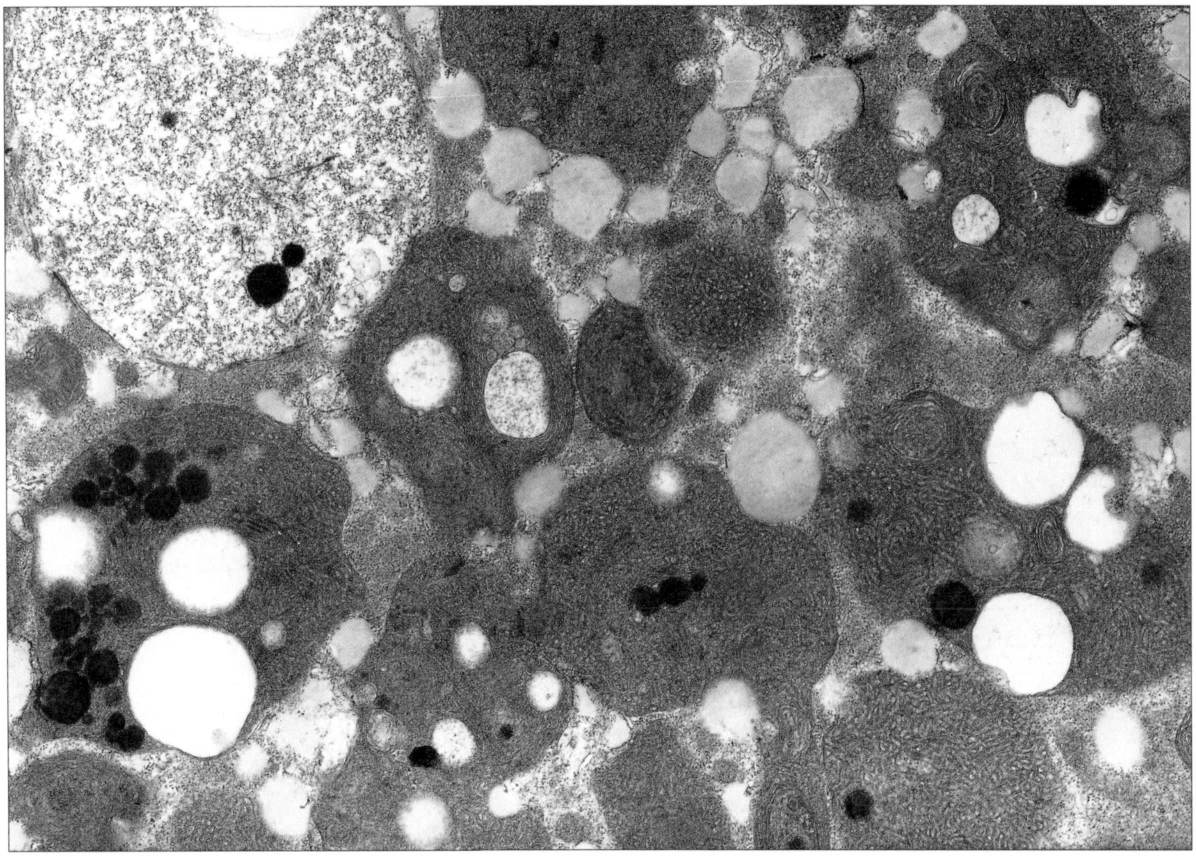

9.44 Other bizarre forms of mitochondria are seen, with both abnormal cristae and multiple inclusions being vacuolar and electron-dense. Many lipid droplets are associated with these abnormal mitochondria. (×17500)

10

Miscellaneous disorders

The illustrations in this volume have all been culled from our own experience. However, we have chosen not to include illustrations for some disorders, not necessarily because they are unavailable, but because we no longer routinely biopsy patients with these conditions.

With particular reference to myasthenia gravis (MyG), none of the authors has any personal experience with the immunocytochemical techniques which have formed the basis of Engel's descriptions of the changes at the motor end-plate (for a detailed review see Engel, 1992).

Patients with unequivocal, clinically identifiable motor neurone disease (MND) are not usually biopsied. However, in some cases it may be difficult to distinguish their presenting signs from those of patients with acid-maltase deficiency (Type II glycogenosis). To this end, (**10.1**) is included to show the classic features of denervation in motor neurone disease.

Likewise, a case of parasitic infestation is included (**10.2**), and in this case an inflammatory response is shown.

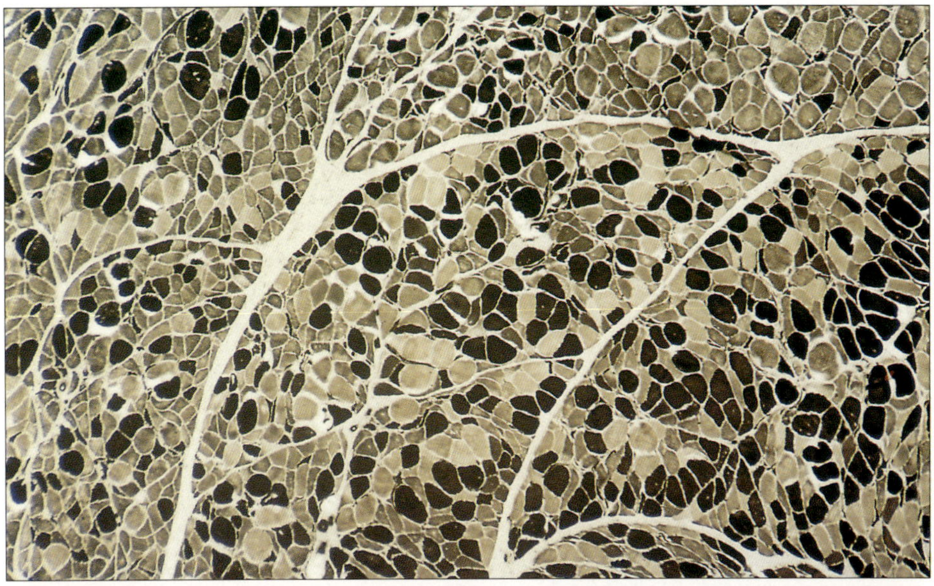

10.1 Motor neurone disease, showing the large variation in muscle fibre size due mainly to the presence of numerous atrophic angulated fibres of both types.
Intermediate staining intensities, and grouping of some fibre types are also evident.
(Myofibrillar ATPase, pH 4.35, ×35)

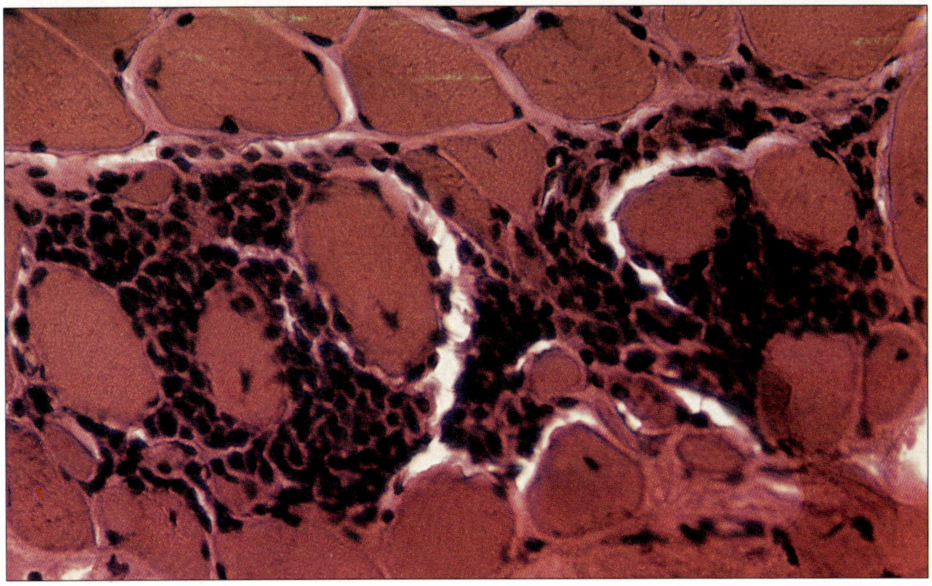

10.2 Massive, localised, inflammatory infiltrate in a muscle with parasitic infestation.
(H&E, ×210)

Myotonic Dystrophy (MyD)

Myotonic dystrophy (MyD) is characterised by Type I fibre atrophy and a marked increase in internal nuclei (**10.3**). In longitudinal section, these may be seen to be arranged in chains. The later stages of the disease are characterised pathologically, by 'ring' fibres (**4.19–4.20**), peripheral sarcoplasmic masses and gradually increasing interstitial fibrosis, although extensive fibre hypertrophy, splitting (**4.14**), patchy oxidative enzyme staining (**10.4**) and tubular aggregates (**10.5–10.7**, **4.73–4.74**) may also occur. It should be emphasised, however, that these abnormalities are non-specific.

In the congenital myotonic disorders, muscle biopsy will usually demonstrate considerable hypertrophy of individual muscle fibres (**4.12**) with occasional internal nuclei but no other evidence of myopathic degeneration. In addition, the histochemical profile of these fibres is usually normal, although tubular aggregates may be seen in Type IIa and IIb fibres (**10.5–10.6**).

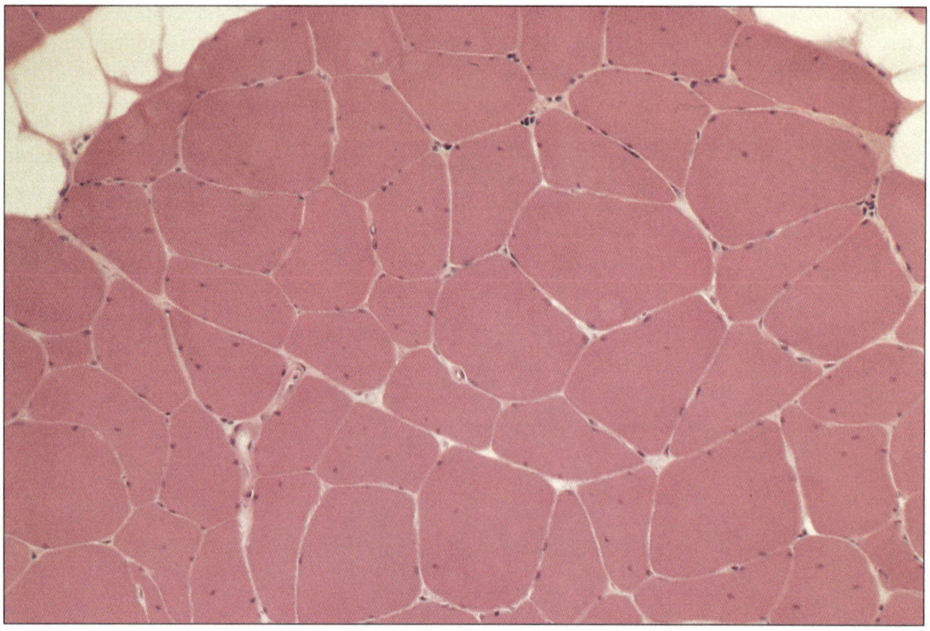

10.3 Myotonic dystrophy, showing large variation in fibre size and the presence of some extremely hypertrophic fibres. Internal nuclei are particularly abundant in this disorder. (H&E, ×140)

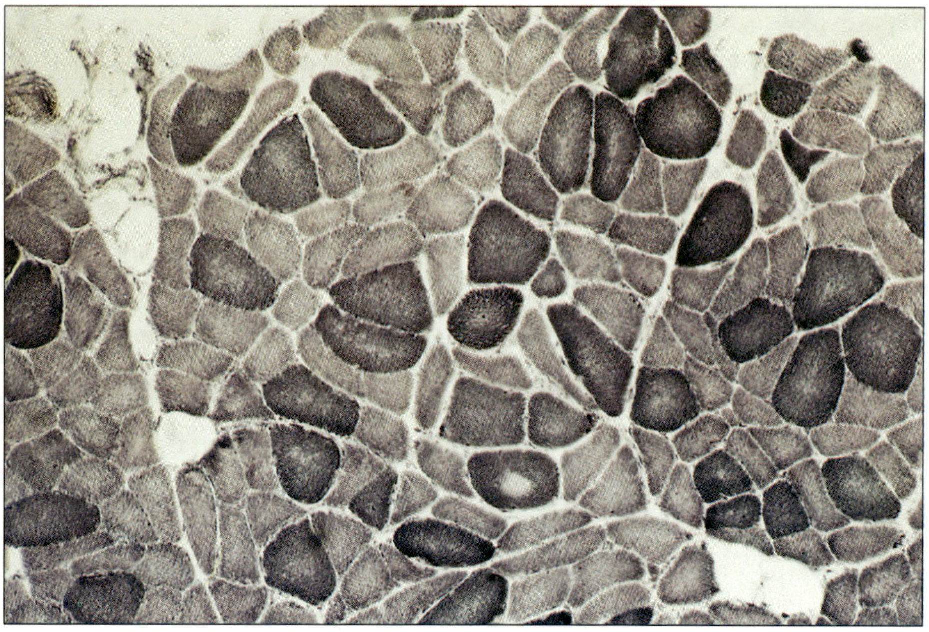

10.4 Myotonic dystrophy (from the same patient as **10.3**), showing Type I hypertrophy and central areas of decreased staining intensity. (NADH-Tr, ×70)

Periodic Paralysis (PP)

The muscle that is taken from patients with Periodic Paralysis (PP) is almost always histologically normal, with vacuolar changes in scattered fibres only being found during the attack in a few cases. Even more rarely, tubular aggregates may be seen (**10.5-10.8, 4.73–4.74**).

In the later stages of the natural history of hypokalaemic periodic paralysis, the patients may develop a fixed and occasionally severe proximal myopathy, which is characterised histologically by the persistence of the vacuolar change described above, with numerous 'groups' of extremely atrophic fibres, often containing nuclear clumps. Their appearance of these is similar to that seen in long standing denervating processes such as SMA (see Chapters 4 and 6).

Patients with hyperkalaemic periodic paralysis rarely develop proximal myopathies.

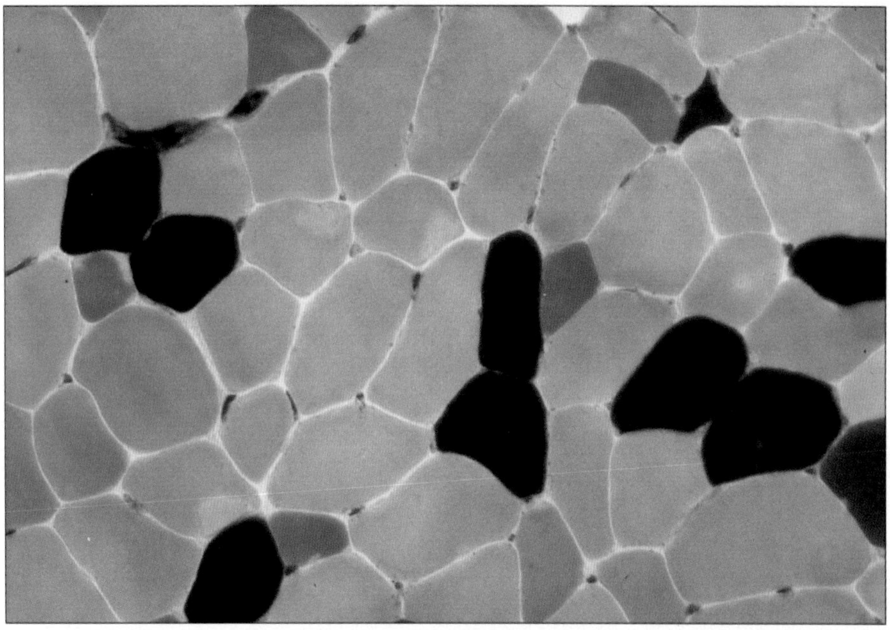

10.5 Pale areas in the subsarcolemmal region of some Type IIa and IIb fibres are due to the presence of tubular aggregates. (See also **10.6**.) (Myofibrillar ATPase, pH 4.6, ×140)

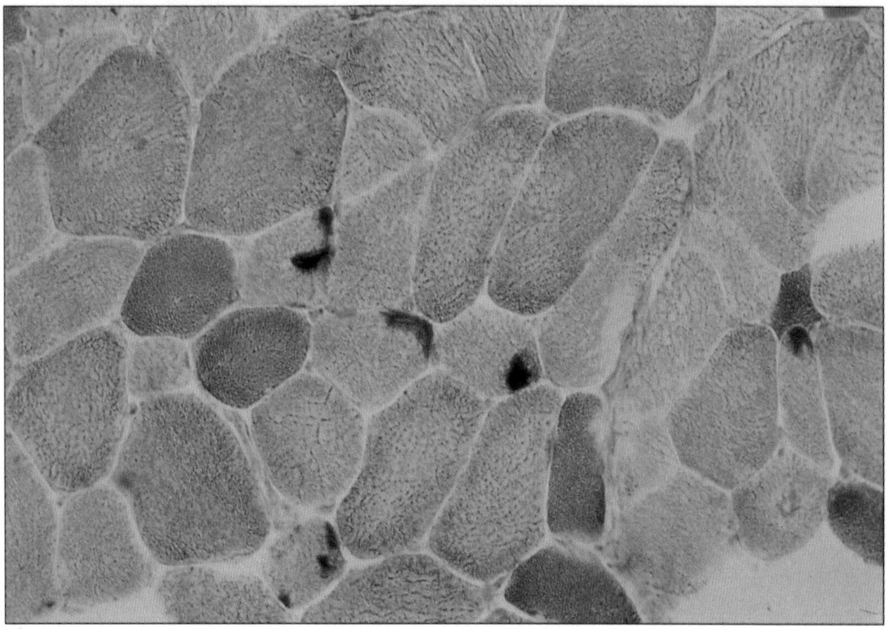

10.6 Tubular aggregates in the subsarcolemmal regions of a number of poorly oxidative fibres stain densely for this enzyme. (Serial section to **10.5**). (NADH-Tr, ×140)

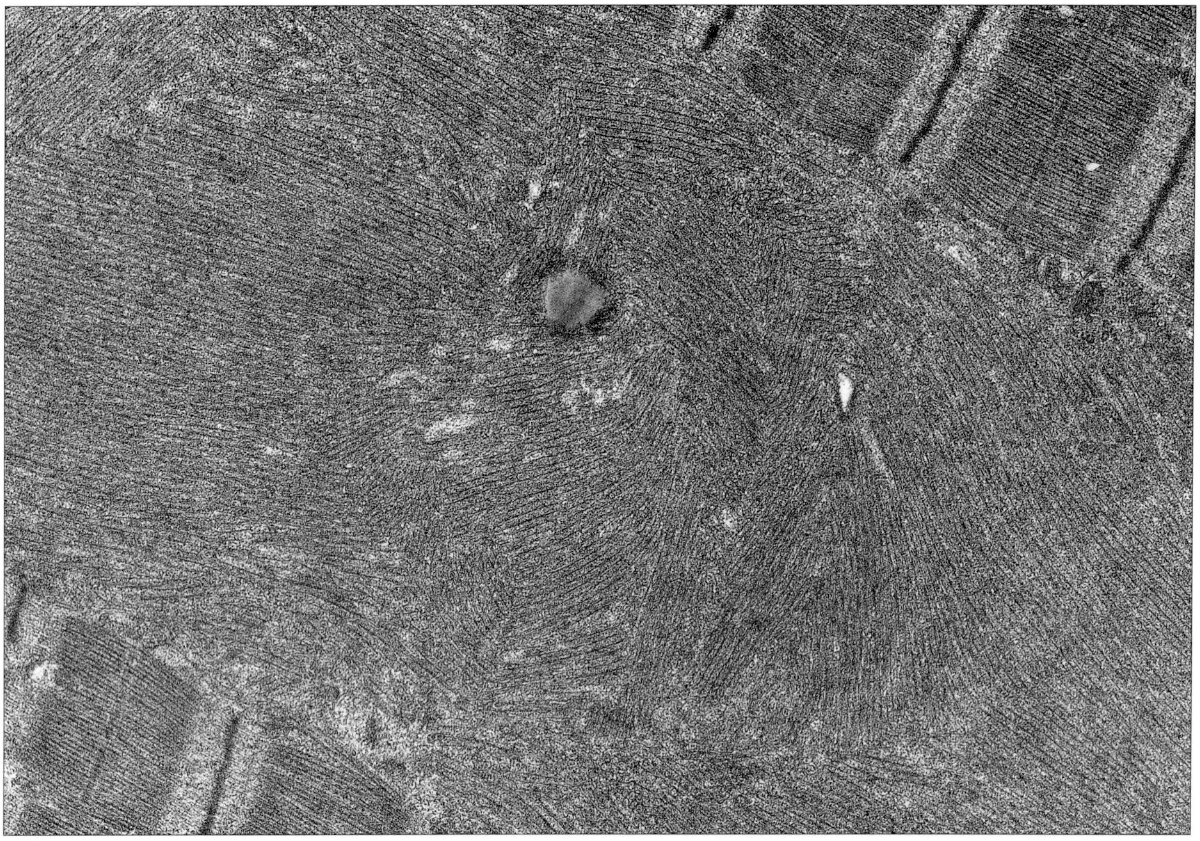

10.7a A large interfibrillar aggregate of randomly orientated tubules. (×20000)

10.7b Cut in transverse section, the internal structure of the tubules in such an aggregate can be seen. (×60000)

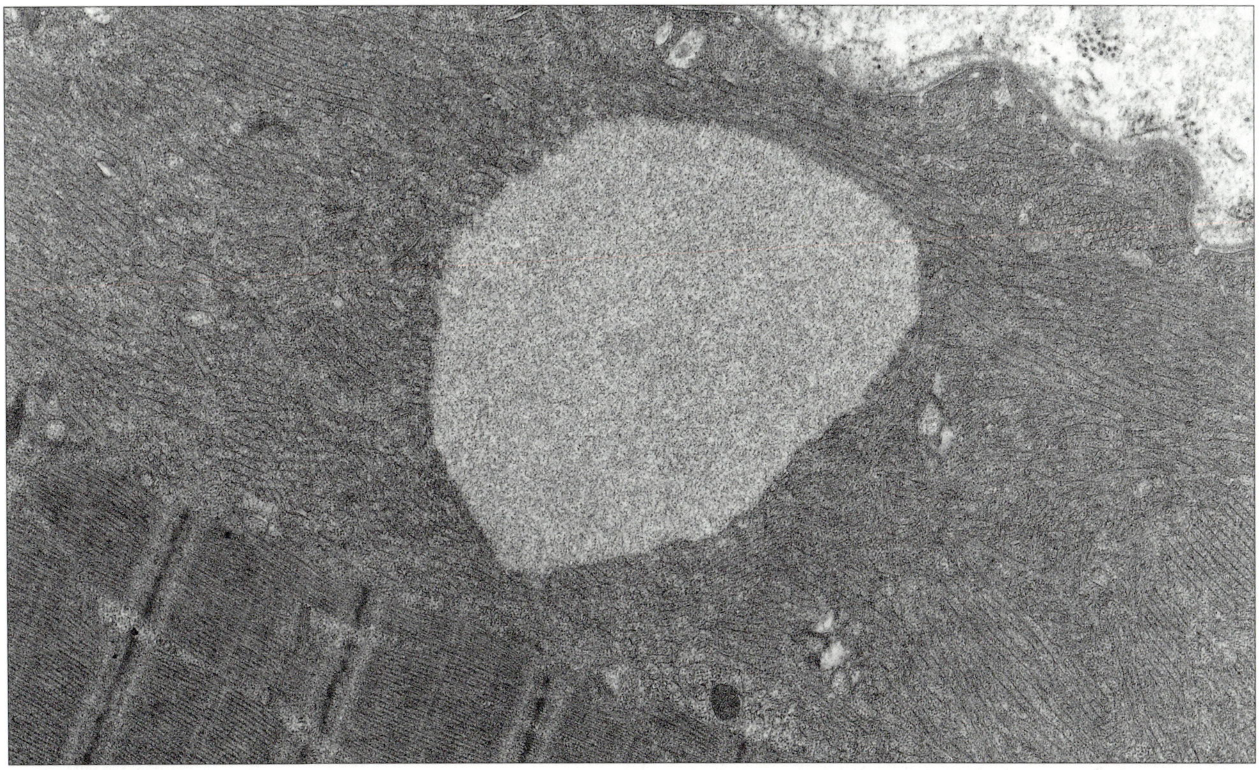

10.8 Vacuoles containing a fine granular material of unknown origin are occasionally seen in association with tubular aggregates. (×18475)

Endocrine disorders

Endocrine disorders, in particular those involving the thyroid, pituitary or adrenal gland, appear to have a direct effect on the neuromuscular system, often causing weakness and occasionally a distinctive myopathology.

Hypothyroid myopathy is almost invariably associated with excess (up to 85%) Type I fibres (**10.9**), often accompanied by Type II atrophy. In hyperthyroidism the muscle biopsy is usually normal.

In cases of pituitary hypersecretion (e.g. acromegaly), a wide variety of non-specific changes are apparent both with the light and electron microscope. Many of these changes involve disturbances in mitochondrial activity (**10.10**).

Adrenal cortex lesions (Cushing's disease) are also associated with mitochondrial abnormalities in skeletal muscle. The most conspicuous change, however, is the marked Type II fibre atrophy attributed to glucocorticoid excess (*see* 'Steroid Myopathy' *below*).

In pancreatic disorders, muscle changes appear to be a consequence of ischaemia (microvascular alterations) or diabetic neuropathy (denervation atrophy), and are not considered further here.

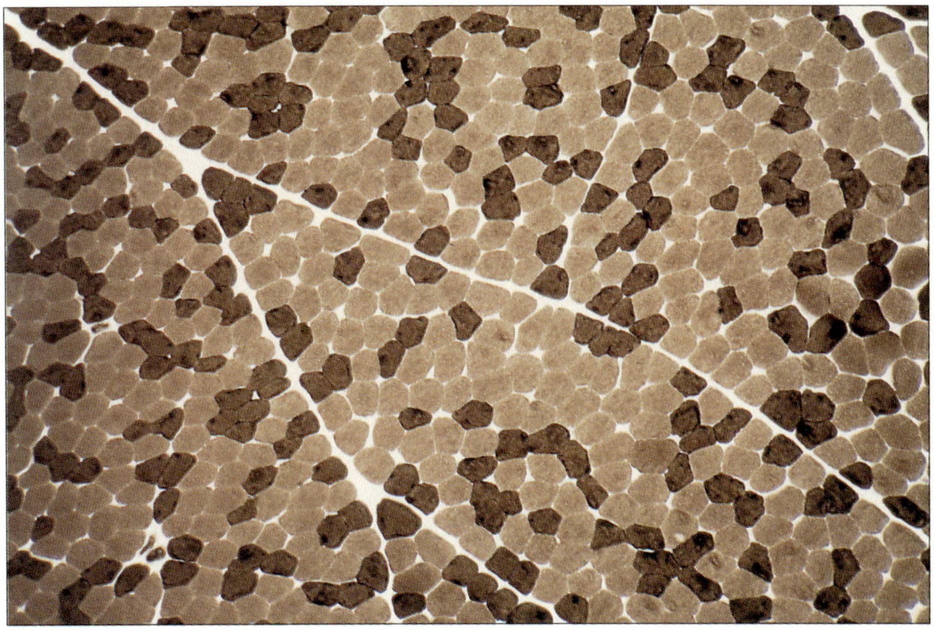

10.9 Excess Type I fibres in a case of hypothyroid myopathy. (Myofibrillar ATPase, pH 10.2, ×175)

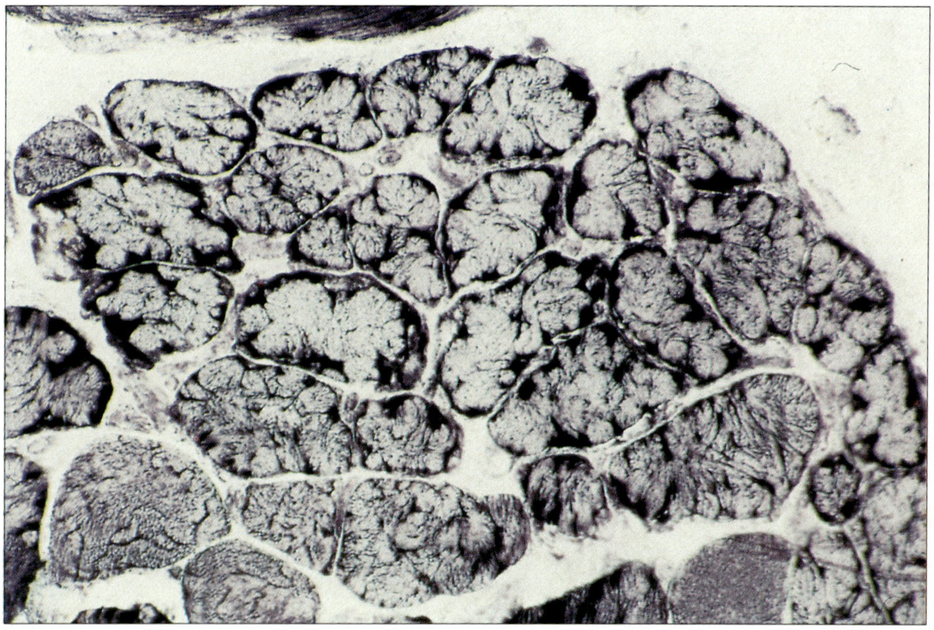

10.10 The disturbed amount and distribution of mitochondrial enzyme activity, producing this lobulated appearance, is one of the non-specific alterations associated with acromegaly. (NADH-Tr, ×263)

Drugs, toxins and trauma

The reactions of the muscle cell to injury by toxins and drugs are limited, and usually take one of four possible forms:

- Myonecrosis, phagocytosis and regeneration.
- Vacuolar myopathy.
- Type II fibre atrophy.
- Disturbance of mitochondrial enzyme distribution and function.

In practice, a specific Type IIb fibre atrophy (identical to that depicted in **4.11**) is by far the commonest fibre atrophy seen, and is an almost universal consequence of prolonged courses of treatment with adrenal corticosteroids ('Steroid Myopathy'). Corticosteroids in very high doses may also induce an acute necrotising myopathy. This may relate to abrupt changes in serum potassium.

Occasionally, combined myotoxic and neurotoxic reactions will be seen to drugs and other toxic substances.

Grossly excessive alcohol abuse may induce massive rhabdomyolytic attacks ('Alcoholic Myopathy'), with a corresponding risk of renal failure. Prolonged, excessive alcohol abuse may be associated with extensive degeneration and regeneration, and an accompanying inflammatory reaction in skeletal muscle (**7.18**).

Chronic alcoholics may also develop a much more banal proximal myopathy, the pathogenesis of which remains uncertain.

Similar episodes may punctuate the natural history of opiate abuse, particularly with heroin, although these are thought to be idiosyncratic reactions rather than the result of a direct toxic effect on the muscle cell.

Patients who abuse diuretics and purgatives, or who have been treated for peptic ulcers with liquorice-derivatives (carbenoxolone), sometimes develop severe vacuolar myopathies due to potassium depletion. Very occasionally, these may culminate in rhabdomyolytic episodes.

Vacuolar myopathies have also been reported in patients taking chloroquine or other 4-amino quinolones. Barbiturate overdose may induce mitochondrial enzyme disturbances and produce core and target-like appearances within the muscle fibres.

Trauma causes extensive muscle degeneration, necrosis and fibrosis (Volkmann's ischaemic contracture). However, this can be minimised if an adequate vascular supply is maintained, and complete regeneration may occur, leaving only internal myonuclei as evidence of previous damage.

One of the most dramatic necrobiotic myopathies seen recently occurred in a young boy who was badly burned in a 'chip-pan fire'. The muscle biopsy revealed widespread necrosis, phagocytosis and early regeneration, with foci of myofibre calcification (**10.11-10.14**). The appearance of this muscle is reminiscent of the changes occurring in skeletal muscle damaged by snake and spider venoms. It is unusual to find such dramatic histological abnormalities in either toxic or drug-induced myopathies, although epsilon aminocaproic acid (EACA) and clofibrate can induce severe muscle necrosis, sometimes culminating in massive rhabdomyolysis and renal failure.

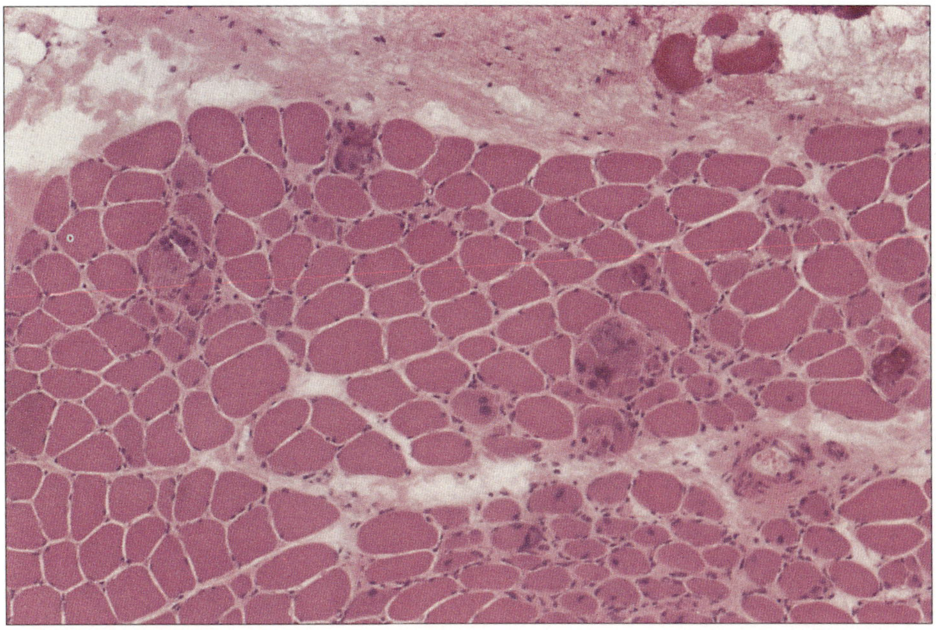

10.11 Muscle tissue damaged by a scalding burn injury (see also **10.12–10.14**). Foci of damage are associated with increased nucleation and variation of fibre size. Increased connective tissue and oedema are also apparent. (H&E, ×35)

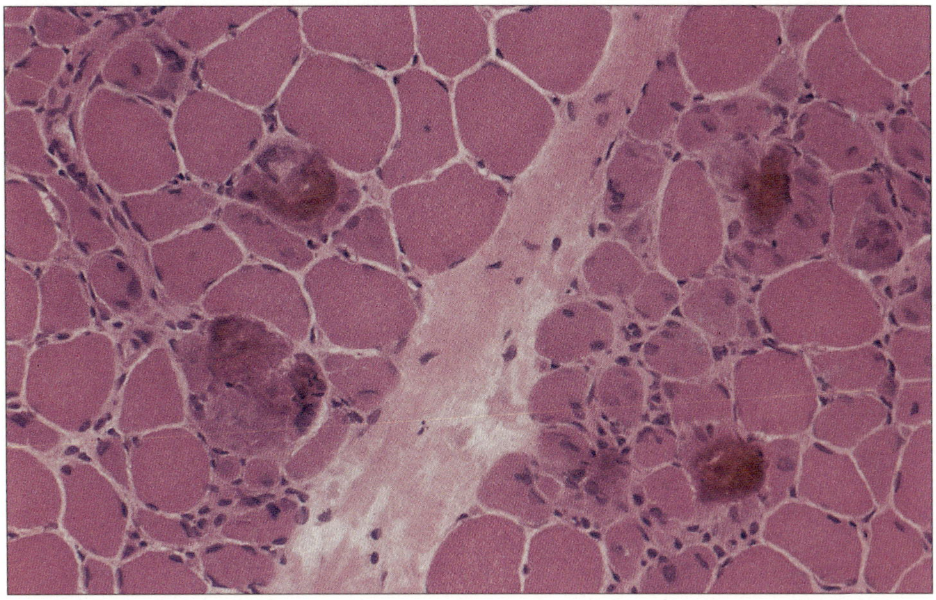

10.12 Higher magnification from the same biopsy as shown in **10.11**. Surrounding the densely stained necrotic regions, many of the smaller muscle fibres contain vesicular nuclei and appear to be regenerating. (H&E, ×140)

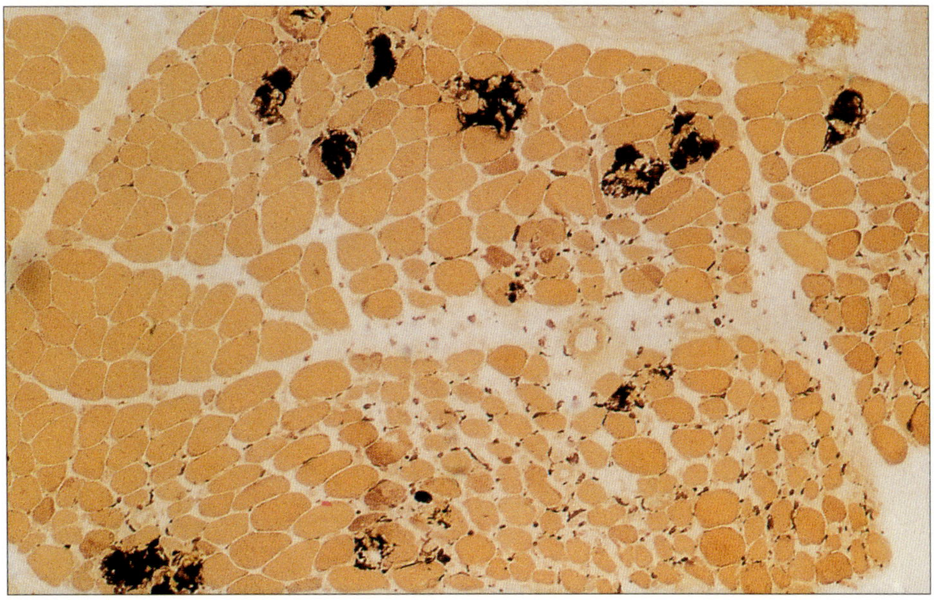

10.13 The damaged areas visible in **10.11** are associated with intense staining for lysosomal enzyme activity. (Nonspecific esterase, ×70)

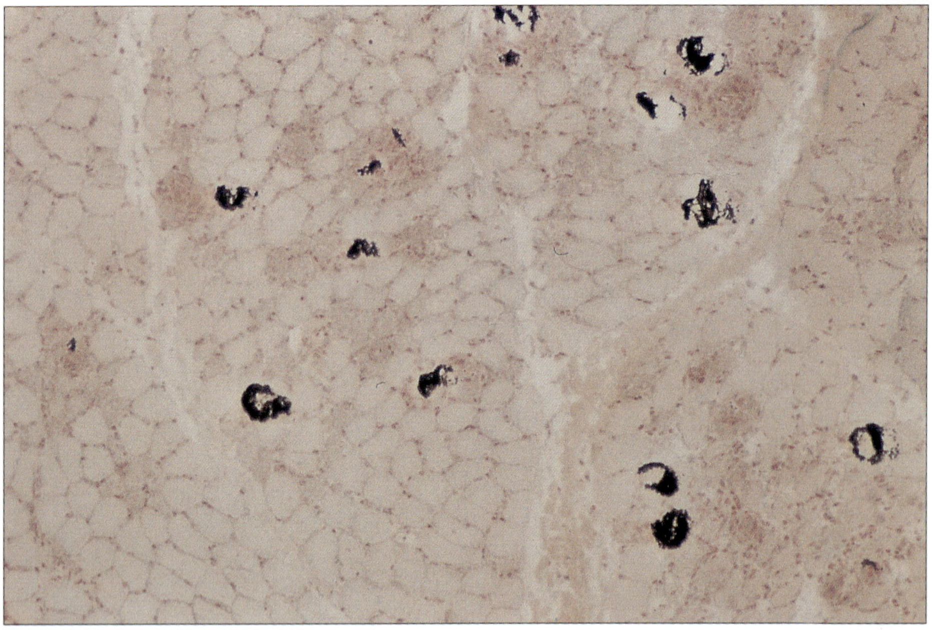

10.14 Note the accumulations of excess calcium in the damaged muscle fibres (see **10.11–10.13**).
(von Kossa, ×70)

11

Interpreting the biopsy

It has been stated several times throughout this book that the pathological features of the muscle biopsy ***must*** be interpreted alongside clinical and other investigations of the patient, in order to come to a decision on the differential diagnosis of the disorder and possible treatment or management of the patient.

This brief chapter underlines this salutation, by first outlining the major features of the clinical and laboratory investigations to be taken into account by the myopathologist, and then summarising the key pathological attributes of the major neuromuscular disorders.

Investigation of the patient with a suspected neuromuscular disorder

The main groups of routine investigations undertaken for patients with suspected neuromuscular disorders are illustrated in **11.1** and described briefly below.

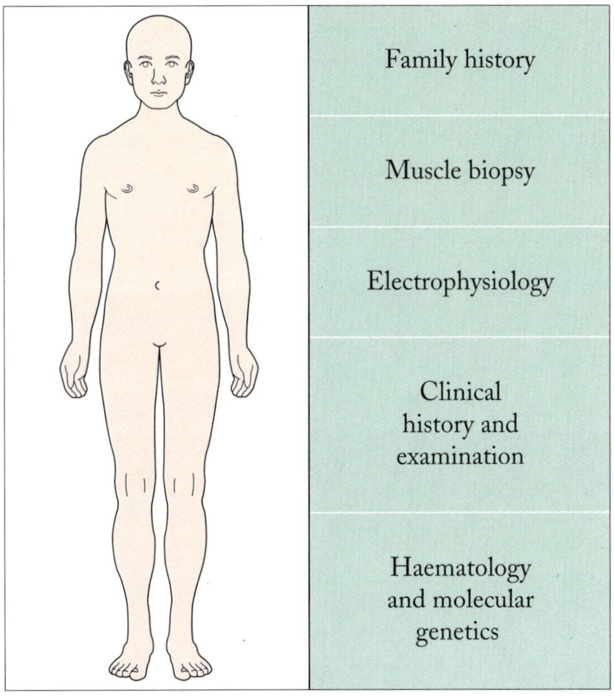

11.1 Routine investigations undertaken for patients with suspected neuromuscular disorders.

Family history

The differential diagnosis may depend on knowing whether the condition appears to have an acquired basis (e.g. in polymyositis and toxic myopathies), or a genetic basis that may be X-linked (e.g. the Duchenne or Becker muscular dystrophies and Kennedy type SMA), autosomal recessive (e.g. severe childhood muscular dystrophy, certain myotonias and metabolic myopathies, and childhood SMA), autosomal dominant (e.g. some FSH dystrophies, some congenital myopathies, ion channel disorders, and a number of neuropathies), or may have a maternal inheritance (e.g. some mitochondrial disorders). The reader is referred to Chapters 5-10, *Table 1.1* and Appendix F for further details.

Clinical history and examination

The age of onset (e.g. delayed motor milestones or decreased activities such as stair climbing), rate of progression (e.g. chronic or acute), anatomical distribution of weakness or wasting (e.g. symmetrical or asymmetrical, the presence or absence of scapular winging and calf hypertrophy, whether the condition is limb-girdle or distal, the presence of Gower's sign and lordosis and other clinical signs or symptoms (e.g. skin rash, myoglobinuria, pain on exercise, hypotonia, fatigue, abnormal reflexes, sensory loss, fasciculations) give strong clues to the aetiology of the clinical condition and pathological appearances.

For example, compare the acute weakness and pain, without muscle wasting, in an adult with polymyositis to the chronic, asymmetric, distal limb-muscle weakness and wasting of a patient with adult-onset spinal muscular atrophy, or the specific pattern of early muscle involvement of FSH.

Likewise, ptosis or diplopia, due to the involvement of external ocular muscles, is often suggestive of a mitochondrial disorder, and dysphagia may be indicative of MND. MRC, Vignos, Quadrant or direct myometric muscle strength ratings are particularly helpful in determining the distribution and rate of progression of muscle weakness, whereas CT scanning or MRI may assist in elucidating the distribution of preclinical muscle wasting. Again, the reader is referred to the specific comments in Chapters 5-10.

Normal **?** Abnormal

Pathological features	Normal	Myopathic		
CT: fibrosis	-	+/+	++/+++	-/+
Inflammation	-	++/++++	-/+	-
Fibre size variation	CV <25%	+/+++	++/++++	-/++
Regeneration	-	++/+++	++/++++	-/+
Morphological abnormalities	-	-	Hyaline fibres Fibrosis	Vacuoles
Enzymes/storage	-	-	-	++/+++
Other features	-	-	Absent dystrophin	Ragged red

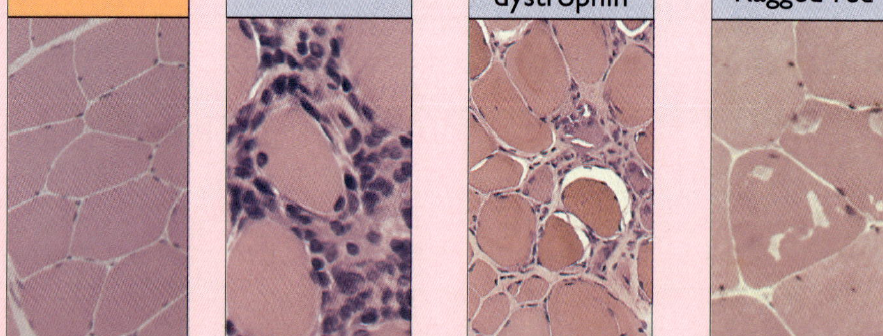

Pathological group	Normal	Inflammatory	Dystrophic	Metabolic
Chapter 4 plus chapter:	3	7	5	9

Clinical history, examination and investigations

Differential diagnoses (examples)	Normal	PM DM IBM	DMD BMD SCARMD	McArdle's Pompe's Mitochondrial

11.2 Characteristic pathological features of the main groups of neuromuscular disorders.

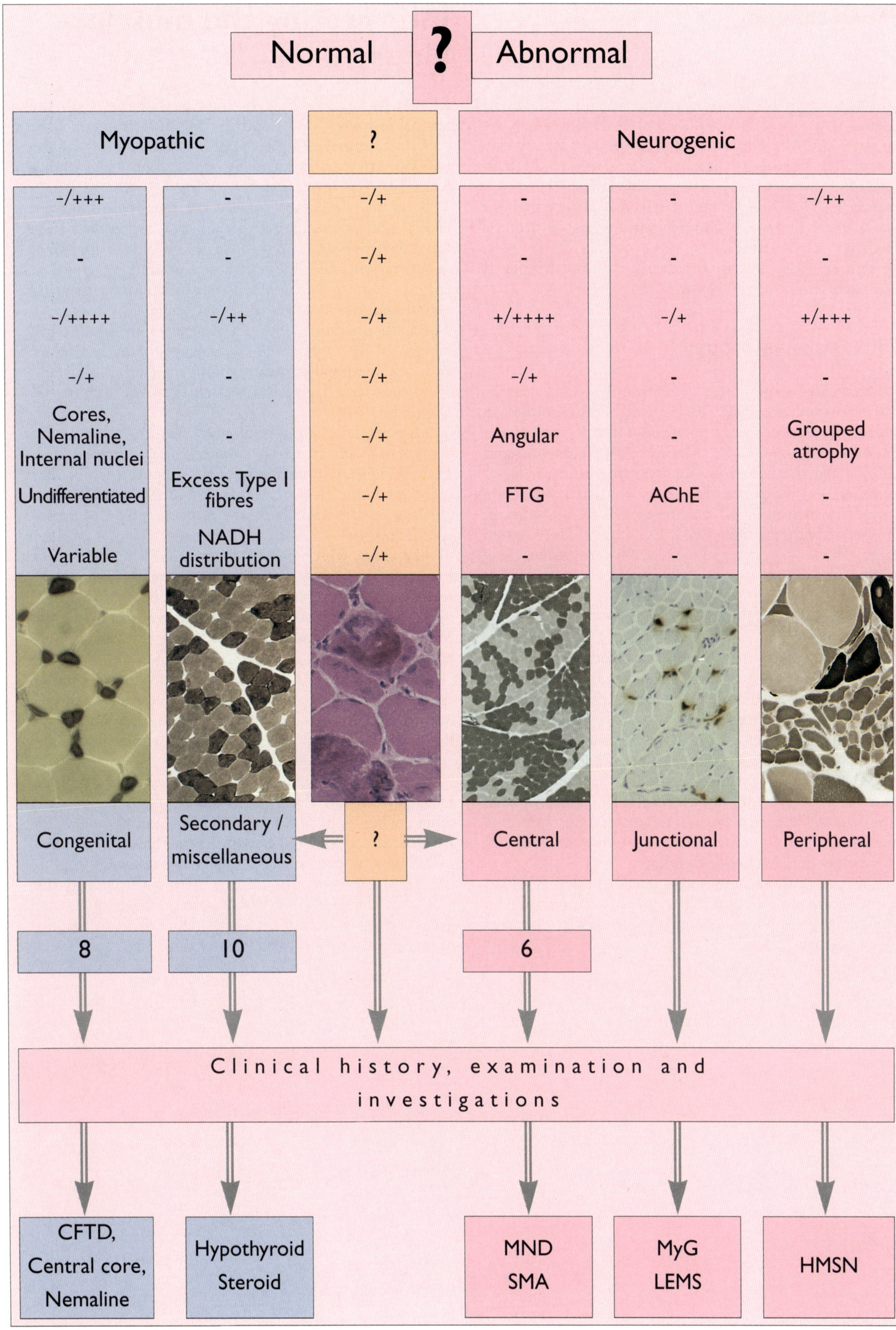
Normal
?
Abnormal
Myopathic
?
Neurogenic
-/+++
-
-/++++
-/+
Cores, Nemaline, Internal nuclei
Undifferentiated
Variable
-
-
-/++
-
-
Excess Type I fibres
NADH distribution
-/+
-/+
-/+
-/+
-/+
-/+
-/+
-
-
+/++++
-/+
Angular
FTG
-
-
-
-/+
-
-
AChE
-
-/++
-
+/+++
-
Grouped atrophy
-
-
Congenital
Secondary / miscellaneous
?
Central
Junctional
Peripheral
8
10
6
Clinical history, examination and investigations
CFTD, Central core, Nemaline
Hypothyroid Steroid
MND SMA
MyG LEMS
HMSN

Muscle biopsy

This should be obtained from a minimally affected muscle that has not been damaged by needle electrodes, and for which there is sufficient information on its normal structure (Chapters 1, 2 and 3). Fixatives must be avoided, and the biopsy must be processed for cryostat sectioning, histology and histochemistry, and ultrastructural investigation where possible (Chapter 2, Appendices C, D and E). Some material may also be used for biochemical analysis or immunoblotting (e.g. dystrophin studies).

The analysis and interpretation of the biopsy is described in more detail below.

Electrophysiology

Primary myopathic and neurogenic disorders may be distinguished by needle EMG and nerve conduction studies. The appearance of giant motor units, polyphasic motor action potentials, or slowed nerve conduction give clues as to the location (i.e. central, junctional, or peripheral) of the motor unit defect. Further information on systemic involvement may be gained from studies of sensory action potentials, EEG, ECG or echocardiogram data.

Haematology

Biochemical data on CK levels are most valuable in indicating myopathies involving rapid muscle destruction (e.g. levels may be raised 10–100 fold in the early to mid-stages of DMD). Abnormally increased blood lactate levels during exercise or abnormal amounts of other enzymes are useful in diagnosing metabolic myopathies.

Other valuable haematological data includes the characterisation of lymphocytic reactions (e.g. in myositic conditions), aberrant leucocyte counts (e.g. infective myopathies), and the identification of abnormal red cell phenotypes (e.g. McLeod syndrome).

Chromosomal and molecular genetic studies of blood samples are now becoming of prime importance in elucidating gene defects in some neuromuscular disorders (Appendix F).

Interpreting the muscle biopsy

First, the myopathologist must be familiar with the normal histology, histochemistry and ultrastructure of the particular muscle biopsied, and must be wary of the possible artefacts induced by preparation and staining techniques (Chapters 2 and 3).

Secondly, the biopsy appearance must be interpreted in the light of the limited pathological repertoire of the muscle fibre (i.e. decrease or increase in size, degeneration and/or regeneration, altered structural or enzyme composition), which may be common to many neuromuscular disorders (end of Chapter 3, Chapter 4). The pathogeneses of these appearances are outlined in Chapter 4. It is often the combination of these features and their temporal correlation with the clinical signs and symptoms which give the real clues to the differential diagnosis.

The characteristic pathological features of the main groups of neuromuscular disorders are summarised in **11.2**, which again emphasises the important link that must be made between muscle biopsy findings and clinical information before attempting to pinpoint a specific disease.

The chart, which depicts the state of a typical biopsy in the early to middle stages of each example, should be seen as a helpful summary. Rather than being read piecemeal, and each part being read dogmatically, it should be read in conjunction with Chapter 4 and with the chapters covering the specific disease (Chapters 5-10).

Obviously, in the coming years, exact categorisation of specific neuromuscular diseases will require additional information from molecular genetic studies. Therefore, the reader is referred to Appendix F for a summary of the current state of knowledge in this field.

However, it is expected that such molecular genetic information will also be extractable from the muscle biopsy when appropriate probes (for example in *in situ* hybridisation studies) become available to the myopathologist, in the same way that staining for dystrophin has revolutionised our approach to interpreting the biopsy of the DMD or BMD patient. Such molecular genetic staining techniques are therefore likely to become prominent in any future editions of this book.

Appendix A

Glossary of abbreviations

AChE	acetylcholine esterase
AD	autosomal dominant
ALS	amyotrophic lateral sclerosis
AR	autosomal recessive
ATPase	adenosine triphosphatase
BMD	Becker muscular dystrophy
CAT	computerised axial tomography
CC	central core
CFTD	congenital fibre type disproportion
CK	creatine kinase
CMD	congenital muscular dystrophy
Co A	co-enzyme A
CPEO	chronic progressive external ophthalmoplegia
CPT	carnitine palmitoyltransferase
CT	connective tissue
CV	coefficient of variation
DAB	diaminobenzidine
DAG	dystrophin associated glycoprotein
DGC	dystrophin glycoprotein complex
DM	dermatomyositis
DMD	Duchenne muscular dystrophy
DNA	deoxyribonucleic acid
EACA	epsilon amino caproic acid
EM	electron microscopy
EMG	electromyography
FSH	facioscapulohumeral dystrophy
FTG	fibre type grouping
H&E	haematoxylin and eosin
HES	hypereosinophilic syndrome
HIV	human immunodeficiency virus
HMSN	hereditary motor sensory neuropathy
IBM	inclusion body myositis
ICI	Imperial Chemical Industries
LEMS	Lambert-Eaton myasthenic syndrome
LG	limb girdle
MAC	membranolytic attack complex
MHC	major histocompatibility complex
MND	motor neurone disease
MRC	Medical Research Council
MRI	magnetic resonance imaging
MyG	myasthenia gravis
MyD	myotonic dystrophy
NADH-Tr	reduced nicotinamide adenine dinucleotide tetrazolium reductase
NCV	nerve conduction velocity
PAN	polyarteritis nodosa
PAS	periodic acid Schiff
PM/DM	polymyositis / dermatomyositis
PP	periodic paralysis
RNA	ribonucleic acid
RT	room temperature
SB	Sudan black B
SCARMD	severe childhood autosomal recessive muscular dystrophy
SDH	succinate dehydrogenase
SMA	spinal muscular atrophy
SP	scapuloperoneal
SPAM	spatial pattern analysis of muscle
SR	sarcoplasmic reticulum
XR	X-linked recessive

Appendix B

Micrograph labels

Abbreviations used to indicate cellular structures in the electron micrographs:

A	A-band of the sarcomere
Av	autophagic vacuole
Ax	myelinated axon terminal
BM	basement membrane
C	connective tissue capsule
Co	collagen fibrils
Cy	cytoplasmic body
D	debris
E	endothelial cell
ER	endoplasmic reticulum
F	muscle fibre extensions
Fb	fibrils
Fl	filaments
G	glycogen
H	H-zone of the sarcomere
Hc	honeycomb structure
I	I-band of the sarcomere
L	lipid
Lc	lymphocyte
Lf	lipofuscin
Lu	capillary lumen
Ly	lysosome
M	M-line of sarcomere
m	mitochondrion
Mf	muscle fibril
Mp	macrophage
My	myelinated axons
N	nucleus
Nm	nuclear membrane
Nn	nucleolus
P	plasma membrane
Ph	phagocyte
R	nemaline rods
RER	rough endoplasmic reticulum
Ri	ribosomes
Sat	satellite cell
SM	sarcoplasmic mass
SR	sarcoplasmic reticulum
T	T-tubule
uMy	unmyelinated axons
V	vacuole
Z	Z-line

Appendix C

Histological and histochemical procedures

Preparation summary (see Chapter 2 for details)

Orientate muscle fibres vertically (for transverse sectioning) and mount supported by agar gel and OCT on labelled cork disc. Freeze muscle tissue in melting Arcton12 ICI (dichlorodifluoromethane) at about -150°C cooled in liquid nitrogen.

Prepare 20 labelled cleaned microscope slides and cut 20 serial sections 8–10 μm thick in a cryostat at -20°C. These slides are used for the stains listed in *Table C.1*. Store blocks or sections in deep freeze at -70°C in sealed bags. Air dry unfixed frozen sections for approximately 30 minutes prior to staining (except toluidine blue and ATPase).

Table C.1 Histological and histochemical stains.

Stain	Slide		Enzyme
1. Toluidine blue	1	11	
2. Haematoxylin & Eosin	4	14	
3 Masson's trichrome	2	12	
4 NADH-Tr	3	13	EC 1.6.4.3
5 Myofibrillar ATPase			
pH 10.2	7	17	EC 3.6.1.3
pH 4.35	8	18	
pH 4.6	9	19	
6 Non-specific esterase	5	15	EC 3.1.1.8
7 Phosphorylase	10	20	EC 2.4.1.1
8 Modified Gomori trichrome	6	16	
9 PAS			
10 SDH			EC 1.3.99.1
11 Sudan black / Oil red O			
12 Methyl green pyronine			
13 Phosphofructokinase			EC 2.7.1.11
14 Acid phosphatase			EC 3.1.3.2
15 Alcian blue			
16 Acetylcholinesterase			EC 3.1.1.7
17 Cytochrome oxidase			EC 1.9.3.1
18 Von Kossa			
19. Acridine orange			

EC: Enzyme commission number
CI: Colour Index
Routine: use procedures: 1-8
Frequently employed for metabolic or other specific queries: procedures: 9-19, cut extra sections if required

1. Toluidine blue

Stains tissue components blue and is used to check orientation and preservation of block during sectioning.

Method

1 Stain unfixed sections for about 1 minute.
2 Wash well in tap water then differentiate in 50% alcohol (ethanol) until excess dye is removed.
3 View specimen.
4 If required, dehydrate, clear and mount in Practamount.

Stain: 1% toluidine blue in 1% borax (disodium tetraborate) in water. Filter before use. Keeps well.

2. Haematoxylin & Eosin

Stains nuclei and basophilic material dark blue and other cytoplasm pink. Used to assess muscle fibre architecture, fibre splitting, position of nuclei, presence of vacuoles, regeneration etc. Hyaline fibres are conspicuous, staining deep pink.

Method

1 Stain unfixed sections in Harris's haematoxylin for 1–3 minutes depending on freshness of stain.
2 Blue by washing in warm tap water for 5 minutes.
3 Counterstain in eosin for 1 minute.
4 Wash briefly in cold tap water.
5 Dehydrate, clear and mount in Practamount.

Harris's haematoxylin: 2.5 litres. Keeps well. Filter before use.

Harris's alum haematoxylin	10 g
Absolute ethanol	100 ml
Ammonium or potassium alum	200 g
Sodium iodate	2 g
Glacial acetic acid	80 ml
Distilled water	make up to 2.5 litres

Dissolve haematoxylin in alcohol and add to alum solution. Bring to boil for 2 minutes, remove from heat and add sodium iodate (Care! – solution may react from thermal shock on adding powder). When cool add glacial acetic acid.

Eosin: 0.5–1% eosin (alcohol soluble) in 70% alcohol, (or 0.5% water soluble eosin in distilled water).

3. Masson's trichrome

Muscle fibres stain red/brown, nuclei blue/black and connective tissue fibres green. Shows muscle fibre architecture and connective tissue proliferation.

Method

1 Stain in Harris's haematoxylin for 2–3 minutes.
2 Blue by rinsing in warm tap water for 5 minutes.
3 Differentiate by immersion in picric acid for 15 seconds.
4 Wash well in cold tap water.
5 Stain cytoplasm red in ponceau for 2.5 minutes.
6 Rinse in cold tap water.
7 Differentiate and mordant by immersion in PMA for 5 minutes.
8 Rinse in cold tap water.
9 Counterstain connective tissue fibres in fast green (FCF) for 15 seconds.
10 Wash in cold tap water.
11 Dehydrate, clear and mount in Practamount.

Nuclear stain: Harris's haematoxylin – see (2) H&E.

Picric acid: 1% picric acid in 70% alcohol (Caution).

Cytoplasmic stain: 1% Ponceau de xylidine (ponceau 2R) in 0.5% acetic acid.

Differentiator & Mordant: 2% phosphomolybdic acid (PMA) in distilled water.

Connective tissue stain: 0.5% fast green (FCF) in distilled water.

4. NADH-Tr

Nicotinamide adenine dinucleotide tetrazolium reductase.

Intermyofibrillar material (mitochondria, sarcoplasmic reticulum, t-tubules) stains dark blue. Used for fibre typing in association with myofibrillar ATPase, (Type I fibres stain darkly, IIa intermediate, and IIb pale). Shows proliferation of mitochondria, cores etc.

Method

1 Incubate unfixed sections for 30 minutes at 37°C.
2 Rinse gently in distilled water.
3 Mount in glycerine jelly or Uvinert.

Incubation medium:	
Tris buffer 0.2M pH 7.4	30 ml
NADH	24 mg
Nitro blue tetrazolium	30 mg

Make fresh, filter before use.

Tris buffer: 0.2M pH 7.4 (Gomori), 2 litres	
Tris(hydroxymethyl)methylamine	12.1 g
HCl (Normal)	82.8 ml
Distilled water	make up to 2 litres

Adjust pH to 7.4 with NaOH/HCl
Add tris to water (1800 ml distilled), gradually add the N HCl whilst adjusting pH.

(Alternative: 25 ml Tris 0.2M plus 20.7 ml 0.2M HCl made up to 100 ml)

Glycerine jelly	
Gelatine	10 g
Distilled water	60 ml
Glycerine	70 ml
Phenol	0.25 g

Melt in a vessel of hot water.

Order:	
NADH	(Boehringer)
NBT	(Sigma N6876)
Tris	(BDH)

5. ATPase

Myofibrillar adenosine triphosphatase.

Used to differentiate muscle fibre types. Serial sections are preincubated at pH 4.35, 4.6 or 10.2 prior to incubation at pH 9.5 and development.

	I	IIa	IIb	IIc
pH 4.35	Dark	Pale	Pale	Intermed.
pH 4.6	Dark	Pale	Medium	Dark
pH 10.2	Pale	Dark/Intermed.	Dark	Dark

Method

1 Air-dry unfixed frozen sections for 2 hours minimum.
2 Preincubate (acids 15 minutes: alkali 20 minutes) at RT.
3 Wash times in tap water; rinse in distilled water.
4 Incubate (acids 45 minutes: alkali 30 minutes) at 37°C.
5 Wash in 2 changes of 1% calcium chloride.
6 Place in 2% cobalt nitrate for 3 minutes.
7 Quickly wash in 3 changes of tap water and 3 changes of distilled water.
8 Develop in dilute (1%) ammonium polysulphide in distilled water for about 5 minutes.
9 Wash well in tap water and then distilled water.
10 Mount in aqueous mountant (e.g. Aquamount, Uvinert).

Incubation solution (per coplin jar):	
ATP (disodium salt)	45 mg
Tris-calcium buffer (0.1M plus 0.018M $CaCl_2$)	30 ml

Adjust pH to 9.5 with N/10 KOH or HCl
Note: incubation solution for acid ATPase should begin at RT, but for alkalis the solution should be brought to 37°C in a microwave oven at low power level.

Acid preincubations (per coplin jar):		
pH 4.6	Sodium acetate 0.2M	15 ml
	Acetic acid 0.2M	15 ml
pH 4.35	Sodium acetate 0.2M	9 ml
	Acetic acid 0.2M	21 ml

Adjust pH to 4.35 or 4.6 with M KOH.

Alkaline pre-incubation (per coplin jar):	
Tris-calcium buffer (0.1M plus 0.018M $CaCl_2$)	30 ml

Carefully adjust pH to 10.2 with N/10 HCl or KOH.
Store at 4°C
1% calcium chloride, 0.2M acetic acid (see 7), 2% cobalt nitrate, 0.2M sodium acetate (see 7).

Tris-calcium buffer (0.1M plus 0.018M $CaCl_2$): should be made up fresh each time.

Tris(hydroxymethyl)methylamine	6.05 g
Calcium chloride (MW 147)	1.32 g

Distilled water make up to 500 ml.

6. Non-specific esterase

Stains motor end-plates and sites of lysosomal activity (muscle or macrophages) dark brown. Sarcoplasmic esterases of type I fibres may also be visualised.

Method
1 Fix sections in formol calcium for 1 hour at 4°C.
2 Wash well in tap water; rinse in distilled water.
3 Incubate for 1 hour at 37°C.
4 Wash well in distilled water.
5 Dehydrate, clear and mount in Practamount.

Fixative: formol calcium

Concentrated formaldehyde	25 ml
Distilled water	90 ml
Calcium chloride (anhydrous)	1 g

(Store over $CaCO_3$ to neutralise.)

Incubation solution: prepare fresh

1% a-naphthyl acetate in acetone	0.5 ml
0.1M disodium hydrogen phosphate	20 ml
Hexazonium pararosanalin	1.6 ml

Adjust pH to 6.5.

Hexazonium pararosanalin:
Mix equal parts solutions A and B and wait about 1 minute for bubbles to appear: add to incubation medium and pH.

Solution A:

Pararosanalin HCl	1 g
Distilled water	20 ml
Heat to near boiling point, cool to RT.	
Add concentrated HCl	5 ml

Filter before use.
Store solution A (after filtration) in a dark bottle at 4°C.

Solution B:
4% aqueous sodium nitrite. Freshly prepared.
Stock: 0.2M disodium hydrogen phosphate (see 10).

7. Phosphorylase

Indicates absence or excess of glycolytic activity and its localisation by staining intermyofibrillar phosphorylase bound to glycogen. Differentiates fibres into enzyme rich (Type IIb – blue/purple), intermediate (Type IIa – brown) and enzyme poor (Type I – yellow).

Method
1 Incubate unfixed sections for 1 hour at 37°C.
2 Wash briefly in distilled water.
3 Develop with Gram's iodine for 1 minute.
4 Wash well in distilled water.
5 Blot dry and allow to air-dry.
6 Clear in Xylene and mount in Practamount.

Note: fades rapidly (especially in the light).

Incubation solution: (per coplin jar)

Glucose-1-phosphate	50 mg
Adenosine monophosphate	15 mg
EDTA	25 mg
Sodium fluoride	20 mg
Dextran	1 g
Acetate buffer 0.2M	20 ml
Absolute ethanol (or Analar alcohol)	5 ml

Adjust to pH 5.9 before use.

Acetate buffer 0.2M pH 5.9:

Sodium acetate 0.2M (stock)	17.5 ml
(0.2M acetate = 27.21 g in 1 litre of water)	
Acetic acid 0.2M (stock)	2.5 ml

(0.2M acetic acid = 11.6 ml glacial acetic acid in 1 litre of water).

Gram's iodine stock:

Potassium iodide (in 3 ml water)	2 g
Iodine crystals	1 g
Distilled water	300 ml

Dilute 1/10 in water before use.

Chemicals:
D-Glucose-1-phosphate dipotassium salt
EDTA - Ethylene diamine tetra acetic acid dipotassium salt (MW 404.7)
Dextran - grade A (MW 200-275,000) BDH No. 38014.

8. Modified Gomori trichrome

Stains mitochondria, nemaline rods, Z-lines and nuclei deep red, other material stains greenish blue. Used to show ragged red fibres with excess mitochondria or nemaline bodies.

Method
1 Stain unfixed sections in Harris's haematoxylin for 3 minutes.
2 Rinse in distilled water.
3 Stain in modified Gomori stain for 1.5 hours.
4 Differentiate with two very quick rinses in 0.2% acetic acid.
5 Dehydrate, clear and mount in Practamount.

Gomori stain	100 ml	400 ml
Chromotrope 2R	0.6 g	2.4 g
Fast green FCF	0.3 g	1.2 g
Phosphotungstic acid	0.6 g	2.4 g
Distilled water	100 ml	400 ml

Make up fresh or store at 4°C for up to 1 month.

9. P.A.S.

Periodic acid Schiff reaction.

Stains glycogen pink. Shows abnormal accumulations of glycogen and related metabolic disorders. Some mucopolysaccharides also stain and glycogen may be confirmed by diastase digestion. Fibre types may be differentiated.

Method
1 Fix sections for 10 minutes in formol calcium (see recipe 6) at 4°C (or coat unfixed sections with 0.25% celloidin in 100% alcohol and rapidly air-dry and move to step 3).
2 Wash thoroughly in tap water.
3 Oxidise sections in 1% periodic acid for 10 minutes.
4 Wash well in running tap water, rinse on distilled water.
5 Immerse in Schiff reagent for 10 minutes (in the dark in the fridge).
6 Wash well in running tap water for 10 minutes.
7 Dehydrate clear and mount in Practamount.

Schiff reagent (de Tomasi method):

Basic fuchsin	1 g
Distilled water (boiled)	200 ml
Sodium metabisulphite 1 g 0.1N HCl	20 ml
Activated charcoal	2 g

Make up in order indicated and mix thoroughly (i.e. dissolve the basic fuchsin in boiling water and shake for 5 minutes, cool to about 50° C before adding acid and to about 25°C before adding metabisulphite). Add charcoal, shake, filter and store in a dark bottle at 4°C. This should be a straw coloured solution which matures and is ready for use in 24 hours. If necessary (i.e. if purple colour remains), add more charcoal, shake and filter.

Diastase digestion: run parallel slides in NaCl/buffer without enzyme

Diastase (or a-amylase)	50 mg
Sodium chloride	400 mg
Phosphate buffer 0.02M pH 6.0	50 ml

After fixation as above, digest for 1 hour at 22°C.

10. SDH (succinate dehydrogenase)

Stains site of mitochondrial enzymes blue. Shows subsarcolemmal and intermyofibrillar distribution. Indicates fibre types and sites of excess mitochondrial activity.

Method
1 Incubate unfixed sections for 30 minutes at 37°C (or overnight at room temperature).
2 Rinse in distilled water.
3 Mount in glycerine jelly (see recipe 4) or Uvinert aqueous mountant.

Incubation solution - (per coplin jar):

Sodium succinate 0.1M	10 ml (0.27 g in 10 ml)
Formdimethylamide	10 ml
Phosphate buffer 0.1M pH 7.6	10 ml
Nitroblue tetrazolium	20 mg

Buffer:

Sodium dihydrogen phosphate 0.2M	3.25 ml
Disodium hydrogen phosphate 0.2M	21.7 ml
Distilled water	25 ml

Stock solutions (0.2M - 1 litre each):

Sodium dihydrogen phosphate	31.20 g
Disodium hydrogen phosphate	28.39 g

11a. Sudan black

Stains neutral and complex lipids and membranes black. Poorly visible in normal muscle. Used to identify storage products.

Method
1 Fix in formol calcium for 5 minutes at 4°C.
2 Wash well in tap water and rinse in distilled water.
3 Stain in filtered Sudan black for 45 minutes.
4 Wash in water.
5 Mount in lukewarm glycerine jelly or Uvinert aqueous mountant.

Fixative (formol calcium):

Formalin (40% formaldehyde)	100 ml
Calcium chloride 2%	600 ml
Distilled water	300 ml

Adjust to pH 7.0 and keep neutralised over calcium carbonate.

Stain:
Saturated solution of Sudan Black B in 70% alcohol (heat in boiling water to dissolve).

Glycerine jelly: see (4) NADH.

11b. Oil red O

Stains lipids bright red, use as for Sudan Black.

Method
1 Fix in formol calcium for 5 minutes at 4°C.
2 Wash in 60% triethylphosphate in distilled water.
3 Stain in 1% Oil red O in triethylphosphate for 15 minutes at RT.
4 Wash rapidly in 60% triethylphosphate.
5 Rinse in distilled water.
6 Mount in Uvinert aqueous mountant.

12. Methyl green pyronine

Stains nuclei, RNA is stained pink to red and DNA bluish green. Used to ascertain nature of basophilic cytoplasm within muscle fibres.

Method
1 Fix in formol calcium for 1 hour.
2 Stain for 10 to 20 minutes.
3 Blot dry and dehydrate rapidly in n-butanol (or acetone).
4 Clear in xylene and mount in Practamount.

Staining solution:

Solution A	10 ml
Solution B	15 ml
Solution C	25 ml

Solution A:
2% aqueous methyl green (CI 42585) washed in chloroform (×5) to extract methyl violet. Keeps for several months.

Solution B:
5% aqueous pyronine Y (CI 45005)

Solution C:
0.2M acetate buffer pH 4.8.

Glacial acetic acid	4.6 ml
Sodium acetate	9.85 g
Distilled water	1000 ml

Note: Method step 3: n-butanol is preferred as acetone or t-butanol may extract too much stain.

13. Phosphofructokinase

The enzyme converts fructose 6 phosphate to fructose 1,6 diphosphate during glycolysis. The presence of the enzyme in normal controls is indicated by dark blue granular deposits similar to the NADH-Tr and SDH reactions and is especially prominent in Type I muscle fibres.

PFK deficiency is indicated by lack of staining in the test section using medium 1 when controls show a positive reaction. Both controls and test sections should also give a positive reaction when incubated with incubation medium 2 (fructose 1,6 diphosphate) and negative reactions with medium 3 (no substrate).

Method
1 Fix sections in 80% ethanol at 4°C for 20 minutes.
2 Air dry for 30 minutes.
3 Incubate at 37°C for 1 hour.
4 Rinse in distilled water.
5 Post-fix in formol calcium.
6 Mount in glycerine jelly.

Incubation medium 1:

Fructose 6 phosphate	18 mg
Nitro blue tetrazolium	12 mg
NAD	19 mg
ATP	17 mg
Magnesium sulphate	7 mg
0.01M arsenate-HCL buffer pH 7.0	30 ml

Note: Incubation medium 3: as above but without F-6-P

Incubation medium 2:

Fructose 1,6 diphosphate	25 mg
Nitro blue tetrazolium	12 mg
NAD	19 mg
0.01M arsenate-HCL buffer pH 8.6	30 ml

0.01M arsenate buffer:

Sodium arsenate (Na_2HAsO_4)	131 mg
Distilled water	100 ml

14. Acid phosphatase

Not apparent in normal muscle. Patches of bright red staining indicate sites of lysosomal activity within the muscle fibre or phagocytic cells.

Method
1 Fix in formol calcium for 1 hour at 4°C.
2 Wash well in tap water, rinse in distilled water.
3 Incubate for 1 hour at 37°C.
4 Wash in distilled water.
5 Dehydrate, clear and mount in Practamount.

Incubation medium: prepare fresh

1% Naphthol ASCL phosphate in DMF	1 ml
0.2M Veronal acetate solution	7 ml
Distilled water	12 ml
Hexazonium pararosanalin (add last)	1.6 ml

Adjust to pH 4.8 and filter.

Hexazonium pararosanalin : see (6) Esterase.

Formol calcium: see (11) Sudan black.

0.2M veronal acetate:
Add equal amounts of:
0.2M Sodium barbitone (41.2 g/l),
0.2M Sodium acetate (27.21 g/l).

15. Alcian blue

Pale blue staining indicates the presence of various mucopolysaccharides. Different types of mucopolysaccharide are detected according to the pH and salt content of the staining solution.

Method
1 Stain in solution A or B for 30 minutes.
2 Wash in running tap water for 3 minutes.
3 Dehydrate, clear and mount in Practamount.

Solution A: Alcian blue pH 1.0

Alcian blue	1 g
0.1N HCl	100 ml

Filter before use.
Stains only sulphated mucosubstances.

Solution B: Alcian blue pH 2.5

Alcian blue	1 g
3% aqueous acetic acid	100 ml

Stains all acid mucosubstances.

16. Acetylcholinesterase

Dense brown staining deposits indicate sites of enzyme activity particularly prominent at sites of neuromuscular junctions or more diffuse in the cytoplasm of denervated muscle fibres. Inhibited control reactions are required to specify the type of cholinesterase activity detected.

Method
1 Fix sections in formol calcium for 20 minutes at 4°C.
2 Wash in tap water.
3 Inhibit in TIPA (10-5M) in 0.075M phosphate buffer pH 7.4 for 30 minutes at RT. (5 ml TIPA solution in 40 ml phosphate buffer in coplin jar).
4 Wash in tap water.
5 Incubate in Gomori stock solution containing 2mg/ml of acetylthiocholine iodide as substrate for 3-18 hours at 37°C.
6 Wash in tap water.
7 Develop in 1% ammonium sulphide for 1 minute at RT.
8 Wash in tap water.
9 Counterstain in haematoxylin for 2 minutes.
10 Blue for 5 minutes.
11 Dehydrate, clear in TCE and mount in Canada Balsam (in TCE).

Gomori stock solution: add chemicals in correct order

Sodium sulphate (anhydrous)	480 g
(filter sodium sulphate after dissolving in distilled water)	
Distilled water	1200 ml
Copper sulphate.$5H_20$	2.1 g
Glycine	2.628 g
Magnesium sulphate	7 g
Maleic acid	12.25 g

Adjust pH to 6.4 with M NaOH.
Place in 37°C oven.

TIPA 10-4M:

Tetra isopropyl pyrophosphoramide	0.0085 g
Phosphate buffer 0.075M	250 ml

TCE (tetrachloroethylene).

Formol Calcium: see (11) Sudan black.

Phosphate buffer:

Sodium dihydrogen orthophosphate	11.65 g/l
Disodium hydrogen orthophosphate	10.65 g/l

Add the disodium salt to the monosodium salt until pH is 7.4.

17. Cytochrome oxidase

Stains sites of mitochondrial activity golden brown. Also to detect disorders with a specific lack of this enzyme.

Method

1 Incubate unfixed sections for 45 minutes at RT.
2 Rinse in distilled water.
3 Fix in formol saline for 15 minutes at RT (optional).
4 Rinse in water, dehydrate, clear and mount in Practamount.

Incubation medium:

A 3/3/diaminobenzidine (DAB)	10 mg

(dissolved in 8 ml of distilled water in a stoppered flask).

Caution- DAB is highly carcinogenic do not incubate in open containers, detoxify in sodium hypochlorite.

B Catalase (20 µg/ml)	2 ml

(2 mg catalase in 100 ml of distilled water in a volumetric flask).

C Cytochrome C (Type II)	12 mg

(dissolved in 10 ml of 0.1M phosphate buffer pH 7.4).

Add A to C carefully, then add 2 ml of solution B giving a total volume of 20 ml, final pH 7.3.
Make up FRESH and DO NOT incubate in open containers or slides.

Phosphate buffer 0.1M pH 7.4:

Sodium dihydrogen phosphate 0.2M	4 ml
Disodium hydrogen phosphate 0.2M	21 ml
Distilled water	25 ml

Formol saline:

10% formaldehyde 40%	100 ml
Sodium chloride	9 g
Distilled water	900 ml

18. von Kossa

Calcium deposits are stained black by the deposition of silver salts.

Method

1 Place in 2% aqueous silver nitrate in a clear glass tube.
2 Expose to bright daylight for 60 minutes or a halogen lamp for 10 minutes.
3 Wash in several changes of distilled water.
4 Treat with 2.5% sodium thiosulphate for 5 minutes.
5 Wash well in tap water.
6 Counterstain as desired.
7 Dehydrate, clear and mount in Practamount.

19. Acridine orange

Autofluoresces orange in association with RNA and yellow-green with DNA and reducing bodies.

Method

1 Use 6–10 µm thick sections fixed in formol calcium for 1 minute. (For paraffin wax sections deparaffinise in xylene and rapidly hydrate prior to step 3.)
2 Rinse well in distilled water.
3 Acidify in 1% acetic acid for 1 minute.
4 Stain for 20-30 seconds in 0.02% acridine orange dissolved in 0.067M phosphate buffer pH 6.
5 Rinse in phosphate buffer.
6 Differentiate for 30–60 seconds in 0.1M calcium chloride.
7 Rinse in buffer and mount wet in buffer.
8 View immediately with fluorescence microscope under blue (about 490 nm) excitation and dichroic mirror and barrier of approximately 515 nm. Acridine orange emission maximum is at 530 nm.

For control sections preincubate with bovine pancreatic ribonuclease (10 mh/ml in phosphate buffer at pH 7.0 for 1 hour prior to step 3).

Phosphate buffer 0.067M pH 6:

Disodium hydrogen phosphate 0.2M	12.3 ml
Sodium dihydrogen phosphate 0.2M	87.7 ml
Distilled water	200 ml

Acridine orange: 2,8,-dimethylaminoacridine (Gurr, CI 46005).

Appendix D

Immunocytochemical procedures

Proteins (antigens) may be accurately localised and visualised in tissue sections using monoclonal antibodies specific to that protein, and may be either directly or indirectly labelled with a marker that is visible via brightfield or fluorescence microscopy.

For brightfield microscopy, either enzymes (e.g. peroxidase, acid phosphatase) or heavy metals (e.g. ferritin, colloidal gold) markers may be used to locate the primary or secondary antibody. Fluorochromes such as FITC (fluoroscein isothyocynate), rhodamine, or Texas Red are used as the final visual step in conjunction with fluorescence microscopy. In the following procedures, one example of each method is given, but the reader is informed that the visualisation procedures may be interchanged providing care is taken in matching the species, immunoglobulin source, and dilutions of primary, secondary and any tertiary markers.

Cautionary note: Enzyme-immunocytochemical procedures allow easy visualisation with a standard brightfield microscope and usually provide permanent preparations, but they may show a relatively weak signal and possible artefacts due to inherent enzyme activity within the tissue. Fluorescent signals are intense and may be useful for detecting small quantities of material, however, they may fade rapidly, require expensive equipment and can be confused by autofluorescence of certain tissue components.

1. Dystrophin – enzyme method

Dystrophin is a 427 kD rod-shaped cytoskeletal protein located immediately beneath the muscle fibre plasma membrane, to which it is attached via a dystrophin associated glycoprotein complex (DGC). Antibodies are available for the central rod (DYS1), C-terminal (DYS2) and N-terminal (DYS3) domains of the dystrophin molecule. For normal muscle, this staining procedure shows a thin continuous rim of labelling (dark brown reaction product) in association with the muscle fibre membrane. Staining is absent in DMD or reduced and/or discontinuous in BMD.

Method

1 Use 6–10 µm thick cryostat sections on slides subbed with chrome alum gelatin. Slides may be stored at -70°C wrapped in cling film and brought to RT before air drying.
2 Air dry sections for 30 minutes. Ring sections with Dako marker pen.
3 Destroy endogenous peroxidase by incubating with hydrogen peroxide (dilution 1:4 in phosphate buffered saline-PBS) for 10 minutes.
4 Wash well in PBS.
5 Incubate with primary antibody (mouse monoclonal anti-dystrophin IgG) for 1 hour at appropriate dilution (e.g. DYS1 1:20; DYS2 1:80; DYS3 1:40 - Novocastra/Europath Labs) in PBS containing 2% normal goat serum.
6 Wash 3 times in PBS.
7 Incubate with secondary antibody (biotinylated rabbit anti-mouse IgG 1:200 {Dakopatts}) in PBS containing 2% normal goat serum for 1 hour.
8 Wash 3 times in PBS.
9 Incubate with Vectastain ABC kit (Vectastain Labs) for 30 minutes. (1 drop A {avidin}, 1 drop B {biotinylated horseradish peroxidase} plus 5 ml 0.05M Tris buffer, activated 30 minutes prior to incubation).
10 Was 3 times in Tris buffer.
11 Develop in diaminobenzidine (DAB) solution for 15 minutes in fume cupboard.
12 Wash in Tris buffer, dehydrate, clear and mount in Practamount.

Chrome Alum Gelatin subbing solution:

Gelatine	2.5 g
Chrome alum (chromium potassium sulphate)	0.25–0.5 g
Thymol crystals	1 crystal
Distilled water	1 litre

Dissolve gelatine in warm water, heat if necessary (55°C), cool and add alum and thymol. Keep refrigerated until needed. Bring up to room temperature for use.
Submerge clean slides in staining pot of subbing solution, drain and leave overnight in hot oven. Store at room temperature.

Phosphate buffered saline (PBS) 0.1M pH 7.2 (dilute 1:10 for use):

Sodium chloride	80 g
Potassium chloride	2 g
Potassium dihydrogen orthophosphate	2 g
Disodium hydrogen orthophosphate (anhydrous)	11.5 g
Distilled water	1 litre

Tris buffer: as for NADH-Tr (Appendix C: recipe 4) diluted 1:4.

DAB solution: (can be stored frozen in 10 mg/ml amounts in distilled water)

3'3' diaminobenzidine (Sigma D5637)	40 mg
Tris buffer	80 ml

Add nickel chloride to enhance stain at a final concentration of 7 mg/10 ml. Immediately before use, add 50 μl of hydrogen peroxide. (In practice the DAB solution is added to half the amount of buffer and the nickel chloride to the other half. Both halves are mixed together followed by the hydrogen peroxide.)

Caution: DAB is highly carcinogenic – incubate in closed containers.

2) Myosin - fluorochrome method

Antibodies are available to a variety of myosin proteins (fast, slow, fetal, embryonic). In normal adult human muscle there should be a clear differentiation into muscle fibres staining homogeneously (red against a black background in this procedure) with either slow or fast myosin thus identifying Type I or Type II fibres respectively and confirming the results of histochemical ATPase procedures. In diseased muscle, fibres may contain mixtures of these and various developmental stage myosins.

Method

1 Use 6–10 μm thick cryostat sections on subbed slides (see Dystrophin procedure above).
2 Air-dry sections for 30 minutes.
3 Ring section with Dako marker pen.
4 Cover section with 100 μl of primary antibody (mouse anti-fast (or slow) myosin IgG {Sigma}) 1:50 diluted in PBS containing 2% normal goat serum (100 μl in 5 ml).
5 Incubate at RT for 1 hour in a humidity chamber.
6 Wash 3 times for 5 minutes each in PBS and carefully blot dry around section.
7 Cover section with biotinylated rabbit anti-mouse IgG (Dakopatts) 1:50 for 30 minutes.
8 Wash slides 3 times for 5 minutes each in PBS.
9 Cover section with Streptavidin Texas red (Amersham) 1:200 for 15 minutes in a humidity chamber.
10 Wash slides 3 times for 5 minutes each in PBS.
11 Mount in aqueous mountant (e.g. Uvinert, glycerine jelly).
12 Examine in fluorescence microscope using green excitation (about 546 nm), dichroic (590 nm) and barrier 610 nm.

PBS solution: as described for Dystrophin staining.

Appendix E

Electron microscopical procedures

The chemicals used to fix and embed tissues for electron microscopy are extremely hazardous. The tissue itself constitutes a biohazard and must be handled accordingly. Prepare the fixatives and perform all of the processing steps in a fume cupboard.

Fixation of skeletal muscle

1 A small bundle of muscle fibres is clamped under slight tension in the modified hair grip described in Chapter 2 and immersed in 5% Glutaraldehyde in 0.1M Phosphate buffer for 1 hour at room temperature.
2 The central portion of the sample is excised and cut into small strips approximately 2mm long by 1mm wide and fixation continued for a second hour.
3 Wash in three changes of phosphate buffer, each of 5 minsute.
4 Fix in 1% osmium tetroxide in phosphate buffer for 1 hour at room temperature.
5 Wash in three changes of phosphate buffer, each of 5 minsute.
6 Wash in distilled water for 10 minutes..
7 Stain with 4% aqueous uranyl acetate for 20 minutes.
8 Wash in distilled water 10 minutes.
9 Dehydrate in 50% alcohol, twice, for 15 minutes each time.
10 70%, 2 × 15 minutes.
11 90%, 2 × 15 minutes.
12 Absolute alcohol.
13 Acetone, 2 × 15 minutes.
14 Immerse in acetone/resin mixed 50:50 for 30 minutes.
15 Place in 100% resin overnight (room temp).
16 Embed in fresh resin and polymerise at 60°C for 24 hours.

Steps 6, 7 and 8 are optional.

Phosphate buffer

Solution A:	
Sodium dihydrogen orthophosphate	31.2 g
Distilled water	1000 ml

Solution B:;	
Disodium hydrogen orthophosphate	28.4 g
Distilled water	1000 ml

For use:	
Solution A	50 ml
Solution B	200 ml
Distilled water	250 ml

Check and adjust to pH 7.4.

Glutaraldehyde fixative

25% Glutaraldehyde (EM grade)	5 ml
0.1M Phosphate buffer pH 7.4	20 ml

Osmium tetroxide fixative

Remove the label from a 1g vial of osmium tetroxide. Wash well to remove any adhesive. Dry with a clean cloth or towel and score around the middle of the tube with a glass cutter. Place in a clean brown glass bottle with a tight fitting stopper. Shake until the vial breaks and add 100 ml 0.1M phosphate buffer. Allow the osmium to dissolve overnight. Store away from light.

Araldite embedding medium

Araldite resin, MY 753	50 g
Hardener, DDSA	50 g
Plasticiser, Dibutylphthalate	2.5 ml
Accelerator, BDMA	2 ml

Measure the components into a disposable beaker and mix thoroughly with a wooden spatula. Allow air bubbles to rise to the surface before use. Store the mixture in the refrigerator until required. Decant the resin/acetone mixture into the toxic waste container provided. Unused resin must be polymerised by placing the beaker and spatula in the oven at 60°C before being discarded.

Caution: the reagents used can cause dermatitis and gloves must be worn whilst preparing and handling the resin.

Uranyl acetate stain for tissue blocks

Uranyl acetate	4 g
Distilled water	100 ml

Store in a brown glass bottle at 4° C.

Uranyl acetate stain for sections

Uranyl acetate	20 g
Methyl alcohol	100 ml

Mix well and stand until the solution is saturated. Filter into a brown glass bottle and store at 4°C.

Lead citrate stain for sections (modified Reynolds stain)

Solution A:	
Trisodium citrate	29.4 g
Distilled water	100 ml
Solution B:	
Lead nitrate	33.1 g
Distilled water	100 ml
Solution C:	
Sodium hydroxide	4 g
Distilled water	100 ml

To prepare the staining solution add 3ml A and 2ml B to 16ml of distilled water, mix well and add 4ml of solution C. Filter through a millipore filter and store at 4°C.

Staining the sections

1. Centrifuge a small volume of each stain.
2. Place a drop of uranyl acetate solution (one for each grid) onto a sheet of dental wax or paraffin film in a covered container.
3. Float the grid with the sections on the underside on the stain for 5 minutes.
4. Wash by dipping several times into small beakers containing 100%, 50%, 25% methanol and two of distilled water.
5. Place a drop of lead citrate solution (one for each grid) onto a sheet of dental wax or paraffin film in a covered container.
6. Float the grid with the sections on the underside on the stain for 3 minutes.
7. Wash by dipping several times into four beakers of fresh distilled water.
8. Remove the excess moisture from the grid by touching the edge with a hard filter paper (Whatman No. 50).
9. Store in a dust free container until required for examination.

Appendix F

Neuromuscular disorders – Gene location

The following table and the accompanying locations on the chromosome map (**C.1**) are adapted from data given in Neuromuscular Disorders (1994), vol. 1, part 4; for full references and updated information consult recent editions of this journal.

Disease	Mode of inheritance	Gene location	Symbol	Gene defect
Muscular dystrophies				
Duchenne/Becker	XR	Xp21.2	DYS	dystrophin
Emery-Dreifuss	XR	Xq28	EMD	
Facioscapulohumeral	AD	4q35	FSHD	
Limb-girdle recessive	AR	15q	LGMD2	
Limb-girdle dominant	AD	5q	LGMD1	
Severe childhood autosomal recessive	AR	13q12	SCARMD	50kD DAG
Congential myopathies				
Myotubular myopathy	XR	Xq28	MTM1	
Central core disease	AD	19q13.1	CCD	ryanodine receptor
Nemaline myopathy	AD	1q21–q23	NEM1	
Fukuyama congenital MD	AR	9q31–q33	FCMD	
Myotonic syndromes				
Myotonic dystrophy	AD	19q13	DM	myotonin-protein kinase
Ion channel muscle diseases				
Myotonia congenita	AD	7q35	CLC-1	muscle chloride channel
Generalised myotonia	AR	7q35	CLC-1	muscle chloride channel
Hyperkalaemic periodic paralysis	AD	17q13.1–q13.3	SCN4A	sodium channel a-subunit
Paramyotonia congenita	AD	17q13.1–q13.1	SCN4A	sodium channel a-subunit
Malignant hypothermia	AD	19q13.1	MH	ryanodine receptor
Metabolic myopathies				
Glycogenoses				
Type II (Pompe)	AR	17q23	GAA	acid maltase
Type V (McArdle)	AR	11q13	PYGM	muscle phosphorylase
Type VII (Tarui)	AR	1cenq32	PFKM	phosphofructokinase
Lipid metabolism				
CPT deficiency	AR	1p11–p13	CPT2	carnitine palmitoyl-transferase

Neurogenic syndromes

Hereditary motor sensory neuropathies (HMSN)				
Hereditary neuropathy	AD	17p11.2	PMP-22	peripheral myelin protein
Charcot-Marie-Tooth				
Type Ia	AD	17p11.2	PMP-22	peripheral myelin protein
Type Ib	AD	1q21-23	CMT1B	
Type II	AD	1p35-36	CMT2A	
X-linked	XD	Xq13	CMTX	
Spinal muscular atrophy (SMA)				
Werdnig-Hoffmann	AR	5q11–q13	SMA	
Kugelberg-Welander	AR	5q11–q13	SMA	
Familial amyotrophic lateral sclerosis	AD	21q22	ALS	Cu/Zn superoxide dismutase
Kennedy disease	XR	Xq21–q22	SBMA	androgen receptor

XR sex linked recessive
AD autosomal dominant
AR autosomal recessive

C.1 Gene location map for the neuromuscular disorders.

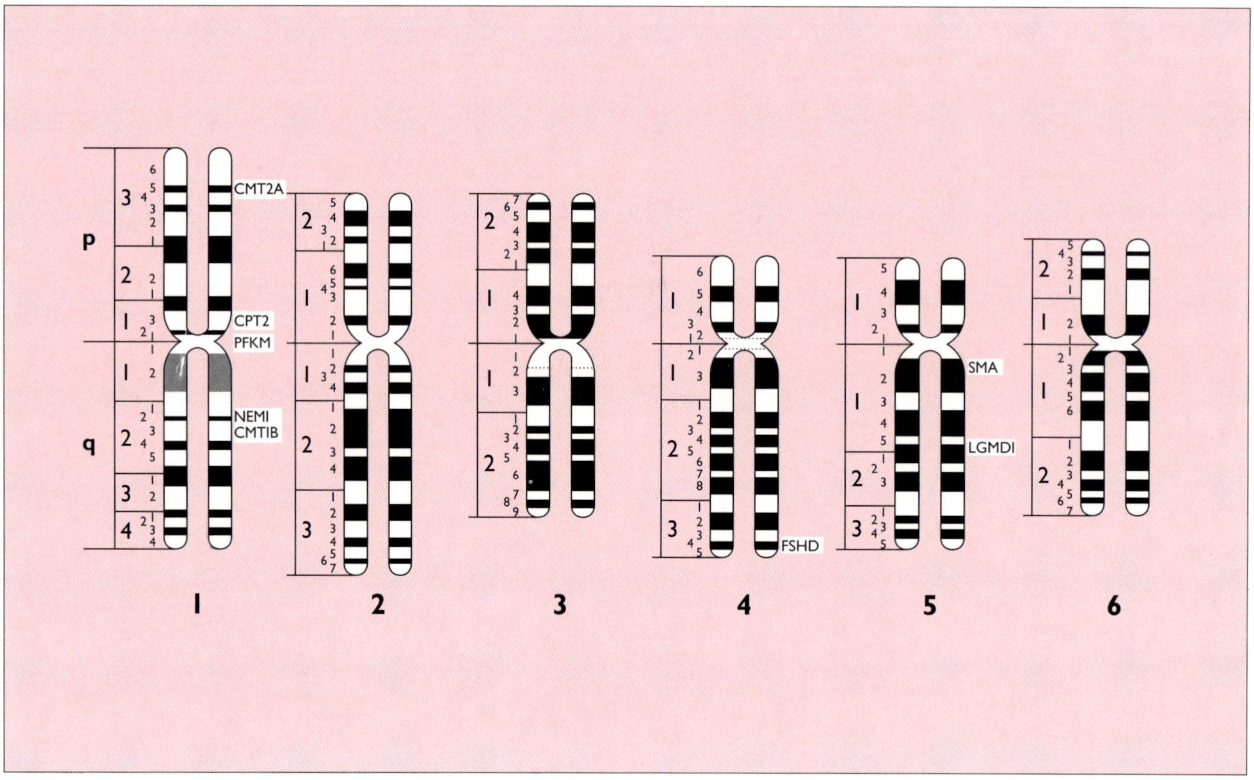

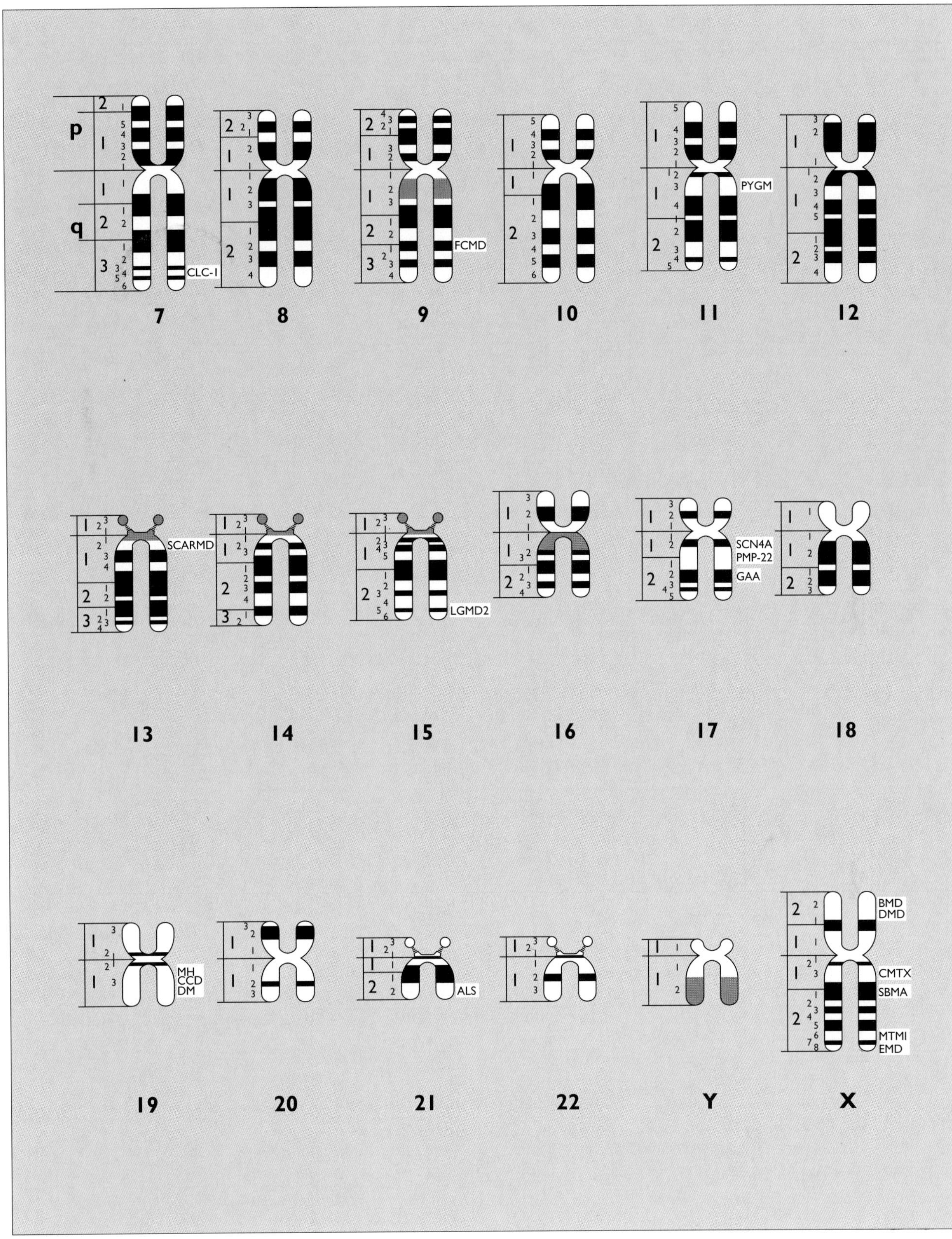
p
q
CLC-1
7
8
FCMD
9
10
PYGM
11
12
SCARMD
13
14
LGMD2
15
16
SCN4A
PMP-22
GAA
17
18
MH
CCD
DM
19
20
ALS
21
22
Y
BMD
DMD
CMTX
SBMA
MTMI
EMD
X

References

Chapter 1

Adams R. D., Denny-Brown D., Pearson C. M. (1953) *Diseases of Muscle: A Study in Pathology* 1st ed. New York Hoeber

Cumming W. J. K., Hudgson P. and Lecky B (1995, in preparation) *Clinical Management of Patients with Neuromuscular Disorders*

Chapter 2

Bergstrom J. (1962) Muscle electrolytes in man determined by neutron activation analysis on needle biopsy specimen: a study in normal subjects, kidney patients, and patients with chronic diarrhoea. Scandinavian Journal of Clinical and Laboratory Investigation. **14** (Supplement 68), 1–110.

Coakley J. H., Smith P. E. M., Dictrichson P., Helliwell T., Edwards R. H. T. (1986) Percutaneous muscle biopsy with conchotome. Clinical Science, **71** (Supplement 15), 23P.

Cumming W. J. K., Mahon M. (1984) Needle biopsies. Muscle & Nerve, **7**, 679–680 (Letter).

Duchenne G. B. A. (1868) Récherches sur la paralysie musculaire pseudohypertrophique ou paralysie myo-sclérosique. Archives Generales de Medécine **11**, 179–209.

Edwards R. H. T. (1971) Percutaneous needle biopsy of skeletal muscle in diagnosis and research. Lancet **ii**, 593–595.

Eriksson P. O., Eriksson A., Ringqvist M., Thornell L. E. (1980) The reliability of histochemical fibre typing of human necropsy muscles. Histochemistry, **65**, 193–205.

Young A., Hughes J. M., Round J. M., Edwards R. H. T. (1978) University College Hospital muscle biopsy needle. Lancet **ii**, 1285.

Chapter 3

Dubowitz (1985) – see Bibliography

Johnson M. A., Polgar J., Weightman D., Appleton D. (1973) Data on the distribution of fibre types in thirty-six human muscles - An autopsy study. Journal of the Neurological Sciences, **18**, 111–129.

Loughlin (1993) – see Bibliography

Mahon M., Toman A., Willan P. L. T., Bagnall K. M. (1984) Variability of histochemical and morphometric data from needle biopsy specimens of human quadriceps femoris muscle. Journal of the Neurological Sciences, **63**, 85–100.

Mahon M., Kristmundsdottir F., Cumming W. J. K. (1987) Spatial pattern analysis of muscle (SPAM). Journal of Anatomy, **152**, 229.

Williams (1977) – see Bibliography

Williams P. L., Warwick R., Dyson M., Bannister L. H. (1989) *Gray's Anatomy*. 37th ed., pp. 546–563, 915–916. Edinburgh, Churchill Livingstone

Chapter 5

Ahn A. H. & Kunkel L. M. (1993) The structural and functional diversity if dystrophin. Nature Genetics, **3** April 283–291

Cullen M. J., Fulthorpe J. J. (1975) Stages in fibre breakdown in Duchenne muscular dystrophy. An electron microscopic study. Journal of the Neurological Sciences, **24**, 179–200.

Emery (1993) – see Bibliography.

Erb W. H. (1884) Über die 'juvenile Form' der progressiven Muskelatrophie und ihre Beziehungen zur sogenannten Pseudohypertrophie der Muskeln. Deutches Archiv fur Klinische Medizin, **34**, 467–519.

Hoffman E. P., Brown B. H., Kunkel L. M. (1987) Dystrophin: the protein product of the Duchenne muscular dystrophy gene. Cell, **51**, 919–928.

Koenig M., Monaco A. P., Kunkel L. M. (1988) The compete sequence of dystrophin predicts a rod-shaped cytoskeletal protein. Cell **53** 219–228.

Meryon E. (1852) On granular and fatty degeneration of the voluntary muscles. Medico-chirurgical Transactions (London), **35**, 73–84.

Lindenbaum M. H. & Carbonetto S. (1993) Dystrophin and partners at the cell surface. Current Biology **3** 109–111.

Love D. R., Byth B. C., Tinsley J. M., Blake D. J., Davies K.(1993) Dystrophies and dystrophin related proteins: a review of protein and RNA studies. Neuromuscular Disorders **3** 5–21.

Matsumura K. & Campbell K. P.. (1993) Deficiency of dystrophin-associated proteins: a common mechanism leading to muscle cell necrosis in severe childhood muscular dystrophies. Neuromuscular Disorders **3** 109–118.

Matsumura K. & Campbell K. P. (1994) Dystrophin-Glycoprotein complex: its role in the molecular pathogenesis of muscular dystrophies. Muscle Nerve **17** 2–15.

Walton J. N., Nattrass F. J. (1954) On the classification, natural history, and treatment of the myopathies. Brain, **77**, 169–231.

Chapter 6

Brzustowicz L. M., Lehner T., Castilla L. H., Penchaszadeh G. K., Wilhelmsen K. C., Daniels R., Davies K. E., Leppert M., Ziter F., Wood D., Dubowitz V., Zerres K., Hausmanowa-Petrusewicz I., Ott J., Munsat T. L., Gilliam T. C. (1990) Genetic mapping of chronic childhood-onset spinal muscular atrophy to chromosome 5q11.2-13.3. Nature, **344**, 540–541.

Dyck et al (1993) – See Bibliography.

Jennekens F. G. I., Tomlinson B. E., Walton J. N. (1971) Histochemical aspects of five limb muscles in old age: an autopsy study. Journal of the Neurological sciences, **14**, 259–276.

Munsat T. L. , Skerry L., Korf B., Pober B., Schapira Y., Gascon G. G., Al-Rajeh S. M., Dubowitz V., Davies K., Brzustowicz L. M., Penchaszadeh G. K., Gilliam T. C. (1990) Phenotypic heterogeneity of spinal muscular atrophy mapping to chromosome 5q11.2-13.3 (SMA 5q). Neurology, **40**, 1831–1836.

Chapter 7

Dalakas M. C. (1992) Inflammatory myopathies. In: *Handbook of Clinical Neurology*, eds. Vinken P. J., Bruyn G. W., Klawans H. L., vol 62, pp. 369–390. Amsterdam, Elsevier

Chapter 8

Fardeau M. (1982) Congenital Myopathies. In: *Skeletal Muscle Pathology*, eds. Mastaglia F. L., Walton J. N., pp. 161–203. Edinburgh, Churchill Livingstone

Chapter 9

DiMauro S., Tonin P., Servidei S. (1992) Metabolic myopathies. In: *Handbook of Clinical Neurology*, eds. Vinken P. J., Bruyn G. W., Klawans H. L., vol 62, pp. 479–526. Amsterdam, Elsevier

Chapter 10

Engel AG. (1992) Myasthenia gravis and myasthenic syndromes. In: *Handbook of Clinical Neurology*, eds. Vinken P. J., Bruyn G. W., Klawans H. L., vol 62, pp. 391–455. Amsterdam, Elsevier

Bibliography

Listed below are some of the major textbooks dealing with human muscle pathology and neuromuscular diseases to which the reader is referred for more detailed accounts and references to individual disorders, both clinical and pathological. In addition, a few contemporary texts describing recent advances in our understanding of molecular, genetic, and cellular aspects of muscle biology are listed along with texts dealing with technical procedures.

Adams R. D. (1975) *Diseases of Muscle – A Study in Pathology*, 3rd ed. Hagerstown, Harper & Row

Anderson J. R. (1985) *Atlas of Skeletal Muscle Pathology*, ed. Gresham G. A., Current Histopathology; 9. Lancaster, MTP Press Ltd.

Angelini C., Danielli G. A., Fontanari D. (1991) eds. *Muscular Dystrophy Research: From Molecular Diagnosis toward Therapy*. Amsterdam, Excerpta Medica

Bethlem J.(1970) *Muscle Pathology - Introduction and Atlas*. Amsterdam, Elsevier/North Holland

Bethlem J. (1977) *Myopathies*. Amsterdam, North-Holland

Brooke M. H. (1986) *A Clinician's View of Neuromuscular Diseases*. 2nd ed. Baltimore, Williams & Wilkins

Brumback R. A., Leech R. W. (1984) *Color Atlas of Muscle Histochemistry*. Littleton, PSG Publishing

Buller A. J., Goodfellow J., Newsom-Davis J. M. (1989) eds. *Molecular Genetics of Muscle Disease – Duchenne and Other Dystrophies*. Edinburgh, Churchill Livingstone; British Medical Bulletin **3**, 45

Busch H. F. M., Jennekens F. G. I., Scholte H. R. (1981) eds. *Mitochondria and Muscular Diseases*. Beetsterzwaag, Mefar b.v.

Carpenter S., Karpati G. (1984) *Pathology of Skeletal Muscle*. Edinburgh, Churchill Livingstone

Cumming W. J. K. (1993) Neuromuscular Disorders. In: *Muir's Textbook of Pathology*, ed. Anderson J. R. 13th ed., vol 21, pp. 69–76. Edinburgh, Churchill Livingstone

Cumming W. J. K. (1992) Aging and Neuromuscular Disease. In: *Textbook of Geriatric Medicine and Gerontology*. eds. Brocklehurst J. C., Tallis R. C., Fillitt H. M., 4th ed. pp. 834-842. Edinburgh, Churchill Livingstone

Drury R. A. B., Wallington E. A. (1980) *Carleton's Histological Technique*, 5th ed. Oxford, Oxford University Press

Dubowitz V. (1978) *Muscle Disorders in Childhood*, Philadelphia, WB Saunders (Schaffer A. J., Markowitz M. eds. Major Problems in Clinical Paediatrics; XVI).

Dubowitz V. (1985) *Muscle Biopsy. A Practical Approach*. 2nd ed. London, Bailliere Tindall

Dubowitz V. (1989) *A Colour Atlas of Muscle Disorders in Childhood*. London, Wolfe Medical Publications Ltd.

Dyck P. J., Thomas P. K., Griffin J. W., Low P. A., Poduslo J. F. (1993) *Peripheral Neuropathy*. 3rd ed. vol. 1–2. Philadelphia, WB Saunders

Ebashi S, Ozawa E. (1983) eds. *Muscular Dystrophy. Biomedical Aspects*. Tokyo/Berlin, Japan Scientific Societies Press/Springer-Verlag

El Haj A. (1992) ed. *Molecular Biology of Muscle*. Cambridge, The Company of Biologists Ltd. (Symposia of The Society for Experimental Biology; XLVI).

Emery A. E. H. (1993) *Duchenne Muscular Dystrophy*. 2nd ed. Oxford, Oxford University Press (Oxford Monographs on Medical Genetics; 24).

Engel A. G., Banker B. Q. (1986) eds. *Myology. Basic and Clinical*. vol. 1–2. New York, McGraw Hill

Evered D., Whelan J. (1988) eds. *Plasticity of the Neuromuscular System*. Chichester, John Wiley & Sons (Ciba Foundation Symposium; 138).

Galdi A. P. (1984) *Diagnosis and Management of Muscle Disease*. Lancaster, MTP Press Ltd.

Griggs R. C., Karpati G. (1990) eds. *Myoblast Transfer Therapy*. New York, Plenum (Advances in Experimental Medicine and Biology; 280).

Gutmann E., Hanzlikova V. (1972) *Age Changes in the Neuromuscular System.* Bristol, Scientechnica Ltd.

Heffner R. R. (1984) ed. *Muscle Pathology.* New York, Churchill Livingstone.

Jones D. A., Round J. M. (1990) *Skeletal Muscle in Health and Disease. A textbook of muscle physiology.* Manchester, Manchester University Press

Kelly A. M., Blau H. M. (1992) eds. *Neuromuscular Development and Disease.* New York, Raven Press (Fox C. F., ed. Molecular and Cellular Biology; 2).

Korenyi-Both A. L. (1983) *Muscle Pathology in Neuromuscular Disease.* Springfield, Charles C Thomas

Loughlin M. (1993) *Muscle Biopsy. A Laboratory Investigation.* Oxford, Butterworth-Heinemann Ltd.

Mastaglia F. L. (1988) ed. *Inflammatory Diseases of Muscle.* Oxford, Blackwell Scientific Publications

Mastaglia F. L., Walton J. N. (1992) eds. *Skeletal Muscle Pathology.* 2nd ed. Edinburgh, Churchill Livingstone

Mauro A. (1979) ed. *Muscle Regeneration.* New York, Raven Press

McComas A. J. (1977) *Neuromuscular Function and Disorders.* London, Butterworths

Partridge T. (1993) ed. *Molecular and Cell Biology of Muscular Dystrophy.* London, Chapman & Hall (Wright D. J. M., Archard L. C., eds. *Molecular and Cell Biology of Human Diseases*; 3).

Peachey L. D. (1983) ed. *Skeletal Muscle.* (Handbook of Physiology; 10). Besthesda, American Physiological Society

Pette D. (1990) ed. *The Dynamic State of Muscle Fibers.* Berlin, Walter de Gruyter

Polak J. M., Van Noorden S. (1983) eds. *Immunocytochemistry* 2nd ed. Bristol, John Wright & Sons

Rowland L.P. (1977) ed. *Pathogenesis of Human Muscular Dystrophies.* Amsterdam, Excerpta Medica

Rowland L. P., DiMauro S. (1992) eds. *Myopathies*, vol. 62 Amsterdam, Elsevier Science Publishers Revised Series (*Handbook of Clinical Neurology*; 18, eds. Vinken PJ, Bruyn GW, Klawans HL)

Rowland L. P., Wood D. S., Schon E. A., DiMauro S. (1989) eds. Molecular Genetics in Diseases of Brain, Nerve, and Muscle. New York, Oxford University Press

Sarnat H. B. (1983) *Muscle Pathology and Histochemistry.* Chicago, American Society of Clinical Pathologists Press

Schmalbruch H. (1985) *Skeletal Muscle.* Berlin, Springer-Verlag

Serratrice G. *et al.* (1984) eds. *Neuromuscular Diseases.* New York, Raven Press

Siegel I. M. (1986) *Muscle and its Diseases.* Chicago, Year Book Medical Publisher

Stoward P. J., Pearse A. G. E. (1992) eds. *Histochemistry. Theoretical and Applied* 4th ed. Edinburgh, Churchill Livingstone

Swash M., Schwartz M. S. (1981) *Neuromuscular Diseases. A practical approach to diagnosis and management.* 1st ed. Berlin, Springer-Verlag

Swash M., Schwartz M. S. (1984) *Biopsy Pathology of Muscle.* London, Chapman & Hall

Uehara Y., Campbell G. R., Burnstock G. R. (1976) *Muscle and its Innervation. An Atlas of Fine Structure.* London, Edward Arnold

Vrbova G., Gordon T., Jones R. (1978) *Nerve-Muscle Interaction.* London, Chapman & Hall

Walton J. N., Canal N., Scarlato G. (1970) eds. *Muscle Diseases.* Amsterdam, Excerpta Medica

Walton J. N., Mastaglia F. L. (1980) eds. *The Muscular Dystrophies.* London, British Council. (British Medical Bulletin; 36).

Walton J. N. (1988) ed. *Disorders of Voluntary Muscle.* 5th ed. Edinburgh, Churchill Livingstone

Williams M. A. (1977) *Quantitative Methods in Biology.* Amsterdam, North-Holland (*Practical Methods in Electron Microscopy*; ed. AM Glauert, 6 pt 1).

The following Journals are excellent sources of primary literature on neuromuscular disorders.

Journal of the Neurological Sciences – Amsterdam, Elsevier
Muscle & Nerve – New York, Wiley Interscience
Neuromuscular Disorders – Oxford, Pergamon

Index

Abbreviations
electron micrograph cellular structure, 182
general, 181
Acetylcholinesterase, 183, 187–188
Acid maltase deficiency (Type II glycogenosis), 147, 148–152, 169
Acid phosphatase, 8, 183, 187
Acridine orange, 8, 183, 188
Acromegaly, 173, 174
Alcian blue, 8, 183, 187
Alcohol abuse, 174
Alligator forceps (chonchotome biopsy), 8
Amyotrophic lateral sclerosis (ALS), 106
Andersen's disease (type IV glycogen storage disorder), 147, 154–155
Anterior horn disorders, 45, see also specific disorders
pathology, 106–111
Araldite embedding medium, 191
Artefacts, 9, 26–27, 180
'Arthrogryposis multiplex congenita', 135
ATPase (myofibrillar adenosine triphosphatase), 8, 10, 183, 184
Atrophy, 2, 25, 40, 44, 45, 49, 50, 76, 105, 107, 110, 116, 120, 130, 137, 170, 173, 174, 177, 181
Autopsy specimens, 10

Barbiturate overdose, 174
Basement membrane, ultrastructural disorders, 83
Becker muscular dystrophy (BMD), 14, 91, 94, 177
differential diagnosis from DMD, 103
dystrophin, 92, 103
female carrier, 103
gene location, 193
pathological features, 101–103
SMA misdiagnosis, 105
see also Xp21.2 myopathies
Benign congenital myopathy syndrome, see Infantile hypotonia
'Bhudda' posture, 135
Biochemical analysis, 14
Brancher glycogenosis, 147
Burn injury, 174–175

Carnitine deficiency, 147, 156–157
Carnitine palmitoyl transferase (CPT) deficiency, 147, 158, 193
Cell culture, 14
Centronuclear (myotubular) myopathies, 76, 137–140, 193
aetiopathogenesis, 135
clinicopathological correlation, 135
pathophysiology, 136
CFTD (congenital fibre type disproportion), 51, 135, 137
Charcot–Marie–Tooth syndrome, 194
Chloroquine, 174
Chonchotome biopsy (alligator forceps), 8
Chronic progressive external ophthalmoplegia (CPEO), 159
Churg–Strauss syndrome, 116
Clinical background, 2, 15, 177
Clinical examination, 177
Clofibrate, 174
Concentric laminar inclusion body myopathy, 146
Concentric laminated bodies, 83
Congenital fibre type disproportion (CFTD), 51, 135, 137
Congenital muscular dystrophy, 135, 137

Congenital myopathies
centronuclear dystrophy, see Centronuclear (myotubular myopathies
clinicopathological correlation, 135
core, see Core myopathies
fibre type disproportion, 137
intra-uterine maturational delay, 135, 145
muscular dystrophy, 137
nemaline myopathy, 135, 141–143
pathology, 135
pathophysiology, 136
Type I fibre hypotrophy, 146
Connective tissue, normal appearance, 15, 17
Control biopsy, 10
Core myopathies, 74, 135, 136, 144, 193
mini-, 145
multi-, 145
multi/mini-, 74, 145
Corticosteroids, 173, 174
CPT (carnitine palmitoyl transferase) deficiency, 147, 158, 193
Cryostat sectioning, 9–10
Cushing's disease, 173
Cylindrical spirals, 83
Cytochrome oxidase, 8, 183, 188
Cytoplasmic bodies, 83

Delta lesions, 57
Dermatomyositis (DM)
immunocytochemistry, 130
pathological reactions, 113, 116, 120
pathology, 83
presentation, 113
Desmin, 135, 139
Diaminobenzidine (DAB), 188
Diuretic abuse, 174
DMD-like protein (utrophin, dystrophin-related protein), 92
Drug-related myopathy, 174
Duchenne muscular dystrophy (DMD), 2, 14, 91–94, 177
clinical features, 91
differential diagnosis from BMD, 103
dystrophin analysis, 101
female carrier state, 94
gene location, 193
haematology, 180
pathological features, 60, 66, 83, 94, 94–104
see also Xp21.2 myopathies
Dystrophin, 14, 91, 92–93
in BMD, 92, 103
and DGC, 92
in DMD, 101
enzyme-immunocytochemistry, 189–190
molecular genetics, 91
SMA misdiagnosis of BMD, 105
Dystrophin-glycoprotein complex (DGC), 91, 92
Dystrophin-related protein (utrophin, DMD-like protein), 92

Electron microscopy, 10, 191–192
cellular structure abbreviations, 182
normal ultrastructure, 28–38
tissue fixation, 191–192

ultrastructural abnormalities, 74–83
ultrastructural screening, 40–41
Emery–Dreifuss syndrome, 94, 193
Endocrine disorders, 173
Enzyme activity, 8
Eosinophilic myositis, 115
Epsilon aminocaproic acid (EACA), 174

Familial amyotrophic lateral sclerosis, 194
Fibre typing, 8
Fingerprint body myopathy, 146
'Floppy infant' syndrome, see Congenital myopathies
FSH syndrome, 74, 177
Fukuyama muscular dystrophy, 91, 137, 193

Gene location, 193–195
Giant-cell myositis, 115
Glutaraldehyde plus osmium tetroxide fixation, 10, 11, 191
Glycogen storage disorders, 67
gene location, 193
type II, 147, 148–152, 169
type IV, 147, 154–155
type V, 147, 153
Granulomatous myositis, 115
Growth and development disorders, 50, see also Myotonic dystrophy

Haematoxylin and eosin (H&E) stain, 2, 8, 9, 183
Hepatitis B, 3
Hereditary neuropathy, 194
Histochemistry
assessment, 40
specimen preparation, 9–10
Histology
assessment, 40
specimen preparation, 9–10
History
clinical, 177
family, 177
HIV infection, 3, 113
Honeycomb inclusions, 83
Hyaline fibres, 60
Hypercontraction bands, 57–60
Hypereosinophilic syndrome (HES), 115
Hyperkalaemic periodic paralysis, 193
Hyperthyroidism, 173
Hyperthermia, malignant, 193
Hypothyroidism, 173

I-band loss, 60
Immunoblotting, 14
Immunocytochemistry, 10, 11, 14, 189–190
assessment, 40
dystrophin, 189–190
myosin, 190
Immunofluorescence, 10, 189
in situ hybridisation, 14, 180
Inclusion body myositis (IBM), 78, 113, 131–133
concentric laminar, 146
Infantile acid maltase deficiency, 147, 148
Infantile hypotonia
spinal muscular atrophy, 105
see also Congenital myopathies
Infections protocol, 3
Infective myopathies, 113, 180
Inflammatory myopathies
antibodies in, 14
dermatomyositis, 83, 113
granulomatous myositis, 115
idiopathic, 113
immunocytochemistry, 113, 130
inclusion body myositis, 78, 113, 131–133, 146
inflammatory cell infiltrates, 116–119
necrosis, 120
parasitic infestation, 169
pathology, 66, 116–129
perifascicular atrophy, 120
phagocytosis, 120
polymyositis, 83, 113
regeneration, 120
Intra-uterine maturational delay, 135, 145
Intromyofibrilla vacuoles, 83

Kearn–Sayre's syndrome, 159
Kennedy syndrome, 105, 194
Kugelberg–Welander syndrome, 105, 194

Lead citrate stain (modified Reynolds stain), 192
Light microscopy, specimen preparation, 8–9
Lipid disorders, 68
carnitine deficiency, 156–157
CPT deficiency, 158
gene location, 193
Lipofuscin deposits, 83
Liquorice derivatives, 174

McArdle's disease (type IV glycogen storage disorder), 147, 153, 193
McLeod syndrome, 180
Malignant hypothermia, 193
Masson's trichome, 8, 183, 184
Metabolic myopathies
clinical patterns, 147
gene location, 193
glycogen storage disorders
type II, 147, 148–152, 169
type IV, 147, 154–155
type V, 147, 153
lipid
carnitine deficiency, 156–157
CPT deficiency, 158
mitochondrial, 159–167
pathological reactions, 147
summary, 147
Methyl green pyronine, 8, 183, 186–187
Mitochondria, 33, 40, 67, 68, 70, 137, 159, 184, 185
Mitochondrial myopathies, 70–74, 159–167
endocrine disorders, 173
lipid disorders, 68
ragged-red fibre, 70
structured core, 73
Modified Gomori trichome, 8, 183, 185
Modified Reynolds stain (lead citrate stain), 192
Molecular genetic studies, 180
Monoclonal antibodies, 14
Morphometry, specimen preparation, 9–10
Motor neurone disease, 2, 105, 106, 169
Motor unit, 15, 18, 22, 45, 105, 106, 108, 110, 137, 146, 180
Multiple acyl CoA deficiency with lipid storage, see Carnitine deficiency myopathy
Muscle biopsy
analysis, 39–43, 180
quantitative, 41–43
clinical cooperation, 2, 3, 180
control, 10
information accompanying, 3, 180
open, 2, 4–5, 9
pathology report form, 39
principles for satisfactory service, 2
procedure, 180
request form, 3–4
site, 2, 4
see also Normal tissue; Tissue preparation
Muscle fibre
growth and development disorders, 50
metabolic disorders
glycogen storage, 67
lipid disorders, 68
mitochondrial disorders, 70–74
myopathic disorders, 50–66
delta lesions, 57
endomysial fibrosis, 50

Muscle fibre (continued)
  fibre atrophy, 50
  hyaline fibres, 60
  hypercontraction bands, 57–60
  hypertrophy, 50
  I-band loss, 60
  phagocytic activity, 61
  regeneration, 61
  ring fibre, 54
  satellite cell, 64
  Z-band streaming, 60
  Z-disc, 62
  neurogenic disorders
    grouped reinnervation, 49
    secondary myopathic degeneration, 49
    nuclearknocken, 76
    ragged-red, 70
  structural abnormalities
    core myopathies, 74
    nemaline rods, 80
    nuclear myopathies, 76–79
    tubular aggregates, 82
  systemic disorders, 50
    Type I fibre hypotrophy, 135, 136, 146
    Type II fibre hypotrophy, 136
  types, 9, 41, 184
  ultrastructural disorders, 83–90
  vascular disorders, 50
Muscle spindles, 25, 34, 40
Muscular dystrophy, 61
Myasthenia gravis, 169
Myelin whorls, 83
Myofibrillar adenosine triphosphatase, see ATPase
Myonuclei, 17, 77, 137, 174
  internal, 25, 40, 43, 103, 110, 137, 170
Myosin, 190
Myotonia, congenita, 193
Myotonia, generalised, 193
Myotonic disorders, 82
Myotonic dystrophy (MyD0, 193
Myotonic dystrophy (MyD), 50, 170, 193

NADH-Tr (nicotinamide adenine dinucleotide tetrazolium reductase), 8, 183, 184
Necrosis, 40, 43, 50, 54, 61, 66, 76, 77, 97, 101, 110, 113, 116, 120, 130, 174
Needle EMG, 180
Needle muscle biopsy, 2, 6–8, 9
Nemaline myopathy, 80, 135, 141–143, 193
Nerve conduction studies, 180
Neurogenic disorders, 45–50
Nicotinamide adenine dinucleotide tetrazolium reductase (NADH-Tr), 8, 183, 184
Non-specific esterase, 8, 183, 185
Normal muscle
  anomalies, 25
  histochemistry, 18–25
  histology, 15–17
  individual variation, 15
  ultrastructure, 28–38
Nuclear myopathies, 76–79
Nuclearknocken, 76

Oil red O, 8, 183, 186
Opiate abuse, 174

Pancreatic disorders, 173
Paraformaldehyde fixation, 11
Paramyotonia congenita, 193
Parasitic infestation, 169
PAS (periodic acid–Schiff reaction), 8, 183, 185–186
Patient investigation
  clinical history, 177
  electrophysiology, 180
  examination, 177
  family history, 177
  haematology, 180
  muscle biopsy, see Muscle biopsy
Periodic acid Schiff reaction (PAS), 8, 183, 185–186
Periodic paralysis (PP), 82, 171–173
Phase contrast microscopy, 10
Phosphofructokinase, 8, 183, 187
Phosphorylase, 8, 183, 185
Photographic records, 10
Pituitary hypersecretion, 173
Polarising microscopy, 10
Polyarteritis nodosa, 115
Polyclonal antibodies, 14
Polymyositis (PM), 66, 113
  misdiagnosis, 2
  pathology, 83
Pompe's disease (type II glycogen storage disorder), 147, 148–152, 169, 193
PP (periodic paralysis), 82, 171–173
Purgative abuse, 174

Ragged-red fibres, 70
Rhabdomyolysis, 2
Ring fibre (ringbinden), 54

Sarcoidosis, 115
Satellite cells, 30, 61, 77, 99, 101
SCARMD (severe childhood autosomal recessive muscular dystrophy), 92, 193
Specimen handling, 8–14
  electron microscopy preparation, 10–13
  immunocytochemistry, 14
Spinal muscular atrophies (SMA), 49, 105–111, 177
  acute infantile, 105
  anterior horn pathology, 106–111
  chronic infantile, 105
  gene location, 194
  late-onset forms, 105–106
  motor neurone disease, 105, 106
Steroid myopathy, 173, 174
Succinate dehydrogenase (SDH), 8, 183, 186
Sudan black B, 8, 183, 186
Systemic disorders, 50

Tarui disease, 193
Tissue preparation, 1–2
  freezing, 9
  handling, 1–2, 4
  and misdiagnosis, 2
  storage, 8, 9, 10
Toluidine blue, 8, 9, 183
Toxin-related damage, 173
Trauma, 174
Trilaminar myopathy, 146

Ultrastructural screening, 40–41
  see also Electron microscopy
Uranyl acetate stain, 191–192
Utrophin (DMD-like protein, dystrophin-related protein), 92

Van Giesson, 8
Vascular disorders, 50
Volkmann's systemic contracture, 174
Von Kossa, 8, 183, 188

Wegener's granulomatosis, 115, 116
Werdnig–Hoffmann syndrome, 105, 194

X-linked recessive myopathies
  pathological features, 94–104
  see also Xp21.2 myopathies
X-linked recessive myopathy, 135, 136
Xp21.2 myopathies, 14, 50, 91–104

Z-band streaming, 60
Z-disc, 62
Zebra body myopathy, 146